REVIEWS IN ENGINEERING GEOLOGY
VOLUME X

CLAY AND SHALE SLOPE INSTABILITY

Edited by

WILLIAM C. HANEBERG
New Mexico Bureau of Mines and Mineral Resources
Campus Station
Socorro, New Mexico 87801

and

SCOTT A. ANDERSON
Department of Civil Engineering
University of Hawaii
Holmes Hall 383
2540 Dole Street
Honolulu, Hawaii 96822

The Geological Society of America
3300 Penrose Place, P.O. Box 9140
Boulder, Colorado 80301
1995

Copyright © 1995, The Geological Society of America, Inc. (GSA). All rights reserved. GSA grants permission to individual scientists to make unlimited photocopies of one or more items from this volume for noncommercial purposes advancing science or education, including classroom use. Permission is granted to individuals to make photocopies of any item in this volume for other noncommercial, nonprofit purposes provided that the appropriate fee ($0.25 per page) is paid directly to the Copyright Clearance Center, 27 Congress Street, Salem, Massachusetts 01970, phone (508) 744-3350 (include title and ISBN when paying). Written permission is required from GSA for all other forms of capture or reproduction of any item in the volume including, but not limited to, all types of electronic or digital scanning or other digital or manual transformation of articles or any portion thereof, such as abstracts, into computer-readable and/or transmittable form for personal or corporate use, either noncommercial or commercial, for-profit or otherwise. Send permission requests to GSA Copyrights.

Copyright is not claimed on any material prepared wholly by government employees within the scope of their employment.

Published by The Geological Society of America, Inc.
3300 Penrose Place, P.O. Box 9140, Boulder, Colorado 80301

Printed in U.S.A.

GSA Books Science Editor Richard A. Hoppin

Library of Congress Cataloging-in-Publication Data

Clay and shale slope instability / edited by William C. Haneberg and
   Scott A. Anderson
      p. cm. — (Reviews in engineering geology ; v. 10)
    Papers presented at a symposium sponsored by the Engineering
Geology Division of the Geological Society of America at the 1992
annual meeting in Cincinnati, Ohio.
    Includes bibliographical references and index.
    ISBN 0-8137-4110-6
    1. Slopes (Soil mechanics). 2. Clay soils—Stability. 3. Shale.
4. Mass-wasting.   I. Haneberg, William C.   II. Anderson, Scott A.
III. Geological Society of America.   Division of Engineering
Geology.   IV. Series.
TA705.R4   vol. 10
[TA711]
624.1'51 s—dc20
[624.1'51]   94-49378
                                                        CIP

Contents

Preface v

1. *Shear Strength and Slope Stability in a Shallow Clayey Soil Regolith* 1
   Scott A. Anderson and Nicholas Sitar

2. *Effect of Test Method and Procedure on Measurements of Residual Shear Strength of Bentonite from the Portuguese Bend Landslide* 13
   Stephen M. Watry and Perry L. Ehlig

3. *Evaluation of Viscoplastic Slope Movement Based on Triaxial Tests* 39
   Wylie W.-H. Wong, Carlton L. Ho, Richard M. Iverson, and Cynthia L. Hovind

4. *Stability of Slopes: Limit Analysis Approach* 51
   Radoslaw L. Michalowski

5. *Groundwater Flow and the Stability of Heterogeneous Infinite Slopes Underlain by Impervious Substrata* 63
   William C. Haneberg

6. *Geology, Hydrology, and Mechanics of a Slow-Moving, Clay-Rich Landslide, Honolulu, Hawaii* 79
   Rex L. Baum and Mark E. Reid

7. *Landslides on Clay and Shale Hillslopes in Tuscany, Italy* 107
   R. Bertocci, P. Canuti, N. Casagli, C. A. Garzonio, and P. Vannocci

8. *Characterizing Durability of Mudrocks for Slope Stability Purposes* 121
   Jeffrey C. Dick and Abdul Shakoor

9. *Slope Stability Considerations in Differentially Weathered Mudrocks* 131
   Abdul Shakoor

10. *Effect of Argillic Alteration on Rock Mass Stability* 139
    Robert J. Watters and Warren D. Delahaut

Index 151

Preface

Clay and Shale Slope Instability grew out of a symposium of the same title sponsored by GSA's Engineering Geology Division at the 1992 annual meeting in Cincinnati, Ohio. The Ohio River valley and surrounding areas are rich with examples of clay and shale slope instability, including colluvium landslides on top of Ordovician shale and limestone bedrock, landslides in weakly indurated Pennsylvanian mudrocks, rockfalls as the result of differential weathering in Pennsylvanian coal measures, and extrusions of plastic Illinoian lake clays into cuts and stream channels. The marginal stability of natural slopes throughout the region also has implications for the stability of engineered slopes. In many cases, the native material responsible for naturally unstable slopes is used for embankment construction. Clay and shale slope instability was therefore a topic geographically well suited to the Cincinnati meeting.

Ten talks were presented at the symposium, the abstracts of which were published by the Society in the 1992 Annual Meeting Abstracts with Programs. Several speakers expressed interest in preparing full-length papers on their topics, and additional contributions were solicited from both the engineering geological and geotechnical engineering communities. The final set of papers is substantially different from the presentations given at the original symposium, as only three of the talks were expanded into papers.

In preparing this volume, we sought papers that integrated both descriptive and analytical aspects of slope instability. Even today, in many consulting firms or university classrooms, one would find qualitative description and quantitative analysis neatly parceled out between geologists and engineers. While we are not suggesting that geologists become engineers and engineers become geologists, we do believe that progress occurs only when integrating, rather than isolating, bodies of knowledge and experience.

William C. Haneberg
Socorro, New Mexico

Scott A. Anderson
Honolulu, Hawaii

Shear strength and slope stability in a shallow clayey soil regolith

Scott A. Anderson
Department of Civil Engineering, University of Hawaii, Honolulu, Hawaii 96822
Nicholas Sitar
Department of Civil Engineering, University of California, Berkeley, California 94720

ABSTRACT

The results of three types of laboratory strength tests on undisturbed samples of a saturated clayey soil regolith are analyzed. Drained-direct shear tests, anisotropically consolidated-undrained triaxial compression tests with pore pressure measurement, and constant-shear-drained unloading test results indicate different modes of failure and different strength parameters. Based on the grain-size distribution and plasticity test results, it appears that the behavior and strength of the regolith would be dominated by the clay content; however, the results of the strength tests illustrate the soil has a structure formed by cemented bonds. This structure has a larger influence on the behavior and strength than the clay content does, at least at the depth of failure. At the in-situ stress level of the shallow regolith, the consequence of failure is found to be a function of the stress path. As a result, some slope failures may move slowly for short distances, and others many mobilize into rapid flow failures because of differences in the stress path leading to failure.

INTRODUCTION

Many natural slopes are covered by a thin soil regolith that consists of residual or colluvial soils. Failures of the regolith are commonly the result of adverse seepage conditions developed as a consequence of heavy rainfall, irrigation, or snowmelt. Failure surfaces are constrained by the underlying bedrock and are consequently subparallel to the slope surface. Failures can occur in large numbers, but because of the thin soil profile they are typically somewhat small in volume. From inspection of failure scars, it appears that some of the failures move slowly for short distances, whereas others mobilize into flow slides which move rapidly for long distances.

In the San Francisco Bay area of California, a particularly large storm occurred in January 1982 and caused thousands of landslides and flow slides in soils of varied gradation, including clayey soils (Ellen, 1988; Howard et al., 1988). The flow slides, which have been generally referred to as debris flows regardless of the apparent grain size, often appeared to have mobilized from soil slips (Ellen, 1988). Following the 1982 storm, a research site was selected on a slope of a clayey soil regolith in Briones Park, about 30 km northeast of San Francisco (Fig. 1), to investigate the hydrologic response and soil behavior leading rainfall-induced landslides and debris flows. The hydrologic response of the slope was measured using recording tensiometers (Johnson and Sitar, 1990; Anderson and Sitar, 1993), and samples were collected for the testing program reported herein.

The purpose of this paper is to illustrate the wide range of stress-strain behavior that can be observed in a clayey soil regolith and to demonstrate the importance of careful selection of strength parameters for accurate stability analysis. This is accomplished by reviewing tests designed and performed to mimic relevant field loading conditions. Observations on the stress-strain behavior of the soil give insight into the modes of slope failure in clayey residual and colluvial soil regoliths and also into why some failures mobilize into debris flows and others do not. Understanding of the conditions for mobilization is important, as the hazard posed by debris flows is much different than that posed by slow-moving landslides. Evidence of the large travel distance that may be associated with flow failure is given in Figure 2. The figure is an aerial photograph

---

Anderson, S. A., and Sitar, N., 1995, Shear strength and slope stability in a shallow clayey soil regolith, *in* Haneberg, W. C., and Anderson, S. A., eds., Clay and Shale Slope Instability: Boulder, Colorado, Geological Society of America Reviews in Engineering Geology, v. X.

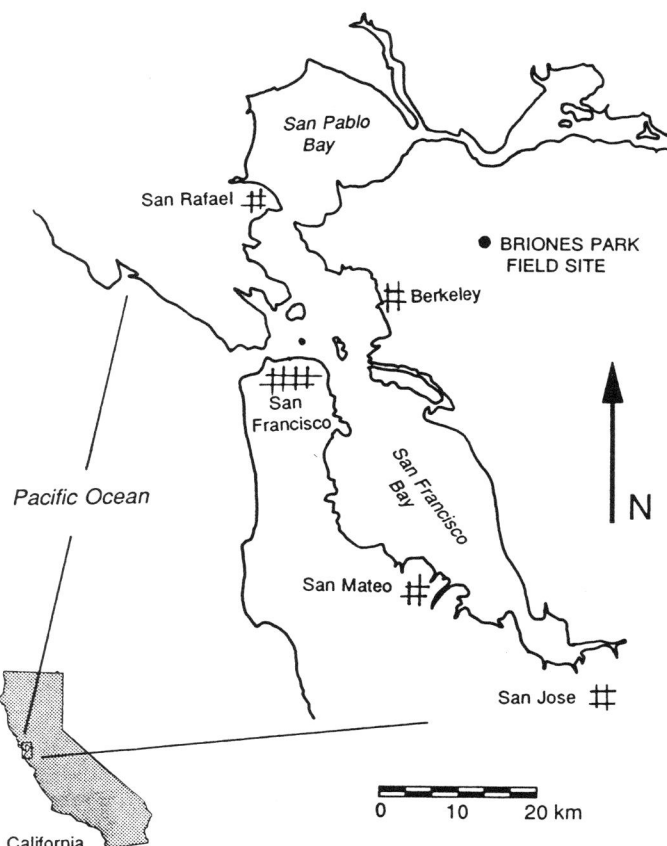

Figure 1. Map of the San Francisco Bay area of California. This research was conducted on soil collected at the Briones Park field site, northeast of the cities of San Francisco and Berkeley.

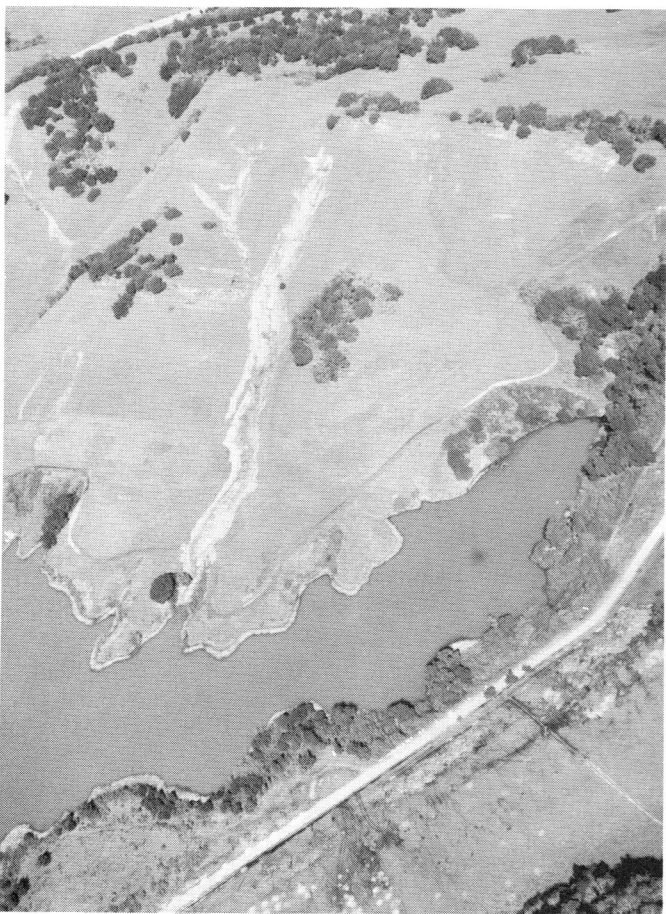

Figure 2. The debris flow shown in this figure occurred at Briones Reservoir, about 2 km from Briones Park, in February 1986. Note the four-lane road at the lower edge of the photograph for scale.

of a large debris flow at Briones Reservoir that occurred in a clayey soil regolith during a February 1986 storm; it is one of many that occurred during the storm.

SITE AND SOIL CHARACTERIZATION

The research site is a hillslope that is typical of many slopes in the Coast Range of northern California: the slope angle varies from about 19° to 34°, and the aspect is southeast. The present-day climate in the area is temperate-Mediterranean, with a mean annual precipitation of about 0.56 m (Rantz, 1971) and a temperature range from 0° to 30°C. The vegetation on the site consists primarily of annual grasses, but also includes some thistle and California poppy. Trees and woody vegetation are common only near channels of the first- and higher-order drainages.

The bedrock consists of Upper Miocene and Lower Pliocene marine and nonmarine siltstones and fine-grained sandstones belonging to the Briones Formation of the San Pablo Group (Wagner, 1978; Johnson and Sitar, 1987). The site is located near the axis of a broad syncline which gives the bedding a regional dip of about 3° to the northeast. Locally, four distinct rock units were identified by inspection of landslide scarps and rock outcrops. The upper unit is a very fine grained, buff-colored, fractured sandstone. It forms the ridge and overlies a highly fractured soft siltstone that, in turn, overlies two distinguishable units of very fine grained sandstone (Johnson and Sitar, 1987).

The depth of the regolith, which at the site includes both colluvial and residual soils, ranges between about 0.6 m and 1.5 m, generally increasing downslope. The soil column shown in Figure 3 indicates a gradual contact with the underlying bedrock and the presence of some gravel-size fragments of rock throughout the soil profile. These fragments become more numerous, less friable, and larger with depth, but their presence is largely unseen in the gradation analysis, which identifies only a slight decrease in clay-size content and an increase in silt content with depth (Johnson and Sitar, 1987). Gradation analyses and Atterberg limit tests reported by

Soil Log: Sampling Trench T1

Topsoil, light to dark brown clayey silt with numerous grass roots and scattered gopher holes.

Light brown, clayey, silty, fine grained sand, moist, firm, with some small tan sandstone and siltstone fragments, and occasional grass roots, root holes and macropores. Sandstone and siltstone fragments are very friable.

Light brown, clayey, silty, fine grained sand, moist, firm, with some tan sandstone and siltstone fragments (up to 7 mm), and macropores. Grass roots and root holes are scarce. Soil has greater structure at depth, with some interlocking of fragments.

Light brown, clayey, silty, fine grained sand, moist, firm, with numerous tan sandstone fragments (up to 13 mm). Soil structure suggests weathering around rock fragments. At depth, fragments are more durable and are typically interlocked.

Severely fractured brown sandstone bedrock. Fracture spacing 15 to 25 mm, rock fragments are weathered so that they may be cut with a knife, and they are supported by a matrix of soil. Some fractures indicate iron staining.

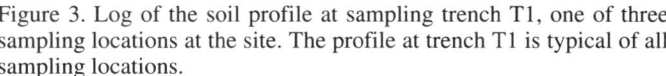

Figure 3. Log of the soil profile at sampling trench T1, one of three sampling locations at the site. The profile at trench T1 is typical of all sampling locations.

Johnson and Sitar (1987) found that there is less than 5% sand, generally between 55% and 70% silt, and 25% to 40% clay-size particles; the average liquid limit is 45; the plasticity index is 19; and the activity, defined by Skempton (1953) as plasticity index/% clay-size particles, varies from 0.4 to 0.9. Individual samples plot as either low plasticity silt (ML) or (CL) according to the Unified Soil Classification system. Lightwood (1988) performed an x-ray diffraction analysis of the clay-size fraction and found smectite to be the only clay mineral present in significant amounts.

TESTING PROCEDURES

The results of three types of tests on the Briones soil are presented here: the direct shear test under drained conditions (DS), the anisotropically consolidated-undrained triaxial compression test with pore pressure measurement (ACU), and a triaxial constant-shear-drained unloading test (CSD)—to model the field stress path generated in the soil regolith by infiltration and increasing pore pressure.

Because of the availability of the equipment and the ease of performing the DS test, it is probably the most commonly performed laboratory strength test for colluvium. This test is used despite recognition that there are difficulties with a rigorous interpretation of the stresses on the failure plane and that there are stress concentrations within the specimen. However, one advantage of the DS test is that it offers a means of determining the strength of a sheared surface after large displacement. DS tests were performed to investigate the loss of strength at large displacement and to determine the "routine" strength parameters as a base level for comparison with the triaxial tests, which are less frequently performed.

The CSD tests were performed so that strength parameters and volume change behavior could be determined using a stress path which stimulates the in-situ stress change as the regolith becomes saturated on a slope with constant shear stress. The ACU tests were performed as an additional means of finding the effective strength parameters and to determine the stress-strain behavior of the soil when subjected to undrained loading. The magnitude of any post-peak loss of strength caused by pore-pressure development during undrained shear is of particular interest in the investigation of the mode of failure and the mechanism of debris-flow mobilization. The anisotropic consolidation state and low confining pressure selected for the CSD and ACU tests are typical of the in-situ stresses at a 1- to 2-m depth on a soil-mantled slope (Anderson and Sitar, 1993).

Sampling

Soil samples were collected from the site at locations shown in Figure 4. Samples were collected from depths of 0.30 m to 0.80 m, a depth below the root zone and above any horizon where rock fragments were too large to be sampled (greater than ~15 mm). At the sampling locations the slope angle is about 22° to 28° and the principal stress ratio, $K = \sigma'_1/\sigma'_3$, is about 2.2 to 2.8 for the soil, when saturated (Anderson and Sitar, 1993).

Disturbance can lead to decreased stiffness and yield strength because of the destruction of the soil structure. Furthermore, disturbance can lead to a reduced tendency for pore pressure development during shear and an increase in undrained strength caused by densification. For these reasons, samples were gathered from trenches that were dug into the hillslope at a time when the soil was moist and least susceptible to disturbance (the typical moisture content was 20% to 30% of the dry weight). Samples for triaxial testing were collected in short sections of 71-mm diameter Shelby tubes. The samples were taken vertically from benches in the wall of the sampling trenches by the repetitive process of pushing a Shelby tube, lightly lubricated with vegetable oil, gently into the soil, then carving the soil away from the outside of the bottom of the tube, and then repeating the process. The Shelby tubes were sealed and stored upright until ready for testing. Undisturbed block samples, approximately 100 mm × 100 mm × 100 mm in size, were trimmed in the field, then encased in paraffin wax and brought to the laboratory where they were kept in a 100% humidity room until they were trimmed into the DS apparatus for testing. All of the DS sampling and some of the DS testing was performed by Lightwood (1988).

Direct shear (DS) test procedure

The DS specimens had a square surface area of 2,580 mm^2, an initial height of 25 mm, and they were tested in accordance with the standard of the American Society for Testing and Materials (ASTM, 1992, D-3080). The specimens were saturated from the bottom up, under a seating pressure of about 10 kPa, and once saturated, they were consolidated in one increment. The initial gap between the halves of the specimen holder was set at 0.6 mm, and the specimens were sheared at a rate of 1 × 10–4 mm/s. Displacement was accumulated in runs of approximately 15 mm, with travel to about 7.5 mm either side of center. At the end of each run the normal stress was removed from the specimen and the displacement direction was reversed. This procedure, whereby all loaded displacement occurred in the same direction, as suggested by Skempton (1985) for obtaining residual strength parameters. By this method, total loaded displacement (in one direction) of up to about 160 mm was achieved along the shear plane.

Constant-shear-drained (CSD) test procedure

The CSD specimens were assembled in a triaxial cell equipped with full-diameter porous stones at each end, and vertical filter-paper strips were placed around the perimeter to ensure rapid pore pressure equilibration and accurate pore pressure measurement. The specimens were enclosed in a latex membrane and saturated by a vacuum extraction/back pressure procedure; B values were generally 0.98 or higher before consolidation. The specimens were then consolidated at K values of 2.4 to 2.8 to restore the in-situ shear stress. Once consolidated, the stress-path loading was applied by reducing the effective confining stress at 0.67 kPa per hour while maintaining constant shear stress. This rate is slow enough to allow equilibration of pressures within the specimen, and it is similar to observed rates of pore-pressure increase caused by heavy rainfall at the site (Johnson and Sitar, 1990).

Anisotropically consolidated-undrained (ACU) test procedure

Specimens for ACU testing were prepared, saturated, and consolidated in the same manner as the CSD tests. B values of 0.99 were achieved before consolidation. Once consolidated to the desired stress state, drainage to the specimen was closed, the pore-pressure transducer was engaged, and the specimen was subjected to undrained loading at an axial strain rate of 0.1% per minute.

TEST RESULTS

Direct shear test results

The yield strength for the drained DS specimens is the peak strength, and it occurs after a displacement of about 2 to 5 mm. With further displacement the specimens soften, and the shear strength decreases to a nearly constant value after about 15 to 20 mm of cumulative displacement. The failure envelopes shown in Figure 5 give the peak (yield) and fully softened strength parameters (as defined by Skempton, 1985) of $\phi'_{pk} = 27.2°$, $c'_{pk} = 16.3$ kPa, and $\phi'_{fs} = 23.7°$, $c'_{fs} = 9.4$ kPa, where ϕ' is the effective stress angle of friction and c' is the effective stress cohesion intercept. In addition, to investigate continued strength loss and to determine the residual strength state, tests on two specimens were continued to 160 mm and 139 mm of displacement. The residual strength of these two specimens is indicated by points R1 and R2, respectively, in Figure 5.

Constant-shear-drained test results

The results of four typical CSD tests are shown in Figure 6 as plots of the mean effective stress, $p' = (\sigma'_1 + \sigma'_3)/2$, axial strain, and volumetric strain versus time. The value of maximum shear stress applied during consolidation (q_c) is shown at the top of the figure for each test, and it is held constant until failure occurs. In the early stage of the test, specimens expand in volume at a rate similar to that which would result simply from the reduction in confining pressure, and there is little or no axial strain. As p' reaches the critical value of p'_y, yield oc-

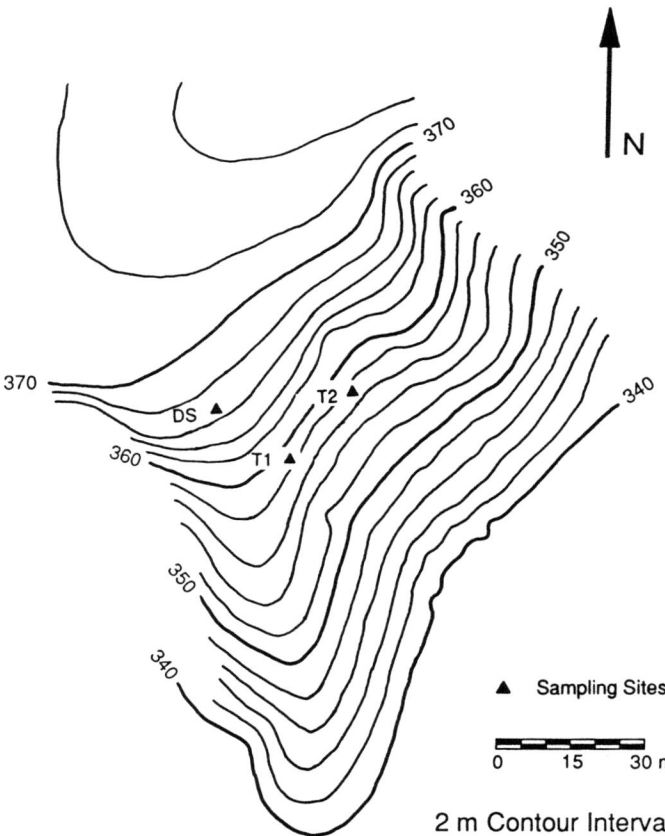

Figure 4. Topographic map of the Briones Park field research site. The three sampling locations are identified as DS, T1, and T2.

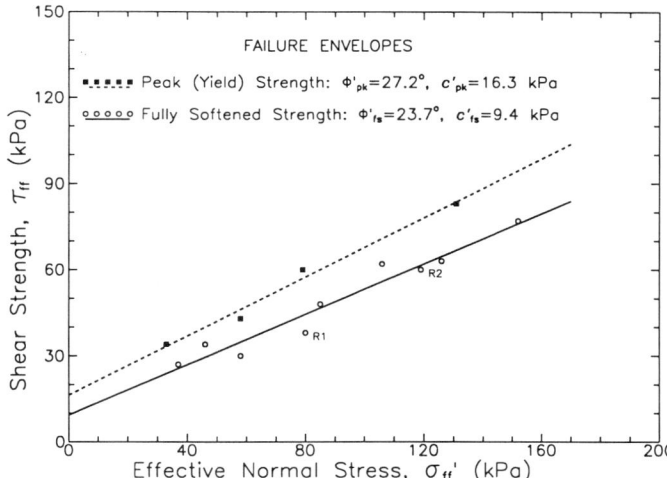

Figure 5. Peak (yield) and large-strain failure envelopes from the DS tests. Points R1 and R2 indicate shear strength after 160 mm and 139 mm cumulative displacement, respectively.

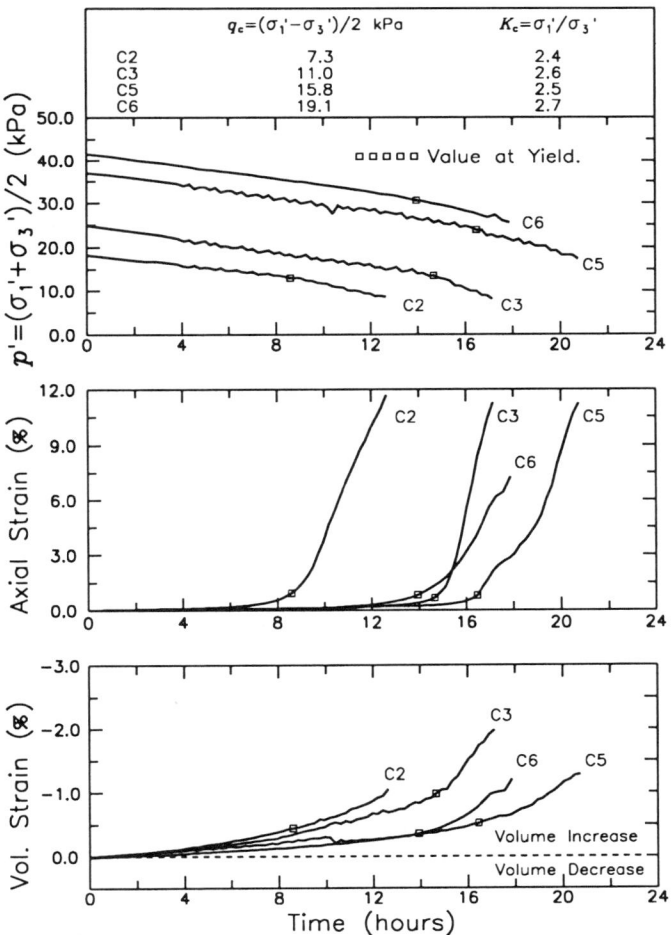

Figure 6. Typical CSD test results. The yield state is identified by the squares and the large-strain state is taken as the end of the test. K_c is the value of K at the end of consolidation and the beginning of the CSD stress path. The constant shear stress is given as q_c.

curs and the axial strain rate accelerates abruptly. The rate of volumetric strain also increases, still in dilation, and further reduction in p' causes the stress state to move down the failure envelope. Failure envelopes determined at yield and large-strain are shown in Figure 7. The yield strength parameters determined from the CSD tests are $\phi'_y = 36.7°$ and $c'_y = 0.7$ kPa, and large-strain strength parameters, determined at the end of the tests, are $\phi'_{ls} = 0.2$ kPa. The difference between the yield and large-strain parameters is not large, considering the variability in the strength of individual specimens, and the material appears to be essentially cohesionless at these low confining pressures. Similar tests at initial confining pressures up to $p' = 220$ kPa show that there is a curvature in the failure envelope and both ϕ'_y and ϕ'_{ls} decrease at higher confining pressures (Anderson and Sitar, 1993). The decrease in ϕ' is accompanied by an increase in the value of c' extrapolated from the data.

Anisotropically consolidated-undrained test results

A typical stress-strain plot of an ACU test is shown in Figure 8. Initially the anisotropically consolidated specimens are very stiff: yield occurs as the peak shear strength is reached at less than 0.5% axial strain. After yielding, the shear strength decreases to a nearly constant value at large strain. Pore pressure increases monotonically during shear, as indicated by the dotted line in Figure 8; approximately half the pore pressure is developed after the peak stress is applied, illustrating that the contractive tendency is a response to shear, rather than an increase in bulk mean stress. The effective stress paths and failure envelopes are shown in Figure 9; the yield strength parameters determined from the ACU tests are $\phi'_y = 30.8°$ and $c'_y = 1.7$ kPa, and the large-strain parameters are $\phi'_{ls} = 31.9°$ and $c'_{ls} = 23$ kPa. As observed in the CSD tests, the difference between the yield and large-strain parameters is small.

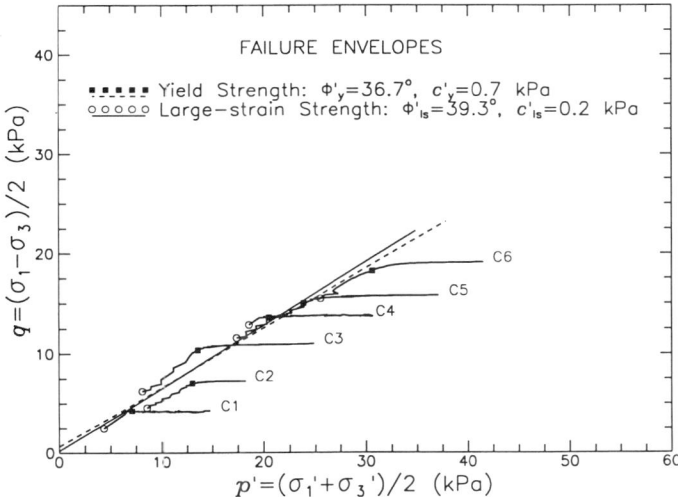

Figure 7. Yield and large-strain failure envelopes and individual stress paths from the CSD test.

DISCUSSION OF TEST RESULTS

Soil characterization procedures

The soil was characterized by conventional procedures such as grain-size analysis, plasticity tests, and even x-ray diffraction, but the observed strength and behavior could not have been predicted based only on these characterization procedures. For example, the ϕ' values found in the CSD tests are much higher than typical values for clayey soils containing smectite. This is apparently the result of the low confining pressure and, also, the rather granular nature of the native material caused by soil-particle aggregation. The siltstone fragments found in the soil are quite friable, especially near the surface. They are readily pulverized into silt and clay-size particles during a standard gradation analysis (ASTM, 1992, D-421), but they are cemented and appear as discrete sand or gravel-size particles in the field. The significance of these cemented fragments is not fully known, but it is clear that they impart some structure to the soil at a depth where many failures originate. In summary, the results of standard grain-size analysis might imply behavior of a clayey soil, whereas the actual behavior is much more like that of a granular material: the soil exhibits a high ϕ', and the potential for collapse and rapid flow failure, especially under the low confining stress of the shallow soil regolith.

Direct shear tests

There is a relatively large scatter in the DS results but there is little difficulty distinguishing peak strength values from fully softened strength values. The high cohesion intercept of the peak strength failure envelope is typical of an overconsolidated clayey soil (e.g., Skempton, 1964). This is because the overconsolidated soil, having once been subjected to a higher "preconsolidation" pressure, σ'_p, is in a dense state with respect to the current confining pressure, σ'_0. Thus, as a result of the stress history, shear strain is accompanied by dilation which, in turn, manifests itself as the cohesion intercept of the failure envelope.

Overconsolidation could be the result of erosion and the subsequent removal of overburden stress on the indurated siltstone; however, a shorter term and perhaps more probable cause of overconsolidation is the stress change caused by seasonal wetting and drying (e.g., d'Appolonia et al., 1967). In the partly saturated Briones soil, negative pore pressures of 80 kPa or more have been measured with tensiometers as the soil dries and becomes desiccated (Johnson and Sitar, 1987; Anderson and Sitar, 1993). Furthermore, Stark and Duncan (1991) found an apparent σ'_p of 144 kPa in a slope wash with a composition similar to the Briones soil and, comparing the desiccated stress state with σ'_0 for a soil at about 1 m depth at saturation, it is easy to arrive at overconsolidation ratios (OCR = σ'_p/σ'_0) of 10 or more. Thus, it appears that wetting and drying alone could produce a highly overconsolidated soil.

Although there is apparently a mechanism for overconsolidation, and thereby an explanation of the high c'_{pk}, the cohesion intercept could actually be unrelated to stress history; it could be the result of structure imparted by cemented bonds within the soil. In a residual soil, the bonds might be relict bonds of the parent material (Vaughan et al., 1988), while in a colluvial soil they might be newly formed bonds, the result of seasonal wetting and drying, or other ongoing processes (D'Appolonia et al., 1967, Allam and Sridharan, 1981, Mitchell and Sitar, 1982). Leroueil and Vaughan (1990) comment that there is no consistent terminology for types of cements and bonds and the resulting soil structure. They use the terms

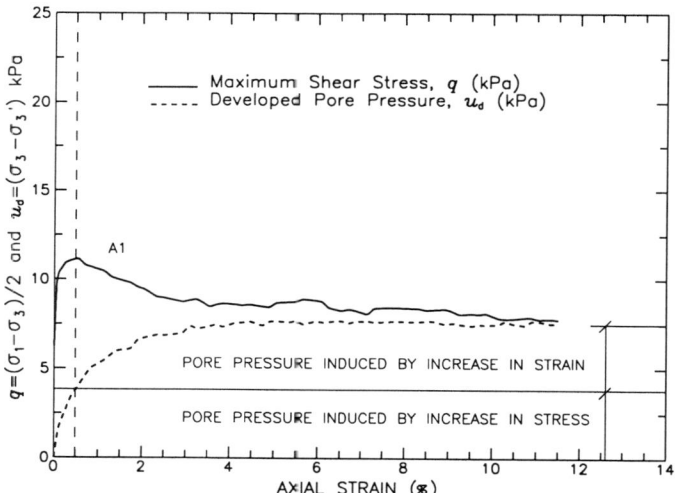

Figure 8. This typical ACU test result, from specimen A1, illustrates that peak strength is reached at very low strain, and that pore pressure increases and strength decreases with continued strain.

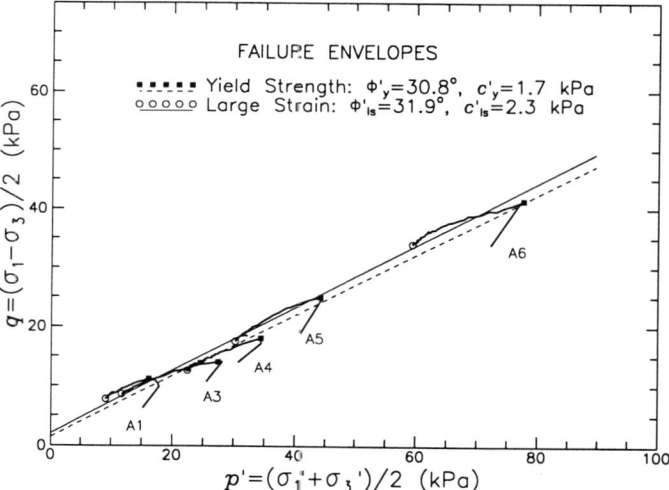

Figure 9. Yield and large-strain failure envelopes and individual stress paths from the ACU tests.

structured and nonstructured which have no implications of origin, but simply indicate whether or not the soil has structure as a result of cementation and bonding of the soil fabric. This terminology is adopted here so that we may interpret the results of each series of tests with reference to models for both structured and nonstructured soils.

The peak strength measured in the DS test can be explained by dilation of an overconsolidated nonstructured soil or by the contribution of bond strength in a structured soil. Unfortunately, neither model can explain the relatively high c'_{fs}, because in the fully softened state a nonstructured soil is no longer dilating, and the bonds of a structured soil would have been destroyed. It appears, therefore, that the shear-force measurement of this test is not being correctly interpreted in terms of the stresses in the soil. This is probably the result of the matrix-supported siltstone fragments which are stiffer than the clayey matrix and, therefore, dragged through the matrix on the forced failure plane, causing local shear and perhaps even developing part of the passive resistance of the matrix before shearing. Additionally, although these specimens were soaked prior to consolidation and testing, there was no way of ensuring saturation on the failure plane, and it is possible that, as a result of partial saturation, the effective confining stress on the failure plane is greater than that measured externally.

There was considerable difficulty in obtaining the residual strength values for R1 and R2 (Fig. 5), because during the accumulation of displacement by the process of multiple reversal of the shear box material was eroded from the failure surface. The residual strength of R1 falls somewhat below the fully softened envelope, but this is not the case for R2, which falls just below the envelope and well within the scatter of the fully softened strengths. Thus, there does not appear to be a consistent, or large, loss of strength from the fully softened to the residual state, when the cumulative displacement is obtained using the multiple reversal method. This seems to suggest that silt and soil particle aggregates of silt and sand size have a significant influence on the strength and behavior of the soil.

In summary, there was difficulty obtaining reliable strength parameters with the DS test, and a residual strength state could not be well defined for the Briones soil. Nevertheless, the relationship of the parameters does suggest overconsolidation and/or structure in intact specimens of undisturbed soil. Both the CSD and ACU tests, where the failure plane is not predetermined and saturated is ensured, defined strength envelopes with a higher ϕ' and a lower c' than the DS tests. Because of fewer uncertainties within the triaxial test specimens at failure, it is believed that the CSD and ACU parameters more accurately represent the strength of the soil, as discussed next.

Constant-shear-drained tests

The CSD results for this clayey soil are readily interpreted within the framework of critical state soil mechanics (Schofield and Wroth, 1968) and with reference to the critical state line (CSL). The critical state is a state of constant density which is defined by the void ratio (e), shear strength (q), and effective normal stress (p') that are attained at large strain. The relationship between these parameters at critical state is unique to a given soil and is the CSL, shown schematically in Figure 10. Strength at the critical state is essentially equivalent to that at the fully; softened state identified previously for the DS tests (Skempton, 1970) and to the large-strain state identified in the triaxial tests. At this state, the effects of initial density and structure are negligible, and strength is a function of the current density and confining pressure only. In Figure 10 the initial, yield, and large-strain states of a CSD test are shown schematically in relation to the CSL. The CSL can be used to predict that a continued decrease in p', beyond the value of p'_{cs}, the value at which the critical state is first reached, will cause a reduction in q and an increase in e as the state moves along the CSL. Hence, it is the soil behavior prior to attaining the critical state which must be evaluated in the CSD tests.

In the early stage of the CSD test, the principal stress ratio, $K = \sigma'_1/\sigma'_3$, is much less than K at yield (K_y) and, despite the fact that K increases through this stage of the test, the specimen swells at a rate very similar to that observed during unloading in an incremental loading consolidation test with a constant $K_c = 2.7$ (Anderson and Sitar, 1993). Swell generally continues at about this rate until the reduction of p' to a value of p'_y leads to an increase in K to K_y, and the soil yields. As the soil yields, the rates of axial and volumetric strain increase, then stabilize, and the continued reduction in p' leads to a reduction in q and an increase in e (negative volumetric strain), as would occur if the soil is already at the critical state and is moving along the CSL, or if the soil is dilating to reach the CSL. Thus, it appears that p'_y and p'_{cs} are reached at approxi-

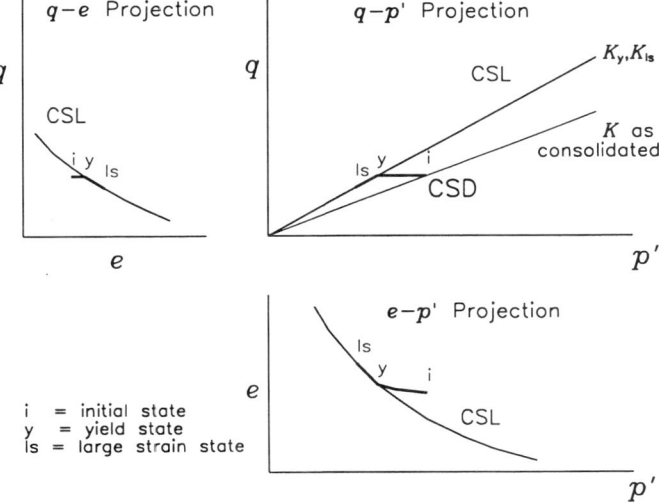

Figure 10. Schematic illustration of the critical state line (CSL) in three different projections. Projections of a schematic stress path for a typical CSD test are shown for reference.

mately the same stress state, an observation supported by the similarity of the yield and large-strain failure envelopes of the CSD tests.

Because of observed debris flow failures near the sampling sites, and the fact that debris flow failure is generally believed to result from rapid strength loss generated by excess pore pressure from shear of a saturated contractive soil (e.g., Johnson and Sitar, 1987; Ellen and Fleming, 1987; Ellen, 1988; Lee et al., 1988; Fleming et al., 1989; Anderson and Sitar, 1994), some evidence of collapse was anticipated in these tests. Collapse could have been indicated either by a contractive volumetric strain, provided there was ample time for drainage, or by a sudden loss of strength, if high positive pore pressure developed before drainage could occur. Collapse could only occur before the CSL is reached because a reduction of p' on the CSL is accompanied only by an increase in volume, as shown in Figure 10. Collapse has been observed in such a manner in very loose uniformly graded cohesionless soils in triaxial specimens (Riemer, 1992) and in a physical model of flow failure (Eckersley, 1990), but no collapse was observed in the CSD tests on the Briones soil.

The constant q of the CSD test precludes the observation of the peak strength, but yield strength is easily identified by the rapid increase in axial strain (Fig. 6). Many structured soils exhibit yield in shear, compression, and swell, and the locus of these yield states is a yield surface of the form shown in Figure 11 (Leroueil and Vaughan, 1990). Because of the strength imparted by the structure, the yield surface for yield in shear may lie outside the CSL, as shown in the figure. This has been observed in CSD-type tests by the Hong Kong Geotechnical Control Office (1982) and Bressani and Vaughan (1989), but the similarity of the yield and large-strain (critical state) failure envelopes for the Briones soil (Fig. 7) suggests that the influence of the soil structure is not particularly significant.

Anisotropically consolidated-undrained tests

In the ACU tests, as in the CSD tests, yield is reached at very small strain; however, the ACU specimens exhibit a tendency for contraction with shear. The tendency to contract causes an increase in pore pressure, a decrease effective stress, and, thereby, a decrease in strength. Thus, the yield strength is the peak strength, and the soil is susceptible to progressive failure and, potentially, flow failure if the peak strength is exceeded. Contractive behavior is typical of normally consolidated clay and loose granular soil, but not soil with an OCR as high as that estimated for the Briones soil. It is surprising then, that the same soil exhibits overconsolidated behavior in the drained DS test and normally consolidated behavior in the ACU test and it leads, once again, to consideration of the contribution of structure to the behavior of the soil. It is probable that the contractive behavior observed in the ACU tests is the result of collapse of a loose bonded structure, occurring as strain accumulates and bonding is destroyed. Only in this manner can the results of the ACU and DS tests on the same soil, with the same stress history, be reconciled. Nevertheless, it can be seen by comparing the large-strain strength, which is not affected by structure, with the yield strength, which is affected by structure, that the structure of this soil does not contribute significantly to the strength. In fact, in terms of the strength parameters for the ACU and CSD specimens, the contribution of structure is nearly lost in the scatter of the data, and the yield and large-strain failure envelopes are very similar (Figs. 7 and 9).

CONSIDERATIONS FOR STABILITY ANALYSIS

Drainage conditions

The shear strength used in most stability analyses of natural slopes is determined by the well-known equation:

$$\tau_{ff} = c' + \sigma'_{ff} \tan \phi'$$

where τ_{ff} and s'_{ff} are the shear stress and effective normal stress on the failure plane at failure. Here, we focus on the determination of the strength parameters, c' and ϕ', but the determination of σ'_{ff} is also not trivial and warrants some discussion. Initially the effective stress on a potential failure plane, σ'_{fc}, can be found by considering the geometry of the problem, the density of the materials involved, and the seepage conditions. The values of σ'_{fc} and σ'_{ff} differ by the value of pore pressure developed during the deformation required to mobilize the strength of the soil. This value is low when the shear rate is low and the hydraulic conductivity is high because a relatively low shear rate and a high hydraulic conduc-

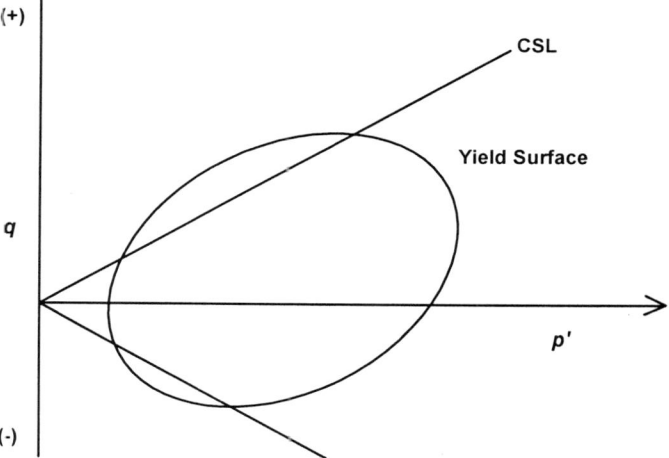

Figure 11. Yield surface for a bonded soil in $p'-q$ stress space. The surface identifies stress states at which yield will occur for this soil, and it shows that the yield surface may coincide with, or even lie above, the CSL (from Leroueil and Vaughan, 1990).

tivity means excess pore pressure is generated slowly and dissipated quickly, before a large pressure gradient is developed. At the extreme, the effective stress change is negligible, and the soil is said to be *drained*.

In contrast, a relatively rapid shear and a low hydraulic conductivity means an excess pore pressure will generate quickly and dissipate slowly, generally only after a large pressure gradient has been established. Thus, for a time, these soils are *undrained* and the σ'_{fc} decreases by the value of pore pressure developed. If the magnitude of pore pressure can be determined at failure, than σ'_{ff} can be evaluated and the effective stress shear strength equation can be used to determine the strength. Pore pressure parameters have been defined with this intent (Moore, 1970; Naylor, 1974; Janbu, 1977; Svano and Nordal, 1987), but at this time they are rarely used in analysis because they are difficult to measure with certainty (Wright and Duncan, 1987; Ladd, 1991). Thus, when the effective stress shear strength parameters are used for analysis, as in this paper, it is implicit that deformation is slow enough for drainage to occur under a very low hydraulic gradient, and $\sigma'_{ff} = \sigma'_{fc}$. When this is not the case, stability analysis must be based on the undrained strength of the soil (Sitar et al., 1992; Anderson and Sitar, 1994).

Modes of failure

The stress-strain behavior observed in the tests gives insight into the different modes of failure for the stress paths generated in the CSD and ACU tests. The CSD tests indicate that when the failure is simply the result of a gradual increase in pore pressure, the soil dilates as yield is reached. In the shearing zone, negative excess pore pressure develops as a result of the dilation, and the effective stress increases until the negative excess pore pressure is dissipated by water flowing into the expanding void volume. Because the negative excess pore pressure leads to higher effective stresses, the soil stiffens and strengthens until this equilibrium takes place. Thus, the failure rate is controlled by the hydraulic conductivity of the soil and the availability of water. This suggests a relatively slow failure rate along the CSD stress path.

The ACU tests indicate that the soil exhibits a very brittle and contractive failure mode when failure is induced by an increase in shear stress that is too rapid for pore pressure dissipation. The DS tests also indicate a brittle type failure, even when the shear stress is increased slowly and drainage does occur, but the positive excess pore pressure developed in the ACU tests indicates a truly contractive behavior. The strength loss measured in the tests is not great, but because of the positive excess pore pressure developed during shear, the shear strength at large-strain is less than the at-rest shear stress applied during consolidation in all but one of the specimens. Thus, when failure is initiated by an increase in shear stress at a rate too fast for excess pore pressure dissipation, the resulting strength loss could lead to a progressive and rapid failure, such as a debris flow.

Strength parameters

The determination of appropriate strength parameters is one of the most difficult aspects of an effective stress stability analysis. This is in part because c' and ϕ' are not simply properties of the soil, they are parameters which include intrinsic soil properties and testing conditions. Consequently, for accurate representation of the strength, testing must mimic field loading conditions. On this basis, the CSD and ACU triaxial tests are far superior to the DS test for determining c' and ϕ' for stability analysis of a soil regolith. A further consideration for analysis is how failure is defined in the tests and how the failure criterion is related to failure of the slope.

The peak strength, or yield strength, is the most difficult to determine because it includes that component of strength which is contributed by the structure of the soil and which is easily destroyed by sampling, transportation, setup, or simply consolidation of the soil during the test (e.g., Brand and Phillipson, 1985; Leroueil and Vaughan, 1990). Furthermore, the shape of the yield surface in Figure 11 indicates that ϕ'_y is not a constant, but a parameter which would tend to be higher for low values of σ'_{ff}, and this is indeed what the CSD and ACU tests show. Yield parameters are suitable for analysis of stability under drained-transient loading, such as the loading which results from the transient hydrologic response to rainfall. Using yield parameters from the CSD tests, measured pore pressure (Johnson and Sitar, 1990), and infinite slope stability analyses, Anderson and Sitar (1993) found a minimum factor of safety (FS) against sliding of 1.06 at the location of Nest 3 during the same storm that generated the debris flow shown in Figure 2. Nest 3 is a group of nested tensiometers located at an elevation of 357 m between sampling trenches T1 and T2 (see Fig. 4). The analyses were performed locally at each nest because it has been shown that the FS varies appreciably along the slope (Johnson and Sitar, 1989); the results of the analyses are summarized in Table 1.

At some time during this storm, local failure occurred on a slope adjacent to the research site and at a location roughly equivalent to Nest 3, where the minimum FS = 1.06. The observation of failure appears to confirm that the FS should have been close to 1.0 at some point during the storm and that parameters determined by careful CSD testing are accurate for predicting the initiation of shallow rainfall-induced landslides and debris flows (Sitar et al., 1992). Of course, the yield parameters say nothing about the nature of post-yield deformation, or whether the failure develops into a debris flow. For that, it is necessary to consider the stress-strain behavior, the relationship between yield and large-strain strength and, once again, the drainage conditions.

The large-strain, or critical state, strength parameters are the most readily determined because they are relatively insensitive to any original structure or the nature of the loading. According to the critical state theory, the cohesion intercept is zero, ϕ' is constant, and the strength is simply a linear function

TABLE 1. FACTORS OF SAFETY CALCULATED WITH THE CSD YIELD STRENGTH PARAMETERS, A TOTAL UNIT WEIGHT OF 18.6 kN/m³, AND PORE PRESSURE VALUES

SLOPE PARAMETERS					
Nest	0	1	2	3	4
Reference Location	Top of slope	8 m downslope	16 m downslope	26 m downslope	33 m downslope
Slope Angle	19	24	34	29	24
Pore Pressure (kPa)*	2.00	0.00	0.50	4.60	5.00
Depth (m)	0.50	0.70	1.50	1.20	1.20
FACTOR OF SAFETY	**1.89**	**1.82**	**1.13**	**1.06**	**1.31**

*From Johnson and Sitar, 1990. Note that the nest numbers refer to locations on the slope where pore pressure was measured.

of σ'_{ff}. The CSD and ACU large-strain parameters are in close agreement with this theory, with the exception that there is a curvature in the large-strain failure envelope, similar to that found at yield, and there is a small cohesion intercept. The intercept could be the result of linear extrapolation of the curved failure envelope, or simply the result of scatter in the data. The large-strain parameters are ideally suited for long-term stability analysis when peak strength cannot be relied upon because creep or progressive failure may cause the peak strength to be destroyed or locally exceeded. Hence, they are often suitable for stability analysis of natural slopes; however, they cannot be used directly to determine the FS against undrained failure because the pore pressure generated during loading causes an unknown change in effective stress.

Finally, despite the high clay content of the Briones soil, there appears to be no well-defined residual-strength state. Even when displacement on a certain failure surface may be quite large, the strength of the soil is essentially equal to its fully softened value and is well represented by the large-strain parameters.

CONCLUSIONS

Results from tests on clayey soil regolith from the Briones site show that different effective strength parameters and different modes of failure are found using three alternate test procedures. This indicates the need for testing on a stress path and stress level close to that which exists in the field, or somehow correcting the strength parameters determined under different stress conditions. More importantly, the DS tests led to uncertain strength parameters, whereas the strength parameters determined from triaxial testing at the in-situ stress level were very consistent. A slope stability analysis using the results of the CSD and triaxial tests predicts a factor of safety near 1 during a storm which initiated a failure on a nearby slope.

Strength and behavior are shown not to depend on the stress history imposed by wetting and drying, as evidenced by the contractive deformation of the ACU tests. Rather, it appears that there is a structure to the soil which, while not strong, leads to a brittle stress-strain response accompanied by contraction when the shear stress is being increased (ACU) and dilation when the confining stress is being decreased (CSD). These different modes of failure in the test specimens suggest different modes of failure for different stress paths in the field. The field stress path that results from infiltration (CSD) leads to a failure rate limited by the hydraulic conductivity of the soil. A failure initiated by increase in shear, particularly if faster than drainage can occur, leads to softening and may well lead to flow failure.

Soil classification can be misleading for shallow residual and colluvial soil regoliths. Desiccation cracks, clay-size content, smectite presence, and a history of wetting and drying would suggest that the Briones soil should behave as an overconsolidated smectitic soil, with a low ϕ' and a dilative tendency in compressional shear. However, because of soil structure and the low in-situ stresses, these expectations are not realized. In fact, ϕ' is, in general, high, and both the yield and large-strain failure envelopes for the Briones soil are found to be curvilinear, with ϕ' increasing and c' decreasing at lower confining pressures.

ACKNOWLEDGMENTS

This research was supported in part by National Science Foundation P.Y.I. Grant No. CEE-8352147, a Jane Lewis Fellowship, and a Mineral Institute Fellowship.

REFERENCES CITED

Allam, M. M., and Sridharan, A., 1981, Effect of wetting and drying on shear strength: Journal of the Geotechnical Engineering Division, American Society of Civil Engineers, v. 107, p. 421–438.

Anderson, S. A., and Sitar, N., 1994, Procedures for the analysis of the mobilization of debris flows, in Proceedings, International Conference on Soil Mechanics and Foundation Engineering, 13th, New Delhi, India: Oxford & IBH, v. 1, p. 255–258.

Anderson, S. A., and Sitar, N., 1993, Debris flow initiation: The role of hydrologic response and soil behavior: Berkeley, University of California, Geotechnical Engineering Report No. UCB/GT/93-02, 220 p.

Brand, E. W., and Phillipson, H. B., 1985, Review of the international practice of the sampling and testing of residual soils, *in* Sampling and testing of residual soils: A review of international practice: Hong Kong, Scorpion Press, p. 194.

Bressani, L. A., and Vaughan, P. R., 1989, Damage to soil structure during triaxial testing, *in* Proceedings, International Conference on Soil Mechanics and Foundation Engineering, 12th, Rio de Janeiro: Rotterdam, A. A. Balkema, v. 1, p. 17–20.

D'Appolonia, E., Alperstein, R., and D'Appolonia, D. J., 1967, Behavior of a colluvial slope: Journal of the Soil Mechanics and Foundations Division, American Society of Civil Engineers, v. 93, p. 447–473.

Eckersley, J. D., 1990, Instrumented laboratory flowslides: Géotechnique, v. 40, p. 489–502.

Ellen, S. D., 1988, Description and mechanics of soil slip/debris flows in the storm, *in* Ellen, S. D., and Wieczorek, G. F., eds., Landslides, floods, and marine effects of the storm of January 3–5, 1982, in the San Francisco Bay region, California: U.S. Geological Survey Professional Paper 1434, p. 63–112.

Ellen, S. D., and Fleming, R. W., 1987, Mobilization of debris flows from soil slips, San Francisco Bay region, California, *in* Costa, J., and Wieczorek, G., eds., Debris flows/avalanches: Process, recognition, and mitigation: Boulder, Colorado, Geological Society of America Reviews in Engineering Geology, Volume VII, p. 31–40.

Fleming, R. W., Ellen, S. D., and Algus, M. A., 1989, Transformation of dilative and contractive landslide debris into debris flows—An example from Marin County, California, *in* Johnson, A. M., Burnham, C. W., Allen, C. R., and Muehlberger, W., eds., Richard H. Jahns Memorial Volume: Engineering Geology, v. 27, p. 201–223.

Hong Kong Geotechnical Control Office, 1982, Mid-levels study: Hong Kong, Geotechnical Control Office report on geology, hydrology, and soil properties.

Howard, T. R., Baldwin, J. E., II, and Donley, H. F., 1988, Landslides in Pacifica, California, caused by the storm, *in* Ellen, S. D., and Wieczorek, G. F., eds., Landslides, floods, and marine effects of the storm of January 3–5, 1982, in the San Francisco Bay region, California: U.S. Geological Survey Professional Paper 1434, p. 163–184.

Janbu, N., 1977, Slopes and excavations in normally and lightly overconsolidated clays: SOA Report, *in* Proceedings, International Conference on Soil Mechanics and Foundation Engineering, 9th, Tokyo: Rotterdam, Balkema, v. 2, p. 549–566.

Johnson, K. A., and Sitar, N., 1987, Debris flow initiation: An investigation of mechanisms: Berkeley, University of California Geotechnical Engineering Report No. UCB/GT/87-02, 179 p.

Johnson, K. A., and Sitar, N., 1989, Significance of transient pore pressures and local slope conditions in debris flow initiation, *in* Proceedings, International Conference on Soil Mechanics and Foundation Engineering, 12th, Rio de Janeiro: Rotterdam, A. A. Balkema, v. 2, p. 1619–1622.

Johnson, K. A., and Sitar, N., 1990, Hydrologic conditions leading to debris flow initiation: Canadian Geotechnical Journal, v. 27, p. 789–801.

Ladd, C. C., 1991, Stability evaluation during staged construction: Journal of the Geotechnical Engineering Division, American Society of Civil Engineers, v. 117, p. 540–608.

Lee, H. J., Ellen, S. D., and Kayen, R. E., 1988, Predicting transformation of shallow landslides into high-speed debris flows, *in* Proceedings, International Symposium on Lakeslides, 5th, Lausanne: Rotterdam, A. A. Balkema, v. 1, p. 713–718.

Leroueil, S., and Vaughan, P. R., 1990, The general and congruent effects of structure in natural soils and weak rocks: Géotechnique, v. 40, p. 467–488.

Lightwood, G. T., 1988, The residual strength characteristics of a weathered residual clay soil [M. Eng. report]: Berkeley, University of California, 35 p.

Mitchell, J. K., and Sitar, N., 1982, Engineering properties of tropical residual soils, *in* Proceedings, American Society of Civil Engineers Specialty Conference: Engineering and Construction in Tropical and Residual Soils, Honolulu: New York, ASCE, p. 30–57.

Moore, P. J., 1970, The factor of safety against undrained failure of a slope: Soils and foundations, v. 10, p. 81–91.

Naylor, D. J., 1974, Stresses in nearly incompressible materials by finite elements with application to the calculation of excess pore pressures: International Journal for Numerical Methods in Engineering, v. 8, p. 443–460.

Rantz, S. E., 1971, Precipitation depth-duration-frequency relations for the San Francisco Bay region, California with isohyetal map of San Francisco Bay region, California, showing mean annual precipitation: U.S. Geological Survey San Francisco Bay Region Environment and Resources Planning Study, Basic Data Contribution 25, 23 p.

Riemer, M., 1992, The effects of test conditions on constitutive behavior of loose sands subject to monotonic loading [Ph.D. dissert.]: Berkeley, University of California, p. 181–203.

Schofield, A. N., and Wroth, C. P., 1968, Critical state soil mechanics: London, McGraw-Hill, 310 p.

Sitar, N., Anderson, S. A., and Johnson, K. J., 1992, Conditions leading to the initiation of rainfall-induced debris flows, *in* Proceedings, American Society of Civil Engineers Geotechnical Division Specialty Conference on the Stability and Performance of Slopes and Embankments—II, Berkeley, California: New York, ASCE, p. 834–848.

Skempton, A. W., 1953, The colloidal activity of clay, *in* Proceedings, International Conference on Soil Mechanics and Foundation Engineering ICOSOMEF, 3rd: Zurich, Organizing Committee ICOSOMEF, v. 1, p. 57–61.

Skempton, A. W., 1964, Long-term stability of clay slopes: Géotechnique, v. 14, p. 77–101.

Skempton, A. W., 1970, First-time slides in overconsolidated clays: Géotechnique, v. 20, p. 320–324.

Skempton, A. W., 1985, Residual strength of clays in landslides, folded strata and the laboratory: Géotechnique, v. 35, p. 3–18.

Stark, T. D., and Duncan, J. M., 1991, Mechanisms of strength loss in stiff clays: Journal of the Geotechnical Engineering Division, American Society of Civil Engineering, v. 117, p. 139–153.

Svano, G., and Nordal, S., 1987, Undrained effective stress stability analysis, *in* Proceedings, European Conference on Soil Mechanics and Foundation Engineering, 9th: Rotterdam, A. A. Balkema, v. 6, p. 871–875.

Vaughan, P. R., Maccarini, M., and Mokhtar, S. M., 1988, Indexing the engineering properties of residual soil: Quarterly Journal of Engineering Geology, v. 21, p. 69–84.

Wagner, J. R., 1978, Late Cenozoic history of the Coast Ranges east of San Francisco Bay [Ph.D. dissert.]: Berkeley, University of California, p. 17–130.

Wright, S. G., and Duncan, J. M., 1987, An examination of slope stability computation procedures for sudden drawdown: Vicksburg, Geotechnical Laboratory, U.S. Army Engineer Waterways Experiment Station, Miscellaneous Paper GL-87-25.

MANUSCRIPT ACCEPTED BY THE SOCIETY MAY 20, 1994

Effect of test method and procedure on measurements of residual shear strength of bentonite from the Portuguese Bend landslide

Stephen M. Watry*
William Lettis & Associates, 1000 Broadway, Suite 612, Oakland, California 94903
Perry L. Ehlig
Department of Geological Sciences, California State University, Los Angeles, California 90032 and City Geologist, City of Rancho Palos Verdes

ABSTRACT

Palos Verdes Peninsula contains large landslides in areas that are otherwise desirable for residential development. The landslides can be developed providing the factor of safety is at least 1.5 or can be raised to 1.5 during development. The greatest uncertainty in the calculated factor of safety is the residual shear strength of bentonite that forms the bases of slides. The use of too high a shear strength inflates the calculated factor of safety and can result in landslide failure after development. Tests by various investigators yield a wide range of residual shear strengths for bentonite samples from Palos Verdes Peninsula. Most residual friction angles (ϕ_r) range between 6° and 14° with one lower value (3.5°) and several higher values reported. Residual cohesions (C_r) range mostly between 0 and 36 kPa (0 to 750 lb/ft^2). Data for various bentonite samples from Palos Verdes Peninsula indicate they have similar compositions and Atterberg limits. Calcium montmorillonite is the principal clay mineral; the liquid limit is generally between 80% and 110% with the plasticity index between 40% and 70%. The data suggest that much of the reported variation in residual shear strength is the result of differences in sample preparation, testing methods, and interpretation of results rather than true differences in strength. Bulk samples of bentonite from the base of the Portuguese Bend landslide were remolded and tested by conventional direct shear, long sample direct shear, and ring shear devices to determine the effect of testing method on residual shear strength measurements. The three devices gave similar results—for conventional direct shear, ϕ_r = 6.9° and C_r = 33.0 kPa (690 lb/ft^2); for long sample shear, ϕ_r = 6.8° and C_r = 23.8 kPa (497 lb/ft^2); for ring shear, ϕ_r = 6.7° and C_r = 7.2 kPa (150 lb/ft^2). Back calculations indicate the ring shear results most closely simulate the residual strength along the base of the Portuguese Bend landslide. For all those devices, the residual shear envelope is a curve whose slope reaches an asymptote near a confining pressure of 200 kPa (approximately 4,000 lb/ft^2). Reported ϕ angles are commonly too high because samples were not tested at sufficient confining pressure to define the asymptote and a best fit straight line was drawn through the data points.

---

*Present address: 820 Idylberry Road, San Rafael, California 94903.

Watry, S. M., and Ehlig, P. L., 1995, Effect of test method and procedure on measurements of residual shear strength of bentonite from the Portuguese Bend landslide, in Haneberg, W. C., and Anderson, S. A., eds., Clay and Shale Slope Instability: Boulder, Colorado, Geological Society of America Reviews in Engineering Geology, v. X.

INTRODUCTION

Its mild climate and aesthetic views makes the Palos Verdes Peninsula one of the most sought after residential locations in the greater Los Angeles area of southern California (Fig. 1). However, the peninsula's development is impacted by large prehistoric landslides, including four that have reactivated since 1956. As is true in most other areas, cities within the peninsula permit land development providing geotechnical studies can demonstrate that the area will have a factor of safety of at least 1.5 (available shear strength divided by mobilized shear strength) upon completion of the development. Thus, the decision of whether or not a landslide can be developed is dependent upon geotechnical studies culminating in stability calculations.

Calculations of the factor of safety of an existing landslide are dependent upon five factors: (1) the configuration of the slide's basal slip surface, (2) the distribution of mass within the slide, (3) the magnitude of fluid pressure along the slide's base, (4) the method used to calculate the factor of safety, and (5) the residual shear strength assigned to material along the slip surface. Field studies can determine the first three factors with satisfactory precision.

Methods used to calculate the factor of safety are standardized and subject to review. There are also standard procedures for laboratory testing of the residual shear strength of material along slip surfaces. But, commonly used procedures may not yield reliable results when testing sheared bentonite obtained from slip surfaces of Palos Verdes landslides.

The large landslides on Palos Verdes Peninsula have failed on bentonite beds in the marine Miocene Monterey Formation that forms the near-surface bedrock. Reported values for the residual shear strengths of bentonite samples from landslides within the peninsula have a broad range of values. Merriam (1960, p. 149) stated that the results of laboratory tests on samples taken from the Portuguese Bend landslide are "both controversial and apparently contradictory and hence are best reported upon by the soils engineers who have done the work." Such an approach has led to unacceptable results. For example, a study of the Flying Triangle landslide indicated its factor of safety exceeded 1.5; however, it started moving while homes were under construction. Many residual shear strengths obtained by laboratory testing appear to be too high based on strengths obtained by back calculations in which sliding is assumed to have occurred when the factor of safety was at or below 1.0.

The senior author undertook this study to determine (1) the extent to which disparities in residual shear strengths of bentonite samples taken from different geographic and stratigraphic locations on the peninsula can be explained in terms of differences in physical properties of the samples, (2) the cause of disparities not attributable to differences in physical properties, (3) the most reliable method of testing the residual shear strength of bentonite, and (4) the most representative value for the residual shear strength of bentonite along the base of the Portuguese Bend landslide. To assess disparities in residual shear strength values caused by differences in physical properties of bentonite samples, tables were compiled listing reported Atterberg limits and residual shear strength parameters. Information on clay fraction, mineralogy, and chemical properties of Palos Verdes bentonite samples was also checked for differences. Results from this study were compared with published correlations between residual friction angle and liquid limit, plasticity index, and percent clay fraction to determine if the results are consistent.

To assess disparities in residual shear strength values caused by differences in testing procedures, technical reports were reviewed for information on shear test set ups and procedures used during testing of Palos Verdes bentonite samples. The procedures and assumptions used to define residual shear strength parameters from the plotted shear stress versus effective normal stress were also reviewed.

To assess potential shear strength disparities resulting from use of different shear testing equipment, three different devices were used to determine residual shear strength parameters for a uniform sample of remolded bentonite obtained from the toe of the Portuguese Bend landslide. The three devices used were (1) a conventional direct shear device requiring reversals of shear direction to have enough deformation to reach a residual shear strength; (2) a direct shear device modified to shear a long sample so that reversals of shear direction would not have to be made; and (3) a torsional ring shear device that can shear a sample continuously in one direction by

Figure 1. Palos Verdes Peninsula vicinity map.

rotation. The residual shear strength values obtained from the three devices were plotted and compared. The residual shear strength parameters determined by each device were used in a stability analysis of the Portuguese Bend landslide in order to determine which test device yielded strength values most representative of the basal slip surface.

GEOLOGIC SETTING

The Palos Verdes Peninsula is a northwest-trending domical ridge, 14.5 km (9 mi) long and up to 8 km (5 mi) wide located 32 km (20 mi) south of downtown Los Angeles, California (Fig. 1). Its crest has gently rolling topography at elevations ranging from 451 to 335 m (1,480 to 1,100 ft) above sea level. Below this upland, remnants of a flight of Pleistocene marine terraces ring the peninsula and demonstrate that it was an island during most of its geomorphic evolution. The peninsula's shape reflects its Quaternary uplift as a doubly plunging anticline within the hanging wall of the southwest dipping Palos Verdes fault. The Mesozoic Catalina Schist forms the core of the Palos Verdes anticline. In early middle Miocene, the schist formed an irregular, southwest-sloping sea floor on which the Monterey Formation was deposited. The Monterey Formation is divided into the middle Miocene Altamira Shale, the late middle to early upper Miocene Valmonte Diatomite, and the late upper Miocene to early Pliocene Malaga Mudstone. The Altamira Shale is further divided into the tuffaceous, cherty, and phosphatic members. The tuffaceous member is of special interest because it contains weak bentonite beds that host the slip surfaces of most large landslides within the peninsula, including the Portuguese Bend landslide complex.

In the vicinity of the Portuguese Bend landslide complex, the Portuguese Tuff is the only easily recognized stratigraphic unit within the tuffaceous member of the Altamira Shale. The Tuff occurs about 137 m (450 ft) below the top of the tuffaceous member and consists of about 18 m (60 ft) of irregularly bedded tuff that appears to have been deposited by turbidity currents during a single eruptive event. Most of it was converted to a nearly pure waxy bentonite. The Portuguese Tuff functions as a zone of low shear strength and as an aquiclude.

During Pleistocene uplift, wave erosion removed the upper part of the Monterey Formation from most of the Palos Verdes Peninsula. Erosion was greatest along the peninsula's seaward-facing south flank. As a result, the tuffaceous member crops out across the entire south flank. Bedding within the tuffaceous member dips seaward subparallel to the slope of the ground surface. This creates conditions suitable for translational sliding along weak bentonite beds. In addition, wave erosion cut notches into the south flank creating steep wave-cut cliffs in which beds were inclined unsupported out of slope, further favoring formation of landslides. For a comprehensive coverage of the geology of the Palos Verdes Peninsula refer to Woodring et al. (1946). Rowell (1982) presents a detailed chronostratigraphy of the Monterey Formation; Conrad and Ehlig (1983) provide details on the lithostratigraphy of the Monterey Formation; and Bryant (1987) provides information on the marine terraces and Quaternary tectonics of the peninsula.

PORTUGUESE BEND LANDSLIDE

The active Portuguese Bend landslide is located in the southeast portion of a 5 km^2 (2 mi^2) late Pleistocene landslide complex which has been termed the Portuguese Bend landslide complex (Fig. 2). The active Portuguese Bend landslide is bounded on the west and the east by other reactivated portions of the Portuguese Bend landslide complex, known as the Abalone Cove and Klondike Canyon landslides.

The Portuguese Bend landslide covers 105 ha (260 acres). The thickness of the active landslide varies from less than a few meters to over 60 m (200 ft) but is generally between 30 to 45 m (100 to 150 ft). The deepest slip surfaces of the landslide are located in close proximity to the Portuguese Tuff.

The Portuguese Bend landslide became active in mid-August 1956 shortly following the placement of a fill embankment for the extension of a roadway across the head of the landslide (Ehlig, 1992). The fill embankment was located at the northeast corner of the ancient and active landslide masses and was up to 21 m (70 ft) high. The first sign of instability was the formation of cracks in a recently built concrete culvert just east of the embankment (Vonder Linden, 1989). Within a month, the cracks propagated westward and seaward to define that part of the Portuguese Bend landslide which failed toward Portuguese Bend. During the next several months, movement progressed westward to include the segment of the Portuguese Bend landslide that failed toward the cove between Inspiration Point and Portuguese Point. The landslide continued to widen to the north and west until the spring of 1957 at which time it attained the approximate dimensions that exist at present. The fully developed slide encompassed about 160 homes. Most of the homes were destroyed by the slide movement, a few were moved to adjacent areas, and a few remain in the slide area.

The rate of movement measured in the first weeks after reactivation of the slide was on the order of 7.5 to 10 cm (3 to 4 in) per day (Vonder Linden, 1989). By October 1956, the rate of movement had slowed to less than 2. cm (1.0 in) per day (Vonder Linden and Lindvall, 1982). Studies to obtain information about the landslide commenced shortly after reactivation. Merriam (1960) reported that 62 borings, using several types of drilling equipment, were drilled in the landslide. Most of the borings were drilled within about three months of reactivation of the landslide, but a few were drilled up to about a year later. About two-thirds of the borings penetrated the basal slip surface of the active landslide (Merriam, 1960). One of the early landslide studies suggested that the landslide may be halted by installing shear pins (reinforced concrete cylinders) through the toe of the landslide. Shear pins were installed over a period of several months in 1957. The shear pins were 1.2 m (4 ft) in diameter, 6 m (20 ft) in length and embedded 3 m (10

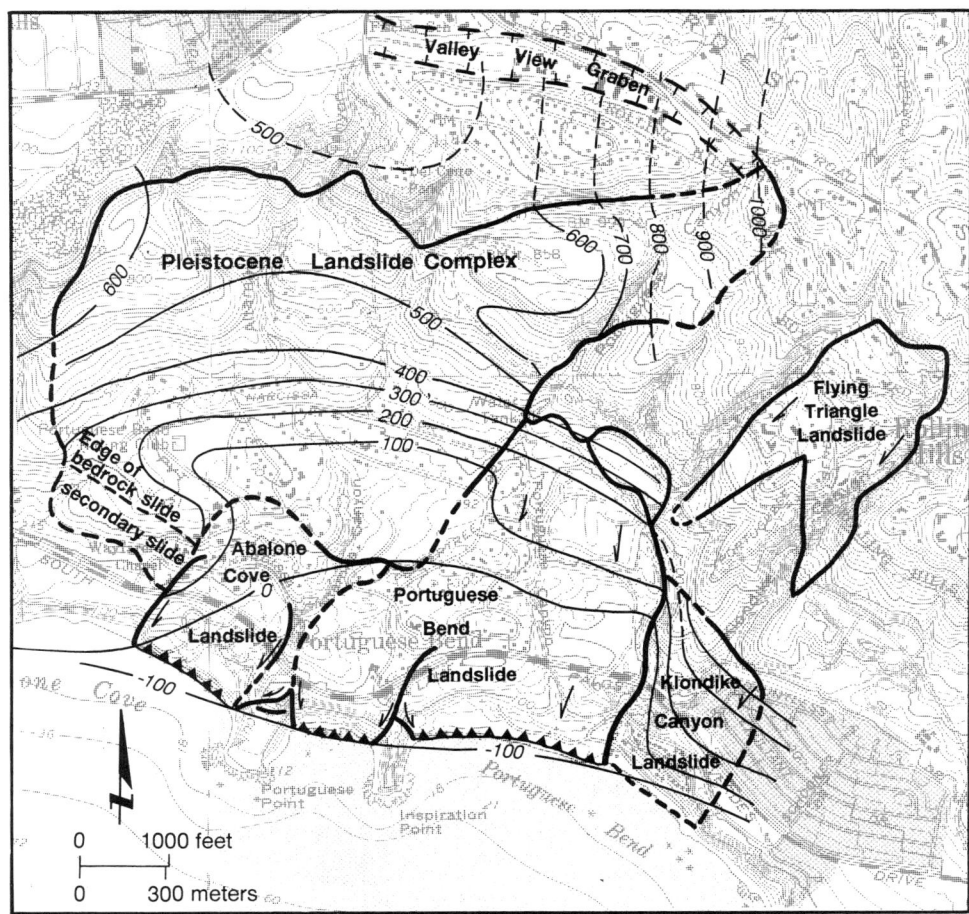

Figure 2. Map showing the Pleistocene landslide complex and the Portuguese Bend, Abalone Cove, Klondike Canyon, and Flying Triangle landslides. Structure contours in feet are on the base of the Portuguese Tuff. From Ehlig (1992); reproduced with permission of Star Publishing Company and the Association of Engineering Geologists.

ft) below the basal slip surface. The first two shear pins installed were instrumented with strain gauges to measure the driving force of the landslide. The velocity of the landslide mass dropped 15% after installation of the first two shear pins (Vonder Linden, 1989) prompting the installation of six more shear pins in an attempt to further decrease the rate of slide movement. Finally an additional fifteen shear pins were installed. The rate of movement decreased 50% after installation of the final fifteen shear pins and remained at this rate for a period of about five months (Vonder Linden, 1989). The rate of movement then swiftly increased until reaching approximately the rate measured before installation of the shear pins. The shear pins failed by plucking, tilting, or shearing.

Between the years 1960 through 1977, with the exception of 1969, the average daily rate of movement varied between 0.75 and 1.0 cm (0.3 to 0.4 in) per day (Ehlig, 1986). In 1969 the average daily rate of movement increased to 1.75 cm (0.7 in) per day in response to heavy rainfall. The average daily rate for the period between 1978 and 1986 was near 2.5 cm (1.0 in) per day except for the year 1983 where the average daily rate was 3.8 cm (1.5 in) per day. The higher rates observed during this period were the result of heavy seasonal rainfall during several years. A comparison of the location of monuments as surveyed in March 1990 to those locations surveyed in November 1991 indicates that the portion of the Portuguese Bend landslide that is approximately north of Palos Verdes Drive South is moving at an average daily rate of 0.13 cm (0.05 in) per day or less while that portion of the landslide approximately south of Palos Verdes Drive South is moving at an average daily rate of 0.5 cm (0.2 in). Distension of the landslide mass by listric normal faults, especially in the extreme south portion of the landslide mass has been noted by Ehlig (1987) just north of the present shoreline. The distension of the landslide mass results in a lowering of the effective normal load on the slip surface. Total horizontal displacement on the active landslide since 1956 varies between less than 20 m

(65 ft) to greater than 250 m (820 ft) near the beach. The greatest amount of movement has occurred in the southern and eastern portions of the slide.

Information about the basal slip surface of the landslide has been directly obtained from borings and interpolated from observations of the slide geomorphology and fissure patterns. Vonder Linden (1972, 1989) and Ehlig (1992) have prepared similar structural contour maps of the active slip surface. Figure 3 is the structural contour map prepared by Ehlig. Cross sections through the Portuguese Bend landslide are shown on Figure 4. Ehlig (1992) divided the landslide into five subslides in order to assess stabilization measures. The boundaries of the subslides are controlled by the basal slip surface geometry and the lateral boundary conditions. The basal slip surface of the active landslide is smooth but contains large-scale undulations which are approximately perpendicular to the direction of movement. Near the head of the Portuguese Bend landslide, the basal slip surface consists of a few centimeters of sheared bentonite over shale which dips 15° to 25° to the south (Ehlig, 1987). Proceeding seaward the active basal shear surface flattens out to an average dip of about 7°, then drops into a trough located just north of Palos Verdes Drive South, then rises slightly again before dropping into depressions south of Palos Verdes Drive South and finally rises sharply to toe out at the shoreline.

In the eastern portion of the active slide, the basal slip surface rests on in-place bedrock and is essentially parallel to bedding in the bedrock. The western half of the active landslide is underlain by bedrock with the dip of the bedding being nearly the same as the basal slip surface (Vonder Linden, 1989). The western half of the active landslide is underlain by inactive ancient landslide deposits (Vonder Linden, 1989). The head of the landslide is defined by a structural ramp in the underlying intact bedrock which brings the basal slip surface to the ground surface. Along most of the east edge of the landslide, bedding is upturned by a west-facing monoclinal flexure. As a result, the basal slip plane intersects the ground surface and forms a sharply defined trace that extends from the head to the toe of the landslide. The western margin of the landslide is poorly defined. Displacement dies out across a

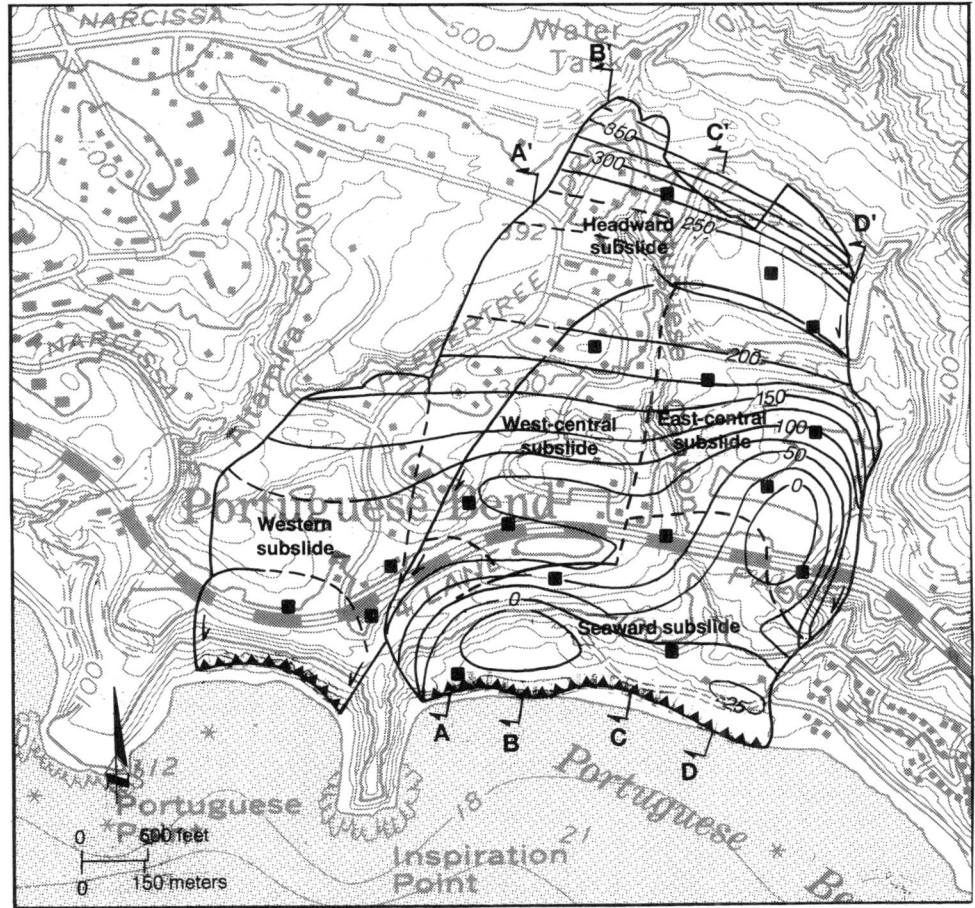

Figure 3. Map of Portuguese Bend landslide showing boundaries of subslides and structure contours in feet on base of slide. From Ehlig (1992); reproduced with permission of Star Publishing Company and the Association of Engineering Geologists.

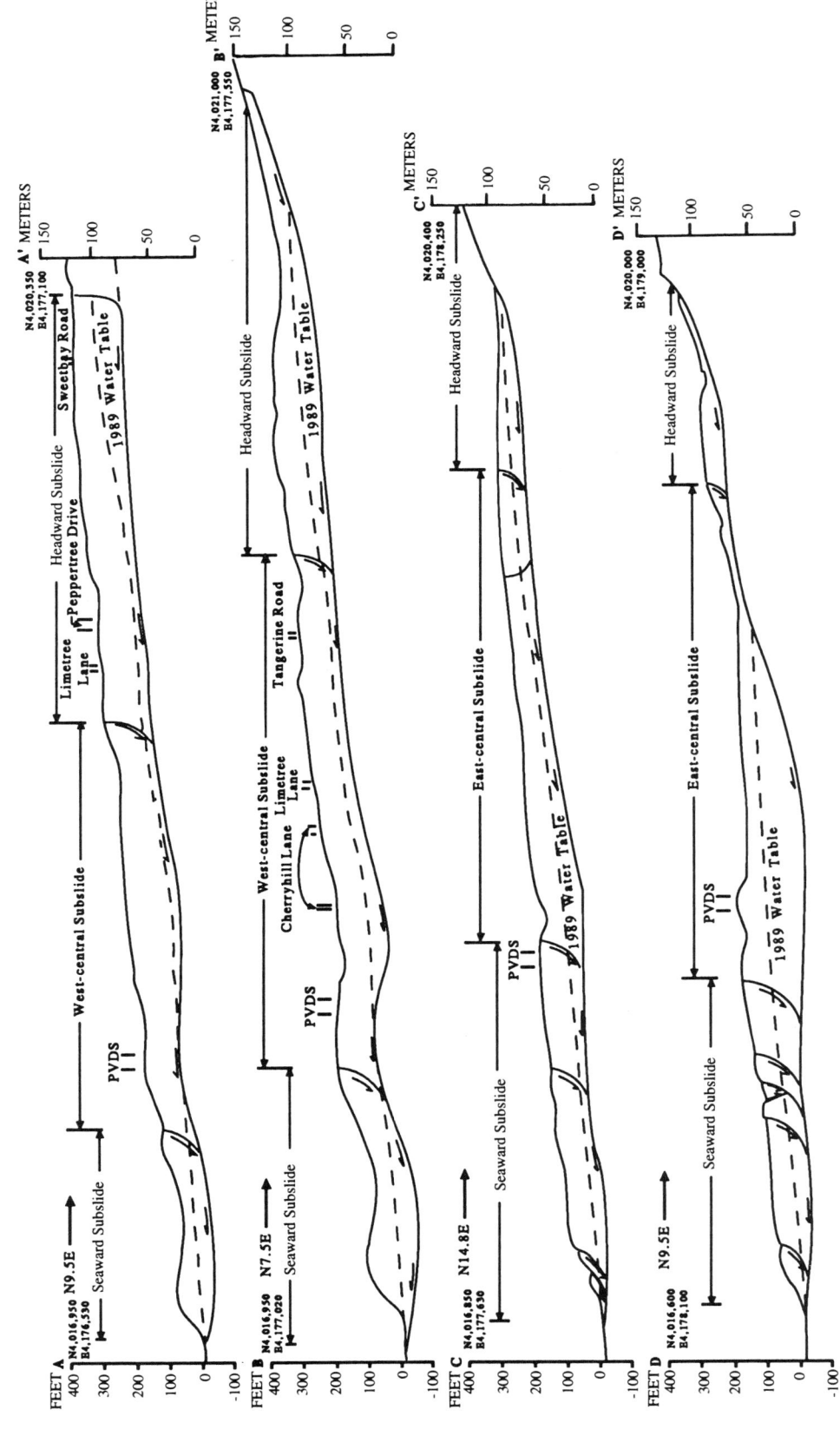

Figure 4. Vertical cross sections through the Portuguese Bend landslide. See Figure 3 for locations of sections. From Ehlig (1992); reproduced with permission of Star Publishing Company and the Association of Engineering Geologists.

broad zone of northeast- to east-trending slices bounded by steeply inclined en echelon faults, fissures, and discontinuous half-grabens (Ehlig, 1992). Vonder Linden (1989) describes the active landslide debris as consisting predominantly of rubble with occasional blocks of bedded slide blocks.

During the summer of 1983, Ehlig analyzed available data on the Portuguese Bend landslide to learn as much as possible about its mechanics and potential methods of stabilizing it. The results of Ehlig's study suggested that stabilization of all but the coastal part of the slide could be achieved at a modest cost by removing groundwater, reestablishing surface drainage, and moving earth from areas where it produces a high driving force to where it would produce a neutral or negative driving force. Beginning in 1984, wells were installed at several locations in the landslide to lower the water table within the landslide. A series of wells were installed in the headward part of the landslide to intercept groundwater flowing into the landslide and deplete groundwater within the headward part of the landslide. Pumping has lowered the groundwater table about 3 m (10 ft) below the groundwater level measured in September 1984 in all but the seaward part of the slide. Grading of the landslide to aid in the stabilization commenced in 1986. The purpose of the grading was to reduce landslide driving forces by removing earth from areas above steeper dipping sections of the landslide basal slip surface and placing it in areas where the landslide basal slip surface is flat, and increase resisting forces by restoring surface drainage to the ocean so as to reduce groundwater recharge.

Based on survey measurements the drainage and regrading of the landslide have significantly improved the stability of the headward and central portions of the landslide. The seaward portion of the landslide remains problematic. Rapid movement and secondary sliding in this area precluded the installation of dewatering wells. A gabion wall was constructed along much of the toe of the active landslide in an attempt to provide temporary protection from wave erosion. However, the gabion wall failed primarily due to overtopping by high surf that allowed erosion of the earth materials behind the wall. The feasibility of constructing a rock revetment to provide permanent shoreline protection is under study by the U.S. Army Corps of Engineers.

For a detailed description of the active Portuguese Bend landslide geomorphology, subsurface conditions, and mechanics, as well as a summary of early observations and studies following reactivation, the reader should review the thesis by Vonder Linden (1972). A paper summarizing the thesis was prepared by Vonder Linden (1989). For information regarding movement of the Portuguese Bend landslide in the 1970s through the earliest part of the 1990s and the details of stabilization measures, the reader is referred to papers by Ehlig (1986, 1987, 1992).

RESIDUAL SHEAR STRENGTH TESTING

A determination of appropriate residual shear strength parameters for the bentonite that forms the slip surface of the Portuguese Bend landslide is critical for a meaningful slope stability analysis. Residual shear strength tests performed on the bentonite from the Portuguese Bend landslide slip surface as well as from other bentonites on the Palos Verdes Peninsula have yielded a wide range of residual shear strengths. Back calculations of several large landslides yielded residual friction angles of 8° or less with cohesion values generally under 7.18 kPa (150 pounds per square foot). These include the ancient South Shores landslide located about 3.5 km (2 mi) east of the Portuguese Bend landslide, portions of the Abalone Cove landslide located adjacent to the west margin of the Portuguese Bend landslide, and the Portuguese Bend landslide itself.

The ancient South Shores landslide is a 65-ha (160-acres) landslide that failed along a seam or seams of bentonite that dip seaward toward coastal bluffs about 16,200 yr B.P. (Ray, 1982). Back calculations of the lower portion of the landslide yielded a residual friction angle of 7.5°, assuming no cohesion, and back calculations of the entire landslide mass yielded a residual friction angle of 9.5° assuming no cohesion (Ray, 1982). The Abalone Cove landslide is a reactivated portion of the Portuguese Bend landslide complex that covers 32 ha (80 acres) and is on the order of 30 to 60 m (100 to 200 ft) thick. The Abalone Cove landslide has failed along seaward dipping bentonitic clay layers both above and below the Portuguese Tuff. The basal slip surface dips very gently (4° to 7°) to a depth about 20 m (60 ft) below sea level before curving steeply upward into the surf zone at the toe of steep bluffs (Ehlig,1982). Back calculations performed on portions of the Abalone Cove landslide yielded residual friction angles of 5.5° to 11.0° when the cohesion was assumed to have a value of 7.18 kPa (150 lb/ft^2) (Lass and Eagen, 1982). When assessing stabilization measures for the Portuguese Bend landslide (Ehlig, 1986), Ehlig assumed the residual shear strength of the bentonite that forms the slip surface of the landslide had a residual friction angle of 6° and a cohesion of 3.59 kPa (75 lb/ft^2). Informal conversations with several investigators suggest that the lower residual friction angles obtained by back calculations are too low and that the back calculations are faulty because of insufficient information on the nature and geometry of the slip surface and the presence of artesian conditions.

Physical, mineralogical, and chemical properties of bentonites from the Palos Verdes Peninsula

Atterberg limits of bentonite samples from the Palos Verdes Peninsula were plotted to evaluate whether significant physical variations occur in the bentonites encountered at different stratigraphic and geographic locations (Fig. 5, Table 1). The values were plotted on a plasticity chart (Fig. 6). The liquid limit of the bentonite samples generally ranged from 80% to 110% with the plasticity index between 40% and 70%. The Atterberg limit values for the three remolded samples used by Watry (1992) in his study are also shown. Some of the variability in the values is caused by physical variations in the bentonite; however, most of the variation is considered to be the result of differences in sample preparation and testing procedures.

Normal preparation for Atterberg plastic and liquid limit tests calls for an air-dried sample to be screened through a 0.425 mm (U.S. Standard Sieve no. 40) sieve prior to the addition of distilled water for the test. Clusters and peds (silt and fine sand–size aggregates of clay particles) will pass through the sieve and if not disaggregated by the addition of distilled water prior to the test will result in lower liquid and plastic limits. Clay samples that are washed to remove salts and/or milled to fully disaggregate the sample will have higher liquid and plastic limits than samples that were only sieved. A sample of clay from the Cucaracha shale tested by La Gatta (1970) yielded a liquid limit value of 49. The clay sample was then slaked four times and crushed in a mill. This sample yielded a liquid limit of 156%. Laboratory testing of the Cucaracha shale by La Gatta yielded a residual friction angle of 6.4°. This value is in good agreement with the residual friction angle determined by a correlation with the liquid limit as presented later in this paper.

The remolded samples utilized by Watry (1992) in his tests were prepared from a slurry that was made by mixing small air-dried pieces of bentonite with distilled water. The bentonite had not been washed or milled prior to preparation of the slurry. The slurry was mixed by hand or with an electric mixer over a period of about a week to fully rehydrate the bentonite. Additional distilled water was occasionally added to the slurry. The addition of the distilled water may have diluted the salt concentration somewhat. The thorough mixing of the slurry broke up any peds that were present.

Sample preparation procedures for Atterberg limit tests on bentonites from the Palos Verdes Peninsula used by other investigators generally followed American Society for Testing and Materials (ASTM Volume 04.08, 1988) guidelines for sample preparation that call for the air-dried sample to be allowed to rehydrate overnight (16 hours). Conversations with two laboratory technicians who have performed many tests on bentonites from the Palos Verdes Peninsula revealed that samples were not milled or intentionally washed to remove salts. The additional time that the remolded samples prepared by Watry (1992) were allowed to rehydrate combined with some inadvertent washing of the soil during preparation of the slurry may have resulted in higher liquid limits. This is based on the

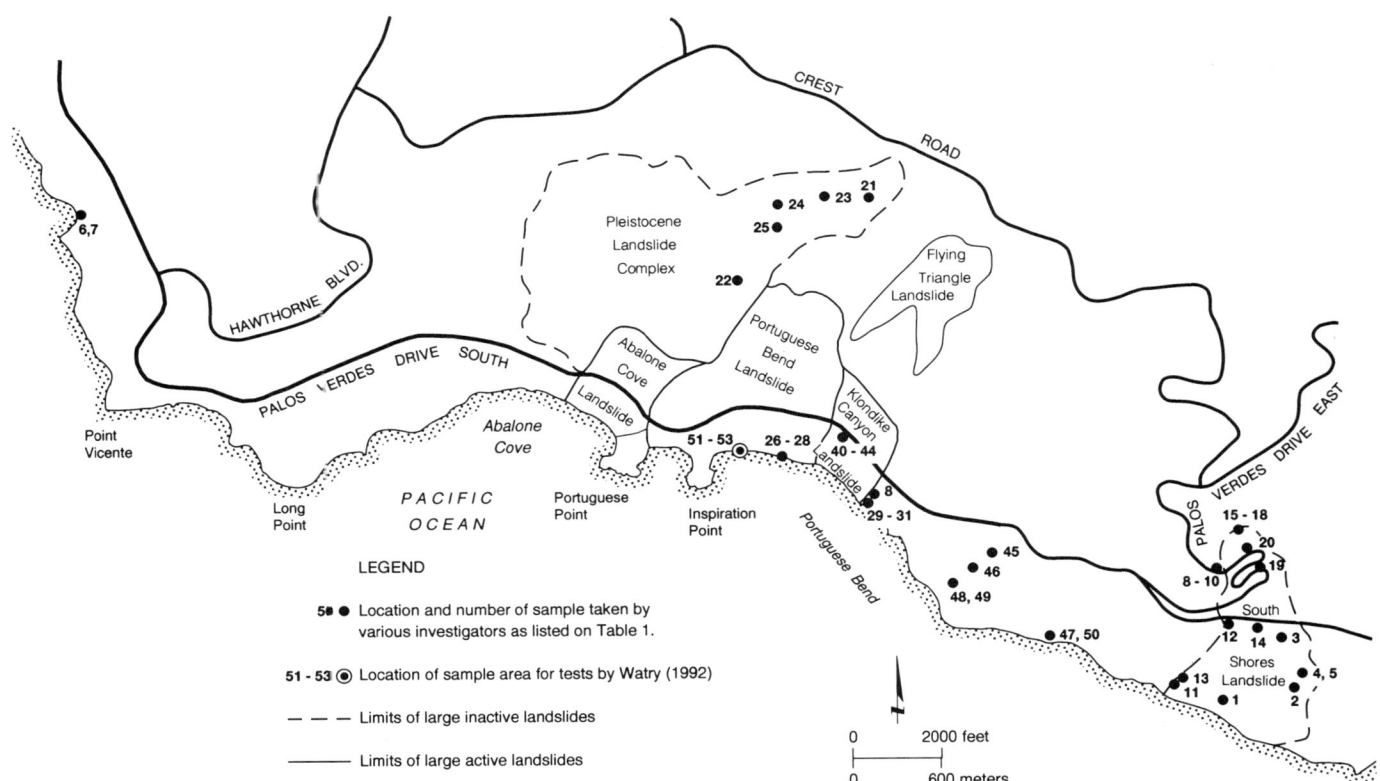

Figure 5. Location map of Atterberg limit samples listed in Table 1.

fact that the liquid limit values for the bentonite exposed at the toe of the Portuguese Bend landslide as determined by Watry are slightly higher than the reported liquid limit values for bentonite samples from near the same location as determined by another investigator (Leighton and Associates, 1990). There is also a matter of reproducibility and repeatability of liquid and plastic limit values. Bowles (1984) reports that the standard deviation for carefully performed tests on the same soil by different technicians is ±3% for liquid limit tests and ±4% for plastic limit tests. It is the authors' opinion that if similar sample preparation procedures were used, there would be even less variability than that shown on Figure 6.

Mineralogical and chemical analyses of bentonites from different stratigraphic and geographic locations in the vicinity of the Portuguese Bend landslide show that the samples are similar. Ten samples of bentonite from the Abalone Cove landslide and Portuguese Bend landslides were analyzed by Novak (1982) for their mineralogical and chemical content. Although the exact location of the samples could not be determined from the referenced paper it is noted that "two samples are from surface exposures near the heads of the slides, three are from borings within the slides, and five are from tuff and gouge along the toes of the slides." The ten samples are described as having similar mineralogies and chemistries. Four of the samples collected at the toes of the Abalone Cove and Portuguese Bend (two from each slide) were analyzed by Novak using atomic absorption spectrometry. These results were compared to chemical analyses of four samples of bentonite that were obtained from bentonite beds located about 0.6 km (2,000 ft) to the east of the Portuguese Bend landslide (Converse Consultants West, 1991). Locations of the later four samples are shown by numbers 45 through 48 on Figure 5. The bentonite seams from which they were obtained are about 80 m (260 ft) above the top of the Portuguese Tuff based on an assessment by Ehlig, whereas the bentonite collected at the toe of the Abalone Cove and the Portuguese Bend landslides are thought to be scrapings from the Portuguese Tuff brought to the surface by the steeply ascending basal slip planes. The chemical analyses of all samples revealed them to be very similar. X-ray diffraction tests by Kerr and Drew (1969), Novak (1982), Leighton and Associates, Inc. (1989a, 1990), and Converse Consultants West (1991) show that the clay mineral in all the samples tested is predominantly a calcium montmorillonite.

Clay fractions reported for bentonite samples from the Palos Verdes Peninsula, as shown on Table 1, are highly variable ranging from 2% to 80%. Kerr and Drew (1969) reported clay fractions of 47% to 71% for five bentonite samples taken within the boundaries or in close proximity to the Portuguese Bend landslide. Leighton and Associates, Inc. (1989b) performed hydrometer analyses on three bentonite samples obtained from the toe of the Portuguese Bend landslide. Clay fractions of 26%, 13%, and 12% were measured. A measured clay fraction of approximately 80% was determined for a remolded sample of bentonite obtained from the toe of the Portuguese Bend landslide and utilized by Watry (1992) in his study. This sample was washed five times with distilled water to remove salts prior to testing. A few investigators (Kerr and Drew, 1969; John W. Byer, personal communication, 1991; Converse Consultants West, 1991; Watry, 1992) have reported difficulty in preventing the bentonite from flocculating in the sedimentation cylinder. Conversations with two laboratory technicians who have performed hydrometer tests on bentonite samples from the Palos Verdes Peninsula revealed that only one had observed flocculation in the sedimentation cylinder. Neither of the technicians washed the samples prior to testing. It is the authors' opinion that many of the lower clay fractions reported may be the result of difficulties in performing the hydrometer tests, primarily caused by flocculation, and that higher and more appropriate clay fractions would be reported if washed samples were tested.

Residual shear strength tests of bentonites from the Palos Verdes Peninsula

The locations of samples of bentonite that have been tested for residual shear strength by various investigators are shown in Figure 7. Table 2 lists the residual shear strengths of the samples along with information on the sample type and test conditions. The table shows that the majority of the tests were performed with a conventional direct shear device. Both remolded and undisturbed samples of bentonite were tested. The typical sample size used has a height of 2.54 cm (1.0 in) and a diameter of 6.12 cm (2.41 in). The samples were repeatedly sheared until a residual strength value was reached. In recent years a few torsional ring shear tests have been performed. Residual friction angles for the bentonite, as determined by the laboratory shear tests, ranged from 3.5° to 27° with residual cohesion of 0 to 191.52 kPa (0 to 4,000 lb/ft^2) (Fig. 8).

The Palos Verdes Peninsula bentonites, like many plastic clays, have a curved failure envelope at low effective normal stresses. The failure envelope then becomes a straight line at higher effective normal stresses. This results in higher, friction angles being measured at low effective normal stresses and lower, constant friction angles being measured over the higher effective normal stresses. Most investigators of bentonites from the Palos Verdes Peninsula ignored or were not aware of the curvature of the failure envelope. Most of the series of residual shear strength tests were performed over a range of effective normal stresses that would include points on the curved portion (lower effective normal stresses), as well as the straight line portion (higher effective normal stresses), of the failure envelope. A straight-line approximation of the entire failure envelope was usually made by either using a linear regression analysis or by a visual approximation. A linear regression analysis would be influenced by the number of points on the curved portion versus the straight-line portion of the envelope. The visual approximation would be subject to engineering judgment by the evaluator.

TABLE 1. ATTERBERG LIMITS AND CLAY FRACTION FOR PALOS VERDES PENINSULA BENTONITES OBTAINED BY VARIOUS INVESTIGATORS

Location Number	Investigator*	Date of Report	Sample	Sample Description	Liquid Limit	Plasticity Index	Clay Fraction
1	Stone Geologic Service, Inc. 66-136	2/24/67	Boring 2-170'	Bentonitic clay	100	63	....
2	Stone Geologic Service, Inc. 66-136	2/24/67	66-R6/76'-78'	Bentonitic clay	108	79	....
3	Stone Geologic Service, Inc. 66-136	2/24/67	66-R8/42'-44'	Bentonitic clay	76	44	....
4	Stone Geologic Service, Inc. 66-136	2/24/67	66-R15/72'-74'	Bentonitic clay	101	64	....
5	Stone Geologic Service, Inc. 66-136	2/24/67	66-R15/86'-88'	Bentonitic clay	142	108	....
6	Byer	6/69	24	Bentonite	110	62	....
7	Byer	6/69	55	Ash-flow tuff bentonite facies	105	62	....
8	Moore and Taber Job No. 386-428	7/17/87	102/4B	Bentonite	130	72	....
9	Moore and Taber Job No. 386-428	7/17/87	102/4B	Bentonite	129	71	....
10	Moore and Taber Job No. 386-428	7/17/87	101/1B	Bentonite	123	72	....
11	Leighton and Associates Project #1870265-10	6/14/89a	C-7 @ 71'	Altered tuff	107	68	....
12	Leighton and Associates Project #1870265-10	6/14/89a	C-13 @ 80'-90'	Bentonite	81	25	....
13	Leighton and Associates Project #1870625-10	6/14/89a	C-16 @ 229'	Altered tuff	111	61	....
14	Leighton and Associates Project #1870265-10	6/14/89a	C-18 @ 158'	Bentonite	84	38	....
15	Leighton and Associates Project #1870265-10	6/14/89a	BA-13 @ 24'	Altered tuff	109	60	....
16	Leighton and Associates Project #1870265-10	6/14/89a	BA-13 @ 27'	Bentonitic tuff	102	51	....
17	Leighton and Associates Project #1870265-10	6/14/89a	BA-13 @ 68'	Bentonitic tuff	103	54	....
18	Leighton and Associates Project #1870265-10	6/14/89a	BA-13 @ 68.5'	....	92	46	....
19	Leighton and Associates Project #1870265-10	6/14/89a	BA-10 @ 20'	....	86	41	....
20	Leighton and Associates Project #1870265-10	6/14/89a	BA-12 @ 105'	....	97	60	....
21	Leighton and Associates Project #1881922-02	11/30/89b	BA-1 @ 80'	....	80	43	....
22	Leighton and Associates Project #1881922-02	11/30/89b	BA-3 @ 70'	....	96	44	....
23	Leighton and Associates Project #1881922-02	11/30/89b	C-2 @ 270'	Bentonitic tuff	89	47	24
24	Leighton and Associates Project #1881922-02	11/30/89b	C-3 @ 175'	Bentonitic tuff	94	51	22

TABLE 1. ATTERBERG LIMITS AND CLAY FRACTION FOR PALOS VERDES PENINSULA BENTONITES OBTAINED BY VARIOUS INVESTIGATORS (continued)

Location Number	Investigator*	Date of Report	Sample	Sample Description	Liquid Limit	Plasticity Index	Clay Fraction
25	Leighton and Associates Project #1881922-02	11/30/89b	C-5 @ 56'-60'	Bentonitic tuff	104	62	26
26	Leighton and Associates Project #1881922-07	8/31/90	Outcrop at beach below Port. Bend landslide	Bentonitic tuff	92	62	26
27	Leighton and Associates Project #1881922-07	8/31/90	Outcrop at beach below Port. Bend landslide	Bentonitic tuff	103	64	13
28	Leighton and Associates Project #1881922-07	8/31/90	Outcrop at beach below Port. Bend landslide	Bentonitic tuff	108	70	12
29	Moore and Taber Job No. 386-428	5/02/89	B-201/A	Bentonitic tuff with sand	61	30	25
30	Moore and Taber Job No. 386-428	5/02/89	B-201/B	Tuffaceous bentonite	105	56	53
31	Moore and Taber Job No. 386-428	5/02/89	B-201/C	Tuffaceous bentonite	92	41	37
32	Moore and Taber Job No. 386-428	5/02/89	B-201/D	Silty bentonitic tuff	92	53	50
33	Moore and Taber Job No. 386-428	5/02/89	B-201/E	Bentonitic tuff	98	53	47
34	Moore and Taber Job No. 386-428	5/02/89	B-201/F	Bentonitic tuff	91	48	34
35	Moore and Taber Job No. 386-428	5/02/89	Pt-A	Bentonitic tuff	113	60	53
36	Moore and Taber Job No. 386-428	5/02/89	Pt-B	Bentonitic tuff	105	51	50
37	Moore and Taber Job No. 386-428	5/02/89	Pt-C	Bentonitic tuff	82	36	43
38	Moore and Taber Job No. 386-428	5/02/89	Pt-D	Bentonitic tuff	91	42	40
39	Moore and Taber Job No. 386-428	5/02/89	Pt-E	Bentonitic tuff	83	40	43
40	Moore and Taber Job No. 386-428	5/02/89	Pt-1	Bentonitic tuff	102	48	36
41	Moore and Taber Job No. 386-428	5/02/89	Pt-2	Bentonitic tuff	90	37	41
42	Moore and Taber Job No. 386-428	5/02/89	Pt-3	Bentonitic tuff	103	48	37
43	Moore and Taber Job No. 386-428	5/02/89	Pt-4	Bentonitic tuff	98	47	38
44	Moore and Taber Job No. 386-428	5/02/89	Pt-5	Bentonitic tuff	108	58	37
45	Converse Consultants West Project No. 88-31-131-04	11/13/91	A	Bentonite	100	65	48

TABLE 1. ATTERBERG LIMITS AND CLAY FRACTION FOR PALOS VERDES
PENINSULA BENTONITES OBTAINED BY VARIOUS INVESTIGATORS (continued)

Location Number	Investigator*	Date of Report	Sample	Sample Description	Liquid Limit	Plasticity Index	Clay Fraction
46	Converse Consultants West Project No. 88-31-131-04	11/13/91	B	Bentonite	85	55	50
47	Converse Consultants West Project No. 88-31-131-04	11/13/91	C	Bentonite	91	49	....
48	Converse Consultants West Project No. 88-31-131-04	11/13/91	D	Bentonite	98	53	46
49	Leighton and Associates Project #1870265-24	11/07/91	1-Upper Bentonite	Bentonite	90	59	2
50	Leighton and Associates Project #1870265-24	11/07/91	1-Lower Bentonite	Bentonite	100	69	3
51	Watry	1992	Remolded Sample 1	Bentonite	108	61	....
52	Watry	1992	Remolded Sample 2	Bentonite	117	72	80
53	Watry	1992	Remolded Sample 3	Bentonite	119	81	....

*See References Cited for more bibliographic information on filed reports.

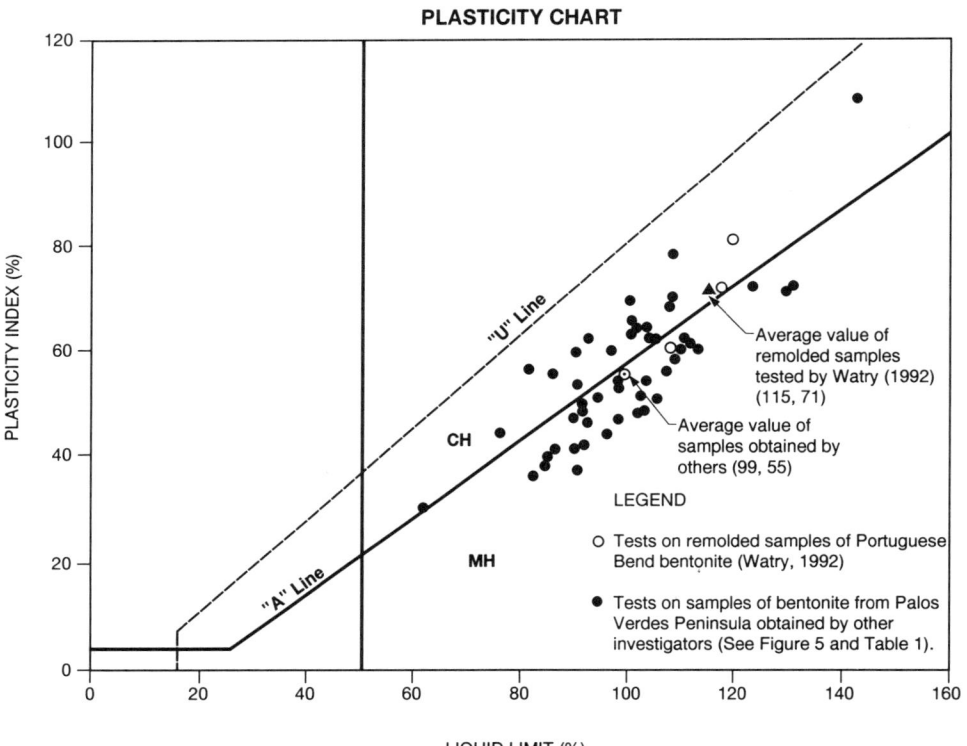

Figure 6. Plasticity chart plot of bentonite samples obtained from Palos Verdes Peninsula by various investigators.

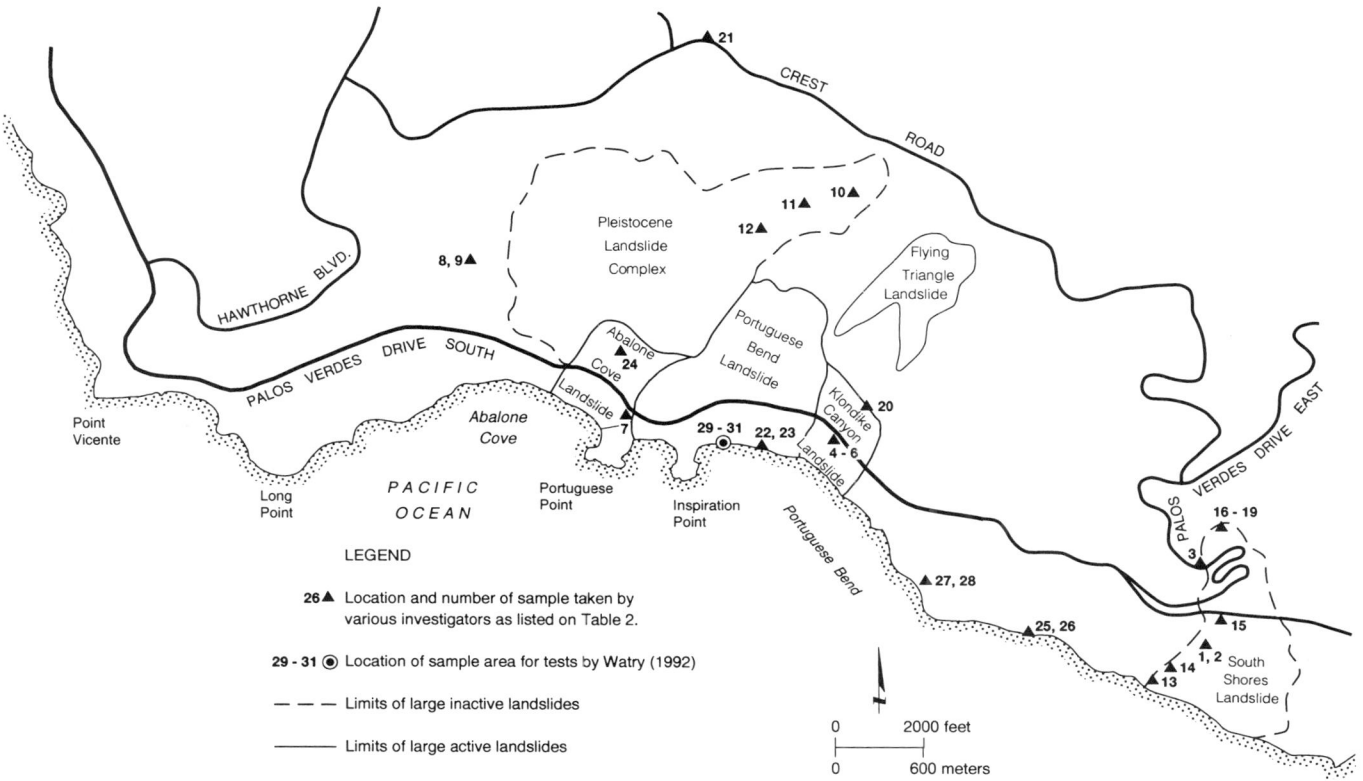

Figure 7. Location map of residual shear strength samples listed on Table 2.

Summary of residual shear strength testing of Palos Verdes Peninsula bentonites

It is the authors' opinion that differences in the physical, mineralogical, and chemical properties of the samples are only a minor cause of the disparities in the measured residual shear strengths of bentonites from the Palos Verdes Peninsula. This is because correlations between the residual friction angle and the plastic and liquid limit of the bentonites, which are discussed later in this paper, define a much narrower range of residual friction angles for the Palos Verdes bentonites than that achieved by the laboratory testing.

The greater causes of the disparities in the laboratory measured residual shear strength of Palos Verdes bentonites are considered to be differences in shear test procedures, errors in shear test procedures, and differences in the interpretation of the shear test results. Differences in the shear test procedures include the use of undisturbed samples versus remolded samples; precutting a sample to establish a slip plane; and varying rates of deformation. Disparities in the residual shear strength due to the aforementioned factors should be very small if the samples are fully consolidated, saturated and deformed at a suitable rate to allow for drainage of the sample as it is being sheared.

Errors in the shear test procedures result from too fast a deformation rate to allow for full drainage of the sample during shear, the testing of samples that are not fully saturated or consolidated when sheared, and insufficient deformation to achieve a residual shear strength. Many of the earlier residual shear strength tests performed on bentonites from the Palos Verdes Peninsula used gear-driven direct shear devices that were unable to run at a slow enough rate to insure full drainage of the standard 2.54 cm (1.0 in) thick sample. Some of the bentonite samples used in the testing were remolded at water contents approximating the optimum moisture content to facilitate compaction and so would possibly need a period of several days to be fully saturated and consolidated. The maximum available deformation in most standard direct shear devices is on the order of 0.64 cm (0.25 in). Numerous reversals of the shear deformation direction are needed to achieve the several centimeters necessary to attain residual shear strength. The bentonite samples are so fine grained and have such a low shear strength that often a significant amount of the sample is squeezed out between the lower and upper halves of the shear box. Shear tests that were halted when sample squeeze became excessive may not have reached residual shear strength.

Differences in the interpretation of the residual shear strength data occurred depending on whether or not a curvilinear failure envelope was considered. Only five of the residual shear tests listed on Table 2 (numbers 8, 9, and 29–31) con-

TABLE 2. RESIDUAL SHEAR STRENGTH FOR PALOS VERDES PENINSULA BENTONITES

Location Number	Investigator	Date of Report	Sample*	Sample Description	Sample Type	Stress Range (kPa)	Strain Rate (mm/min)	Amount of Strain per Pass (mm)	Total Amount of Strain (mm)	Residual Friction Angle (f_r)	Residual Cohesion (C_r)(kPa)
1	Robert Stone and Associates, Inc.	1/16/68	66-136C	Bentonitic clays	Remolded precut	43.86-217.21	0.064	......	......	7	10.23
2	Robert Stone and Associates, Inc.	1/16/68	66-136C	Bentonitic clays	Undistrubed	50.13-196.32	0.064	......	......	7.5	10.37
3	Moore and Taber Job No. 386-428	7/17/87	102/4B 104/1B	Bentonite	Undisturbed	83.54-167.08	0.035	6.4	......	16	2.05
4	Moore and Taber Job No. 377-417	5/23/77	13/2	Tuffaceous claystone	Undisturbed	......	1.02	......	......	27	2.05
5	Moore and Taber Job No. 377-417	5/23/77	14/4	Sheared tuffaceous claystone	Undisturbed	......	1.02	......	......	18	2.05
6	Moore and Taber Job No. 377-417	5/23/77	14/4	Sheared tuffaceous claystone	Remolded	......	1.02	5.1	......	8	3.07
7	F. Beach Leighton Proj. No. G3395	3/27/74	B7 @ 95'	Bentonitic tuff	Remolded precut	11.49-45.95	1.02	5.1	......	6-7	1.53-3.07
8	Leighton and Assoc. Proj. No. 1830817-03	7/27/85	T2 @ 4'-1	Bentonitic tuff	Remolded	45.95-183.79	0.0127	5.1	......	9-21	4.09-26-60
9	Leighton and Assoc. Proj. No. 1830817-03	7/27/85	T2 @ 4'-2	Bentonitic tuff	Remolded	45.95-183.79	0.0127	5.1	......	10-20.5	14.32-35.80
10	Leighton and Assoc. Proj. No. 1881922-02	1/30/89b	C-2 @ 270'	Bentonitic tuff	Remolded	45.95-183.79	0.0127	5.1	......	19.4	10.64
11	Leighton and Assoc. Proj. No. 1881922-02	1/30/89b	C-3 @ 175'	Bentonitic tuff	Remolded	45.95-183.79	0.0127	5.1	......	13.0	8.69
12	Leighton and Assoc. Proj. No. 1881922-02	1/30/89b	C-5 @ 56-60'	Bentonitic tuff	Remolded	45.95-183.79	0.0127	5.1	......	14.6	13.83
13	Leighton and Assoc. Proj. No. 1870265-10	6/14/89a	C-7 @ 71'	Altered tuff	Remolded	45.95-183.79	0.0127	6.4	76.0	9.7	5.85
14	Leighton and Assoc. Proj. No. 1870265-10	6/14/89a	C-16 @ 229'	Altered tuff	Remolded	45.95-183.79	0.0127	6.4	76.0	13.8	8.55
15	Leighton and Assoc. Proj. No. 1870265-10	6/14/89a	C-18 @ 185'	Bentonite	Remolded	45.95-183.79	0.0127	6.4	76.0	14.2	16.78
16	Leighton and Assoc. Proj. No. 1870265-10	6/14/89a	BA-13 @ 24'	Altered tuff	Remolded	45.95-183.79	0.0127	5.1	......	14.9	6.67

TABLE 2. RESIDUAL SHEAR STRENGTH FOR PALOS VERDES PENINSULA BENTONITES (continued)

Location Number	Investigator	Date of Report	Sample*	Sample Description	Sample Type	Stress Range (kPa)	Strain Rate (mm/min)	Amount of Strain per Pass (mm)	Total Amount of Strain (mm)	Residual Friction Angle (f_r)	Residual Cohesion (C_r)(kPa)
17	Leighton and Assoc. Proj. No. 1870265-10	6/14/89a	BA-13 @ 27'	Bentonitic tuff	Remolded	45.95-183.79	0.0127	5.1	......	9.8	4.37
18	Leighton and Assoc. Proj. No. 1870265-10	6/14/89a	BA-13 @ 68'	Bentonite tuff	Remolded	45.95-183.79	0.0127	5.1	......	16	4.18
19	Leighton and Assoc. Proj. No. 1870265-10	6/14/89a	BA-13 @ 68.5'	Bentonite tuff	Remolded	45.95-183.79	0.0127	5.1	......	22	4.18
20	Converse Ward Davis Dixon Proj. No. 81-02167-01	9/8/81	F-5 30.5'	Bentonite	Remolded precut	10.44-41.77	......	......	......	14	0
21	Lockwood-Singh Proj. Ref. 743-42	5/30/84	Road cut	Bentonite	Undisturbed	10.44-83.54	0.0127	6.4	......	20	3.13
22	Leighton and Assoc. Proj. No. 1881922-07	8/31/90	Palos Verdes beach	Bentonite clay	Remolded	45.95-183.79	0.0127	5.1	......	10.9	8.60
23	Leighton and Assoc. Proj. No. 1881922-07	8/31/90	Palos Verdes beach	Bentonite clay	Remolded	45.95-183.79	0.0127	5.1	......	10	6.81
24	Robert Stone and Assoc. Job. No. 1372-98	8/16/79	Test well 2 @ 146'	Bentonite	Remolded precut	20.89-83.54	......	......	......	8	3.13
25†	Leighton and Assoc. Proj. No. 1870265-24	11/7/91	Lower bentonite (sea cliff outcrop)	Bentonite	Remolded ring shear	57.44-459.48	......	......	......	8.1	7.56
26†	Leighton and Assoc. Proj. No. 1870265-24	11/7/91	Lower bentonite (sea cliff outcrop)	Bentonite	Remolded ring shear	87.72-416.67	......	......	......	6.8	11.66
27	Leighton and Assoc. Proj. No. 1870265-24	11/7/91	Upper bentonite (sea cliff outcrop)	Bentonite	Remolded direct shear	45.45-321.64	......	......	......	8.3	8.10
28†	Leighton and Assoc. Proj. No. 1870265-24	11/7/91	Upper bentonite (sea cliff outcrop)	Bentonite	Remolded ring shear	87.72-263.16	......	......	......	7.7	15.25
29†	Watry	1992	Remolded sample 1	Bentonite	Remolded direct shear	150.82-606.74	0.00127 -0.000508	6.4	41-56	6.9	33.04

TABLE 2. RESIDUAL SHEAR STRENGTH FOR PALOS VERDES PENINSULA BENTONITES (continued)

Location Number	Investigator	Date of Report	Sample*	Sample Description	Sample Type	Stress Range (kPa)	Strain Rate (mm/min)	Amount of Strain Per Pass (mm)	Total Amount of Strain (mm)	Residual Friction Angle (f_r)	Residual Cohesion (C_r)(kPa)
30[†]	Watry	1992	Remolded sample 2	Bentonite	Remolded modified direct shear	310.69-471.62	0.0224	203	106-194	6.8	23.80
31[†]	Watry	1992	Remolded sample 3	Bentonite	Remolded ring shear	50-800	0.0178	......	25-207	6.7	7.18
32[‡]	Eben Vey	1961	11 samples from borehole near head of Portuguese Bend slide, 1 sample from toe	......	Direct shear	......	......	......	......	3.5 to 10	5.99 to 3.26
33[‡]	Converse, Davis, and Assoc.	1970	26 samples of various materials from Portuguese Bend slide	......	Direct shear	......	......	......	......	10	191.52

*Samples taken from the south side of the Palos Verdes Peninsula.
[†]Residual friction angle and residual cohesion values determined by Watry by linear regression analysis of data points sheared at effective normal pressures of 201.10 kPa or more.
[‡]Actual shear test data and reports not seen. Data obtained from Table 4 in Vonder Linden, 1972.
47.88 kPa = 1,000 lb/ft².
25.4 mm = 1.0 in.

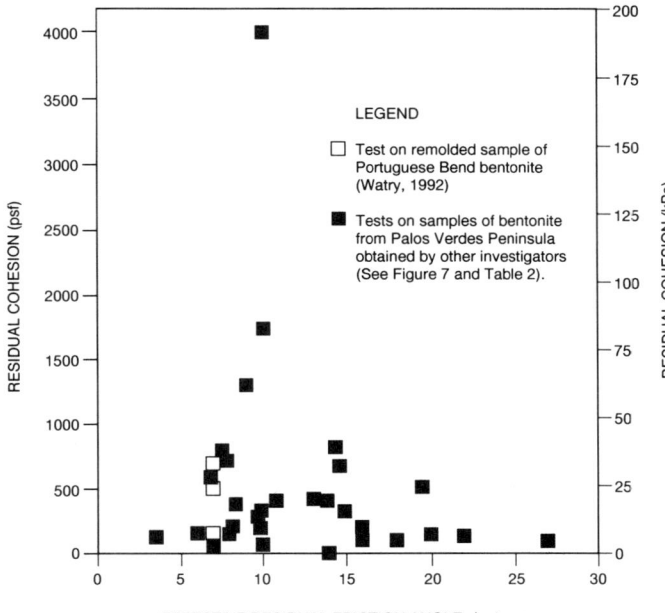

Figure 8. Residual shear strength of bentonite samples from the Palos Verdes Peninsula as determined by various investigators.

sidered the failure envelope to be curvilinear and accounted for it in their determination of the residual friction angle and cohesion. A visual approximation or a linear regression analysis was used by the investigators to model the failure envelope in terms of a residual friction angle and cohesion. Either method is appropriate if sufficient tests have been performed to delineate the failure envelope. However, more than one set of residual friction angle and cohesion shear strength parameters can be developed from a curved failure envelope. In these cases, emphasis should be placed on those shear test points that fall within the range of effective normal stress that will be considered in the stability analyses. A straight line approximation of even a slightly curvilinear failure envelope can be highly erroneous when used in a slope stability analysis if the shear test data points given the most weight in the visual approximation or linear regression analysis do not coincide with the effective normal stress range considered in the analysis.

Portuguese Bend landslide shear strength study

A study of the shear strength of a remolded sample of bentonite obtained from the toe of the Portuguese Bend landslide was performed by Watry (1992). The purpose of the study was twofold: (1) to determine a residual shear strength for the bentonite that forms the slip surface of the Portuguese Bend landslide; and (2) to measure what variation, if any, there is in the residual shear strength measured by different shear test devices. The location where the bentonite sample used in the study was collected is shown on Figures 5 and 7.

Bentonite from the steeply rising slip surface of the Portuguese Bend landslide is exposed along a narrow beach at the toe of the landslide. The beach is only accessible during low tide. The bentonite exposed is blue gray, stiff, moist and highly sheared. Locally, sand and gravel fragments of hard rock are enveloped by the bentonite clay. Approximately 36 kg (80 lb) of bentonite was excavated with a shovel, placed in plastic bags, and transported to the laboratory. The samples that were in large chunks were trimmed into small pieces and air dried. The bentonite was placed in stainless steel bowls and distilled water added to make a slurry. The slurry was then poured into a 25.4-cm (10-in) diameter consolidation tank to create firm remolded samples that were trimmed to accommodate the various shear devices.

A series of consolidated undrained (CU) triaxial tests were first performed to determine various properties of the bentonite such as stress-strain characteristics, pore-pressure development during shear, the peak shear strength, the coefficient of consolidation, and permeability. Figure 9 shows the peak shear strength for the remolded bentonite determined by the triaxial shear tests and the peak shear strength values determined by the other shear test methods.

Three series of residual shear strength tests were performed on the remolded samples of bentonite using three different shear test devices: (1) a conventional direct shear device, (2) a modified direct shear device, and (3) a torsional ring shear device. Schematic diagrams of the shear devices and the samples used in the devices are shown on Figure 10. For full descriptions of the shear devices the reader is referred to Head (1982) for the direct shear device, Watry (1992) for the modified direct shear device, and Bromhead (1979) and Lupini et al. (1981) for the torsional ring shear device.

The conventional direct shear tests were performed similarly to direct shear tests performed by other investigators except that a thinner sample of 1.27 cm (0.5 in) in height was used to allow for quicker drainage of the specimen. The conventional direct shear devices used were driven by a stepper motor capable of very slow rates of deformation (0.000254 mm/min or 0.00001 in/min). The remolded bentonite samples were deformed (sheared) in one direction about 0.64 cm (0.25 in), which is almost the full amount of deformation available before reversing direction. The sample was repeatedly sheared by reversing the direction of deformation until a residual shear strength value was approached. Total deformations were between 28 and 56 mm (1.1 and 2.2 in).

Dull striated, shear surfaces were observed in all of the conventional direct shear tests with the surface being formed near the interface of the filter stone and the upper surface of the sample in all but one test. In the one test the shear surface formed in the middle of the sample. With the one exception noted, the shear surface was formed entirely within the clay just below the roughened surface of the filter stone. The shear surface development here is considered to be analogous to the formation of the slip surface at the soil/polished metal or

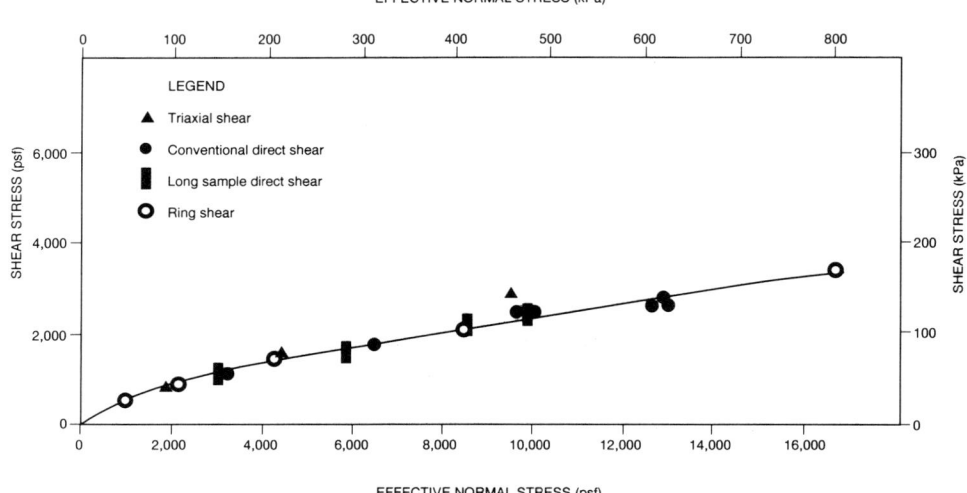

Figure 9. Drained peak failure envelope for shear tests performed on remolded bentonite from Portuguese Bend landslide.

soil/polished stone interface as noted in tests performed by Kanji (1974) and Kanji and Wolle (1977). The polished metal or polished stone surface is thought to have quickened and enhanced the development of a preferred orientation of clay particles along a shear plane in the soil immediately below the polished metal or stone so that a lesser amount of total deformation is required to attain a residual shear strength. The residual failure envelope determined by the conventional direct shear test is shown in Figure 11.

A second series of residual shear tests were performed utilizing a modified direct shear device which can shear rectangular samples 5.08 cm by 25.40 cm (2.00 in by 10.00 in). The prototype device, informally called the Long Sample Direct Shear Device, was developed at the UCLA Soil Mechanics Laboratory. The device was designed to be able to continuously shear a sample in one direction over 20 cm (8 in) while maintaining a constant normal stress centered over the portion of the sample being sheared. In a test on the remolded bentonite, the residual shear strength value was reached after a deformation of the sample about 10.7 cm (4.2 in). The normal stress was then adjusted to increase very slightly with continued deformation so that a failure envelope could be plotted over a range of normal stresses. Figure 11 shows the residual shear strength failure envelope determined by the Long Sample Direct Shear Device.

A third series of residual shear strength tests were performed using a Bromhead torsional ring shear apparatus at the University of Illinois at Urbana-Champaign. The ring shear apparatus utilizes annular specimens contained by concentric confining rings. The inner diameter of the sample is 7.00 cm (2.756 in) and the outer diameter 10.00 cm (3.937 in). The nominal thickness of the sample is 0.5 cm (0.196 in). Porous bronze stones are placed below and above the sample. The upper porous stone also serves as a loading platen. The sample is consolidated in the apparatus and sheared by rotating the base plate and lower bronze porous stone. In the ring shear tests performed on the remolded bentonite sample, the shear surface was induced at the top of the sample just below the porous bronze stone. The residual shear strength failure envelope determined by the ring shear tests is shown in Figure 11.

As determined by the three methods, the residual shear strength of the remolded bentonite had similar friction angles but delineated three separate failure envelopes (Table 3 and Fig. 11). The ring shear tests produced the lowest failure envelope; the conventional direct shear test produced the highest envelope; and the modified long sample direct shear tests produced an intermediate envelope. The lower shear strength value determined by the ring shear test is most likely the result of the ability of the ring shear device to shear the sample in one direction over large shear displacement while maintaining a constant effective normal stress. The prototype long sample shear device, although capable of large displacements in one direction, is not as capable of maintaining a constant effective normal stress as the ring shear device. Also, the shear surface in the long sample shear device is formed within the center of the sample and so the relatively rapid development of a preferred orientation of clay particles by shearing the sample along a hard smooth surface as suggested by Kanji and Wolle (1977) may not have been realized. The conventional direct shear device was limited to a very small amount of displacement that was not able to develop the degree of preferred orientation of the clay particles obtained in the ring shear tests and long sample direct shear tests. The reversal of the shear direction has been shown to disturb the orientation of clay particles developed by the preceding shear displacement (Kenney, 1967). Conventional direct shear tests also are ham-

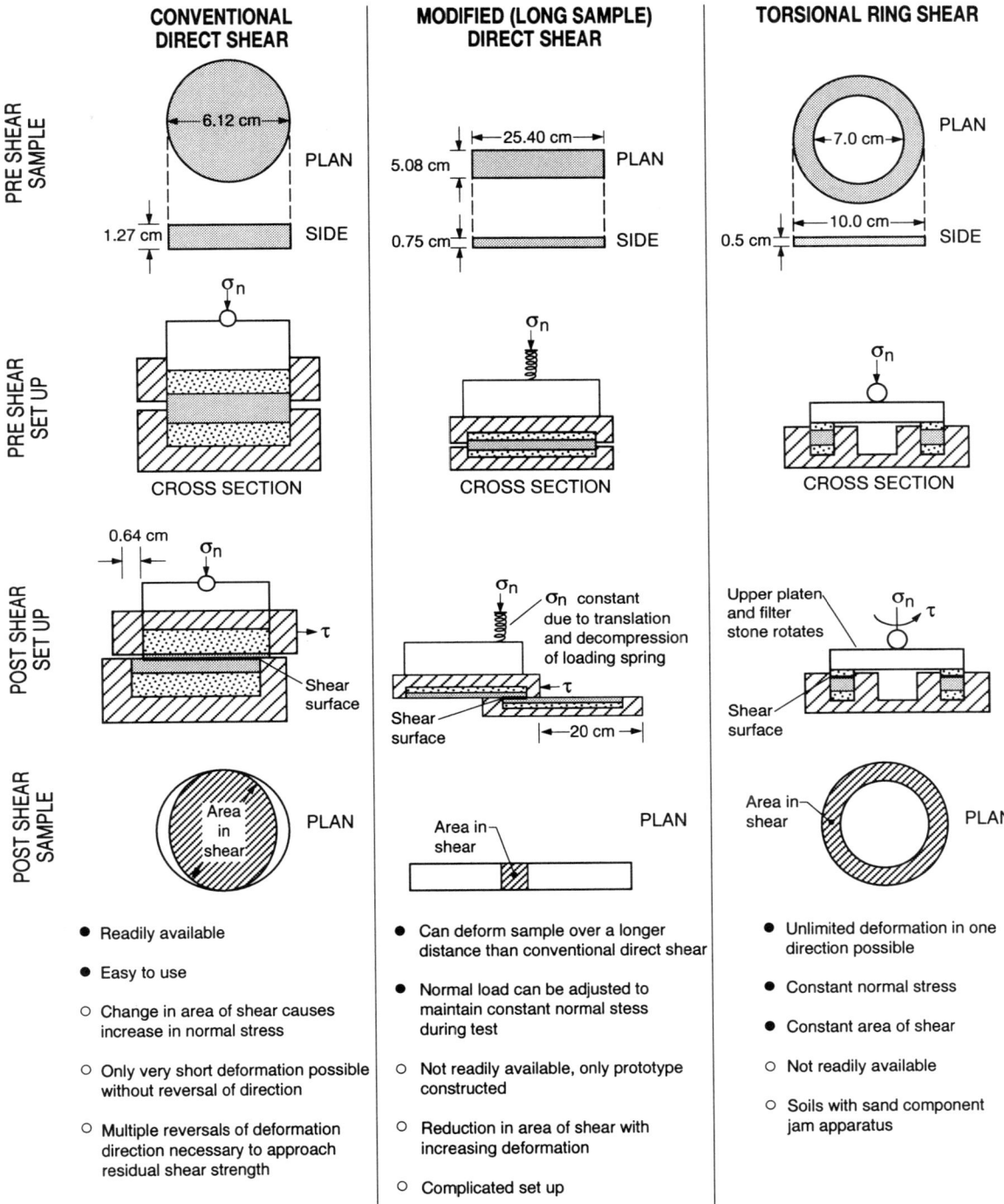

Figure 10. Schematic diagram of shear test devices used to determine residual shear strength of remolded bentonite sample in tests by Watry (1992).

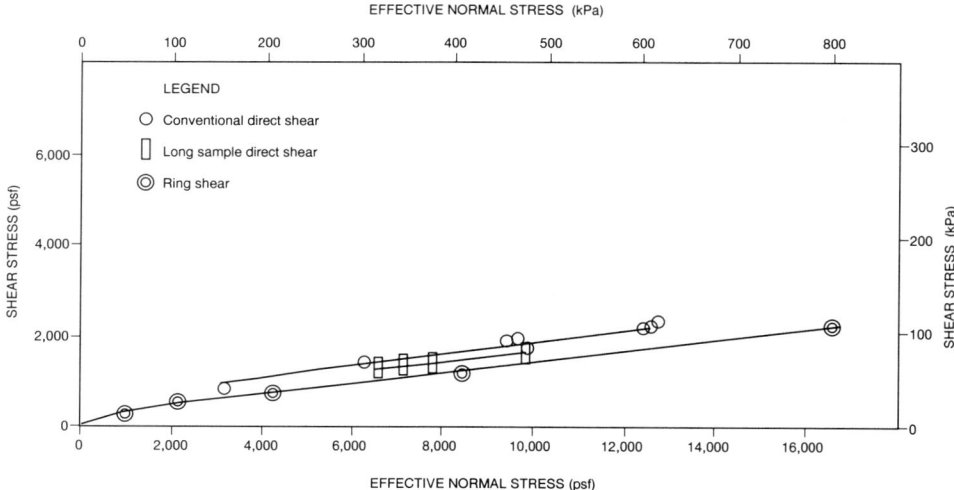

Figure 11. Drained residual failure envelope for combined shear tests on remolded bentonite from Portuguese Bend landslide.

TABLE 3. RESIDUAL SHEAR STRENGTH PARAMETERS DETERMINED BY SHEAR TESTS OF REMOLDED SAMPLES OF BENTONITE OBTAINED FROM THE TOE OF THE PORTUGUESE BEND LANDSLIDE

Testing Device	Residual Friction Angle (ϕ_r)	Residual Cohesion (C_r) (kPa)	Residual Cohesion (C_r) (psf)
Ring shear	6.7	7.18	150
Long sample shear	6.8	23.80	497
Con. direct shear	6.9	33.04	690

pered by the loss of sample by squeezing of the material through the gap between the upper and lower halves of the shear box.

Correlations between soil properties and residual shear strength

The residual friction angle of a soil has been correlated with its plasticity index (Voight, 1973; Kanji and Wolle, 1977; Lupini et al., 1981; Chandler, 1984), liquid limit (Cancelli, 1977; Mesri and Cepeda-Diaz, 1986), and clay fraction (Skempton, 1985). One goal of this study was to determine if similar correlations existed for bentonites from the Palos Verdes Peninsula. The residual friction angle of the remolded bentonite as determined by the laboratory shear tests performed by Watry (1992) was compared to the residual friction angle derived from correlations with the average plasticity index (71%) and the average liquid limit (115%) of the remolded bentonite samples. A correlative residual friction angle for the bentonite also was determined using an average value of plasticity index (55%) and liquid limit (99%) of all the bentonite samples obtained by other investigators. The correlative residual friction angle was compared to the residual friction angle determined by laboratory shear tests by various investigators.

A good correlation exists between the laboratory and correlative residual friction angles for the remolded bentonite samples tested by Watry (Figs. 12–15). The residual friction angle derived from the correlations using the average plasticity index and liquid limit of the three remolded samples ranges from 6.7° to 8.5°. The residual friction angle derived from the correlations using average values of plasticity index and liquid limit of the bentonite samples obtained by other investigators ranges from 7.8° to 11°, which is lower than many of the residual friction angles determined by laboratory testing by these investigators. This suggests that many of the laboratory shear test values determined by other investigators are too high.

A correlation between the clay fraction of the remolded bentonite sample used in the study by Watry (1992) and residual friction angle is shown in Figure 16. This correlation also considers the activity of the clay and the normal stress range. Correlations of clay fraction with residual friction are not accurate when the clay fraction contains minerals that are not platy. Rod, tubular, and needle shaped clay minerals do not develop the degree of preferred orientation that a platy mineral does and so will have higher residual friction angles.

SLOPE STABILITY ANALYSES

Stability analyses were performed along cross section B as shown on Figure 4. The analyses were performed using Spencer's Method as contained on the program TSLOPE© (TAGA, 1985). The analyses were performed utilizing the residual shear strength parameters determined by the conventional direct shear, long sample direct shear, and ring shear devices.

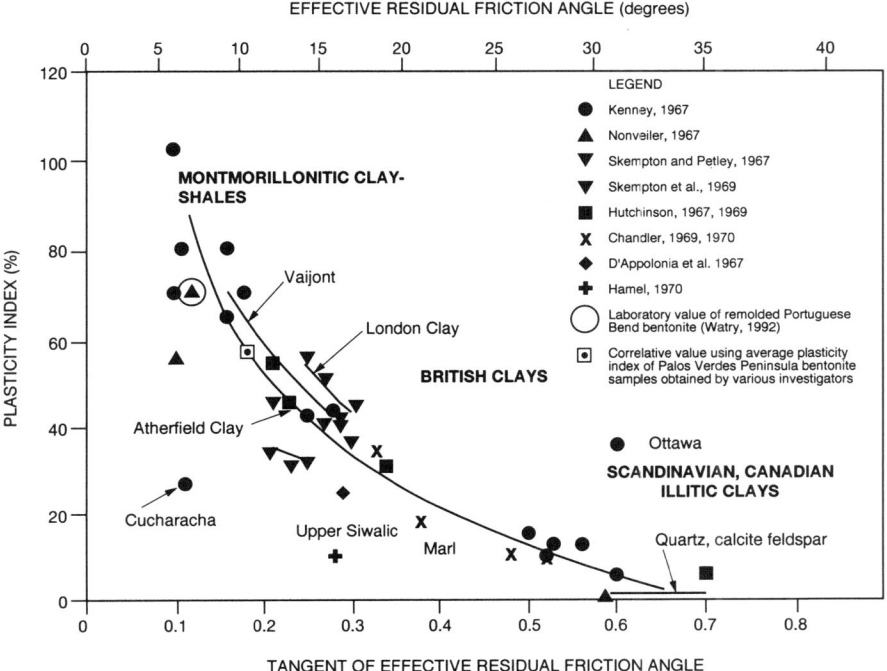

Figure 12. Correlation between residual friction angle and plasticity index. Modified from Voight, 1973.

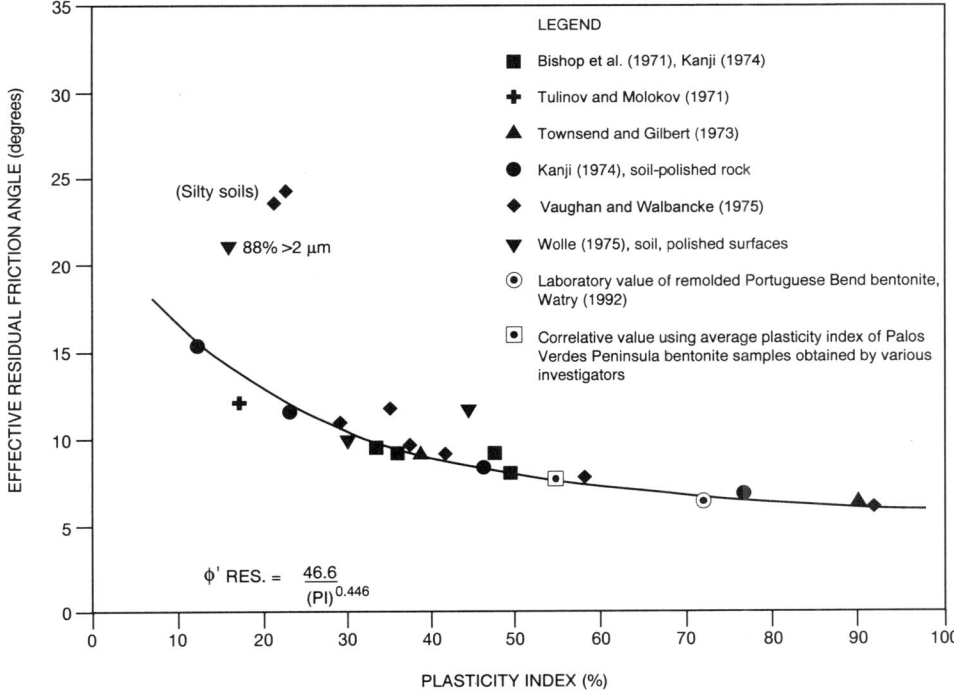

Figure 13. Correlation between residual friction angle and plasticity index. Modified from Kanji and Wolle, 1977.

The analyses were performed utilizing the historic high groundwater level (September 1984) and the current lowered groundwater level which is maintained nearly constant by pumping. Because of the complex nature of the landslide mass, it was considered judicious to analyze segments of the landslide mass as well as the entire landslide mass. Scarp locations defined by listric normal faults as shown on cross section B of Figure 4 were considered to be good locations to segment the landslide mass.

Trial surfaces 1 and 1a, as noted in Table 4, consider the entire landslide segment, trial surfaces 2 and 2a, the west-central and seaward subslides, and trial surface 3 only the seaward subslide.

The residual friction angle and cohesion values used in the analyses, as listed in Table 3, were derived from the linear portion of the failure envelopes. The linear portion of the failure envelopes is located at effective normal stresses above about 191.52 kPa (4,000 lb/ft^2). The effective normal stress acting on the Portuguese Bend landslide slip surface is in excess of 191.52 kPa (4,000 lb/ft^2) except for very thin margins near the toe and head scarp of the slide.

It should be recognized that the cohesion of a clay under no load approaches zero (Mitchell, 1993). The cohesion values listed are the intercept of the straight-line portion of the failure envelope projected back to the ordinate (shear stress) of a shear stress versus effective normal stress plot. The straight-line portion of the failure envelope was determined by a linear regression analysis of those data points above 191.52 kPa (4,000 lb/ft^2) effective normal stress. The correlation coefficients were in excess of 0.99.

Trial slip surfaces 1 through 3 were calculated with the groundwater surface at the historic measured high (September 1984). The historic groundwater high is about 3 m (10 ft) above the 1989 groundwater table shown in Figure 4. Trial slip surfaces 1a and 2a were calculated using the 1989 groundwater level, which is equivalent to the present groundwater level, and is maintained by pumping. The groundwater level in the seaward subslide was not changed since pumps were not placed in this area. A uniform density of 17.43 kN/m^3 (111 lb/ft^3) was assumed for all earth materials.

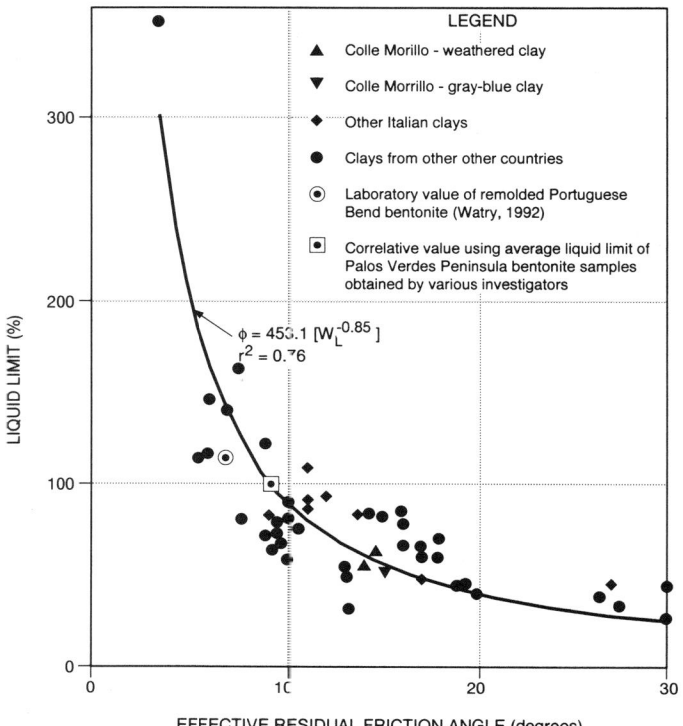

Figure 14. Correlation between liquid limit and residual friction angle. Modified from Cancelli, 1977.

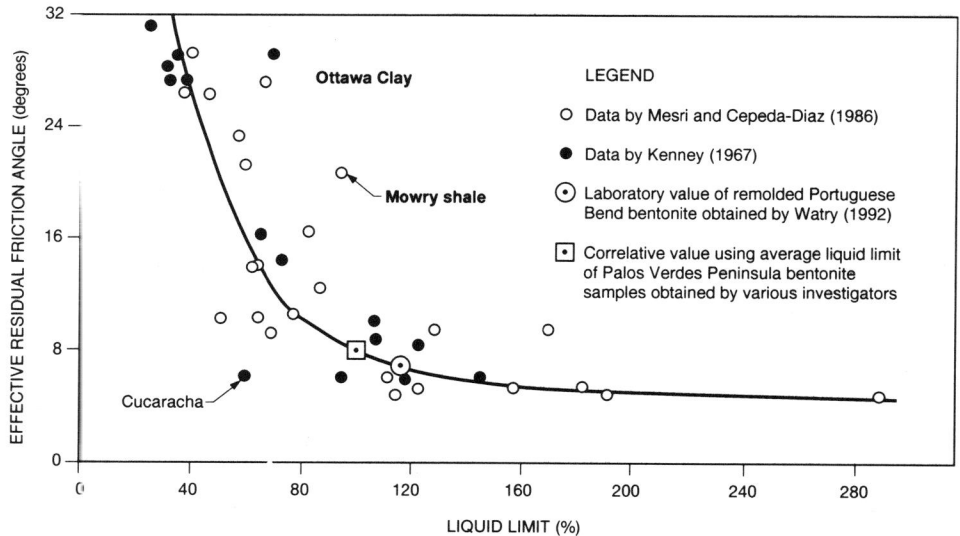

Figure 15. Correlation between liquid limit and residual friction angle. Modified from Mesri and Cepeda-Diaz, 1986.

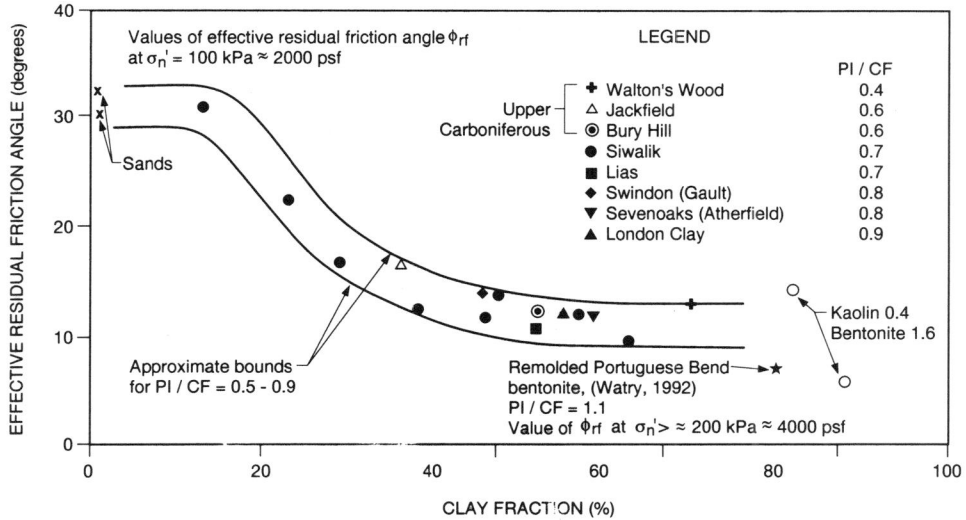

Figure 16. Correlation between clay fraction and residual friction angle. Ring shear apparatus used for all tests plotted. Modified from Skempton, 1985.

The results of the slope stability analyses reveal that even seemingly small differences in shear strength can make a large difference in factor of safety. A factor of safety of near unity for the entire section was obtained using the shear strength parameters measured by the ring shear device and the historic high groundwater surface. Factors of safety for the same section and same groundwater condition obtained using the shear strength parameters measured by the conventional direct shear device and the long sample direct shear device were well above unity suggesting that the residual shear strength parameters determined by these devices are too high.

An assessment of the stability of the Portuguese Bend landslide (excluding the subslide failing between Portuguese and Inspiration Points of which there is little subsurface data) was made utilizing a three-dimensional slope stability analysis, TSLOPE3© (TAGA, 1991). The three-dimensional analysis was performed to determine if the factor of safety determined by the two-dimensional analysis along Section B is representative of the entire landslide mass. The analysis utilized the shear strength parameters determined by the ring shear tests and the 1984 historic high groundwater level. A factor of safety of 1.11 was calculated. This factor of safety is considered to be reasonably consistent with the value determined by the two-dimensional analysis (1.016) considering assumptions and simplifications had to be made regarding subsurface and boundary conditions in areas where information was sparse. A more detailed description of the three-dimensional analysis used is proposed for a future paper.

SUMMARY

Laboratory residual shear strength tests of bentonites from the Palos Verdes Peninsula provide a wide range of values. Back calculations of active and inactive landslides on the Palos Verdes Peninsula that have bentonite clay failure surfaces yielded lower residual shear strengths than those determined by the laboratory testing. Correlations of the effective residual friction angle of the Palos Verdes Peninsula bentonites with the liquid limit or plasticity index generally yielded lower friction angles than those determined by the laboratory shear tests.

Conventional direct shear tests were used by most investigators to determine the residual shear strength of Palos Verdes Peninsula bentonites. Differences and errors in shear test procedures and data interpretation account for much of the variability in residual shear strengths of the bentonites. An inherent limitation of the conventional direct shear test device is that the amount of shear deformation available on a single pass in on direction is often not sufficient to achieve a residual shear strength of a highly plastic clay, even when special measures such as precutting the sample, shearing the sample against a hard, polished surface, and shearing the sample at high effective normal stresses are used. There are also slight changes in the effective normal stress and the area in shear during shear

TABLE 4. RESULTS OF TWO-DIMENSIONAL SLOPE STABILITY ANALYSES

Section B Trial Slip Surface	Factor of Safety Ring Shear	Factor of Safety Long Sample Direct Shear	Factor of Safety Conventional Direct Shear
1	1.016	1.335	1.518
2	0.928	1.214	1.379
3	0.870	0.967	1.109
1a	1.06		
2a	0.965		

deformation. The torsional ring shear apparatus is considered to be the best available device for determining the residual shear strength of highly plastic clays. The torsional ring shear is capable of an unlimited amount of shear deformation in one direction while maintaining a constant area of shear and effective normal stress.

Residual shear strength tests should be performed over a range of effective normal stress, covering the range of effective normal stress that acts on the landslide slip surface. A curvilinear failure envelope should be anticipated when shear tests are performed on highly plastic clays at low effective normal stresses.

A residual friction angle of 6.7° and a residual cohesion of 7.18 kPa (150 lb/ft^2) were assigned to the bentonite forming the slip surface of the Portuguese Bend landslide on the basis of ring shear tests performed on the remolded bentonite obtained from the toe of the Portuguese Bend landslide. These shear strength parameters are considered representative of the slip surface based on the good agreement of the laboratory shear test derived residual friction angle with those residual friction angles determined by correlations with the liquid limit and plasticity index of the bentonite. Back calculations of the Portuguese Bend landslide also indicate that the ring shear test derived residual shear strength parameters are reasonably representative of the slip surface.

ACKNOWLEDGMENTS

We thank Karl Vonder Linden, Martin Stout, William Lettis, and Robert Hollingsworth for reviewing drafts of the manuscript and Carolyn Mosher for preparing the figures.

REFERENCES CITED

American Society for Testing and Materials, 1988, Annual book of ASTM standards, Section 4, construction, Volume 04.08, Soil and rock, building stones; geotextiles, p. 573–583.

Bishop, A. W., Green, G. E., Garga, V. K., Andresen, A., and Brown, J. D., 1971, A new ring shear apparatus and its application to the measurement of residual strength: Géotechnique, v. 21, p. 273–328.

Bowles, J. E., 1984, Physical and geotechnical properties of soils, 2nd edition: New York, New York, McGraw-Hill, Inc., 578 p.

Bromhead, E. N., 1979, A simple ring shear apparatus: Ground Engineering, v. 12, no. 5, p. 40–44.

Bryant, M. E., 1987, Emergent marine terraces and Quaternary tectonics, Palos Verdes Peninsula and San Pedro Bay: Los Angeles, Economic Paleontologists and Mineralogists and American Association of Petroleum Geologists, Pacific Section Volume and Guidebook, Part I, p. 63–78.

Byer, J. W., 1969, Geology and engineering properties of Portuguese Tuff, Palos Verdes Hills, California [M.S. thesis]: Los Angeles, University of Southern California, 88 p.

Cancelli, A., 1977, Residual shear strength and stability analysis of a landslide in fissured overconsolidated clays: Bulletin of the International Association of Engineering Geology, no. 16, p. 193–197.

Chandler, R. J., 1969, The effect of weathering on the shear strength properties of Keuper Marl: Géotechnique, v. 19, p. 321–334.

Chandler, R. J., 1970, A shallow slab slide in the Lias clay near Uppingham, Rutland: Géotechnique, v. 20, p. 253–260.

Chandler, R. J., 1984, Recent European experience of landslides in over-consolidated clays and soft rocks, in Proceedings International Symposium on Landslides, 4th, Volume 1: Toronto, International Symposium on Landslides, p. 61–81.

Conrad, C. L., and Ehlig, P. L., 1983, The Monterey Formation of the Palos Verdes Peninsula, California—An example of sedimentation in a tectonically active basin within the California Continental Borderland, in Larue, D. K., and Steel, R. J., eds., Cenozoic marine sedimentation, Pacific margin, U.S.A.: Society of Economic Paleontologists and Mineralogists, Pacific Section, p. 103–116.

Converse Consultants West, 1991, Response to RPV geotechnical review South Shores parcels 1 and 1A tentative tract 49470, Rancho Palos Verdes, California [unpublished consulting report]: Pasadena, Converse Consultants West Project No. 88-31-131-04, 34 p.

Converse Ward Davis Dixon, Inc., 1981, Geotechnical investigation, Parcel 12, Tentative Tract 37885, Rancho Palos Verdes, California [unpublished consulting report]: Anaheim, California, Converse Ward Davis Dixon, Inc., Project No. 81-02167-01.

D'Appolonia, E., Alperstein, R., and D'Appolonia, D. J., 1967, Behavior of a colluvial slope: Journal of Soil Mechanics, American Society of Civil Engineers, v. 93, p. 447–473.

Ehlig, P. L., 1982, Mechanics of the Abalone cove landslide including the role of ground water in landslide stability and a model for development of large landslides in the Palos Verdes Hills, in Cooper, J. D., comp., Landslides and landslide abatement, Palos Verdes Peninsula, southern California, Geological Society of America Cordilleran Section annual meeting, Anaheim, California: Association of Engineering Geologists, Southern California Section, p. 57–66.

Ehlig, P. L., 1986, The Portuguese Bend landslide: Its mechanics and a plan for stabilization, in Ehlig, P. L., comp., Landslides and landslide mitigation in southern California: Geological Society of America Cordilleran Section annual meeting, Los Angeles, guidebook, field trips 3, 13, and 16, p. 181-190.

Ehlig, P. L., 1987, The Portuguese Bend landslide stabilization project, in Fischer, P. J., Geology of the Palos Verdes Peninsula and San Pedro Bay: Los Angeles, California, Society of Economic Paleontologists and Mineralogists and American Association of Petroleum Geologists, Pacific Section, Volume and Guidebook, part II, p. 2-17-2-24.

Ehlig, P. L., 1992, Evolution, mechanics, and mitigation of the Portuguese Bend landslide, Palos Verdes Peninsula, California, in Pipkin, B. W., and Proctor, R. J., eds., Engineering geology practice in southern California: Belmont, California, Star Publishing Co., Association of Engineering Geologists, southern California section, Special Publication No. 4, p. 531–553.

F. Beach Leighton and Associates, 1974, Report of geotechnical investigation, Abalone Cove, Palos Verdes Peninsula, Los Angeles County, California [unpublished consulting report]: Irvine, California, F. Beach Leighton and Associates Project No. G3395-12-1.

Hamel, J. V, 1970, Stability of slopes in soft, altered rocks [Ph.D. thesis]: Pittsburgh, University of Pittsburgh.

Head, K. H., 1982, Manual of soil laboratory testing, Volume 2: Permeability, shear strength and compressibility tests, London: Plymouth, England,

Pentech Press, 747 p.

Hutchinson, J. N., 1967, Written discussion, in Proceedings, Geotechnical Conference, Oslo: Oslo, Norwegian Technical Institute, v. 2, p. 183–184.

Hutchinson, J. N., 1969, A reconsideration of the coastal landslides at Folkestone Warren, Kent: Géotechnique, v. 19, p. 6–38.

Kanji, M. A., 1974, The relationship between drained friction angles and Atterberg limits of natural soils: Géotechnique, vol. 24, p. 671–674.

Kanji, M. A., and Wolle, C. M., 1977, Residual strength—New testing and microstructure, in Proceedings, International Conference on Soil Mechanics and Foundation Engineering, 9th: Tokyo, International Conference on Soil Mechanics and Foundation Engineering, v. 1, p. 153–154.

Kenney, T. C., 1967, The influence of mineral composition on the residual strength of natural soils, in Proceedings, Geotechnical Conference, Oslo: Oslo, Norwegian Geotechnical Institute, v. 1, p. 123–129.

Kerr, P. F., and Drew, I. M., 1969, Clay mobility, Portuguese Bend, California: California Division of Mines and Geology Special Report 100, p. 3–16.

La Gatta, D. P., 1970, Residual strength of clays and clay-shales by rotation shear tests [Ph.D. thesis]: Cambridge, Massachusetts, Harvard University, Harvard Soil Mechanics Series No. 86, 204 p.

Lass, G. L. and Eagen, J. T., 1982, Introduction to the ancient Abalone Cove landslides, in Cooper, J. D., comp., Landslides and landslide abatement, Palos Verdes Peninsula southern California, Geological Society of America Cordilleran Section annual meeting, Anaheim, California: Association of Engineering Geologists, Southern California Section, p. 81–87.

Leighton and Associates, Inc., 1985, Preliminary geotechnical analysis of concept plan buttress design for parcel 14, City of Rancho Palos Verdes, California [unpublished consulting report]: Irvine, California, Project No. 1830817-03.

Leighton and Associates, Inc., 1989a, Geotechnical feasibility investigation of proposed golf course, golf course clubhouse, and 25 Street widening, South Shores Landslide, City of Rancho Palos Verdes, California [unpublished consulting report]: Santa Ana, California, Leighton and Associates Project No. 1870265-10.

Leighton and Associates, 1989b, Geotechnical feasibility investigation of Peacock Hill, a portion of parcel 10, City of Rancho Palos Verdes, California [unpublished consulting report]: Santa Ana, California, Leighton and Associates Project No. 1881922-02.

Leighton and Associates, 1990, Supplementary feasibility investigation of Peacock Hill, a portion of parcel 10, City of Rancho Palos Verdes, California [unpublished consulting report]: Santa Ana, California, Leighton and Associates Project No. 1881922-07, v. I.

Leighton and Associates, 1991, Geotechnical analysis of gross stability of tentative tract no. 50667, between Paseo Del Mar Avenue, Palos Verdes Drive South, and La Rotunda Drive, west of the former hotel site, subregions 7 and 8, City of Rancho Palos Verdes, California [unpublished consulting report]: Irvine, California, Leighton and Associates Project No. 1870265-24, v. I.

Lockwood-Singh and Associates, 1984, Addendum to report of geotechnical investigation, lot 1, tract 32110, 29131 Crenshaw Boulevard, City of Rancho Palos Verdes, California [unpublished consulting report]: Los Angeles, Lockwood-Singh and Associates Project Ref. 743-42.

Lupini, J. F., Skinner, A. E., and Vaughan, P. R., 1981, The drained residual strength of cohesive soils: Géotechnique, v. 31, p. 181–213.

Merriam, R., 1960, Portuguese Bend landslide, Palos Verdes Hills, California: Journal of Geology, v. 68, p. 140–153.

Mesri, G., and Cepeda-Diaz, A. F., 1986, Residual shear strength of clays and shales: Géotechnique, v. 36, p. 269–274.

Mitchell, J. K., 1993, Fundamentals of soil behavior, 2nd edition: New York, John Wiley and Sons, 437 p.

Moore and Taber, 1977, Landslide investigation, Portuguese Bend Club, Rancho Palos Verdes, California [unpublished consulting report]: Job No. 377-417.

Moore and Taber, 1987, Response to review of geotechnical reports, easterly portion of parcel 15, Portuguese Bend Club, Rancho Palos Verdes, California [unpublished consulting report]: Job No. 386-428.

Moore and Taber, 1989, Response to geotechnical review, supplemental geologic and geotechnical study, easterly portion of parcel 15, Portuguese Bend Club, Rancho Palos Verdes, California [unpublished consulting report]: Anaheim, California, Moore and Taber Job No. 386-428.

Nonveiler, E., 1967, Shear strength of bedded and jointed rock as determined from the Zalesina and Vajont slides, in Proceedings, Geotechnical Conference, Oslo: Oslo, Norwegian Geotechnical Institute, v. 1, p. 289–294.

Novak, G. A., 1982, Mineralogy and chemistry of bentonite from the Abalone Cove and Portuguese Bend landslides southern California, in Cooper, J. D., comp., Landslides and landslide abatement, Palos Verdes Peninsula southern California, Geological Society of America Cordilleran Section annual meeting, Anaheim, California: Association of Engineering Geologists, Southern California Section, p. 27.

Ray, M. E., 1982, Geologic investigation, grading stabilization measures, and development of the South Shores landslide, in Cooper, J. D., comp., Landslides and landslide abatement, Palos Verdes Peninsula, southern California, Geological Society of America Cordilleran Section annual meeting, Anaheim, California: Association of Engineering Geologists, Southern California Section, p. 29–38.

Robert Stone and Associates, Inc., 1968, Supplementary soils engineering study of stability of South Shores Mobile Home Park, 25th Street—Paseo del Mar, Palos Verdes Hills [unpublished consulting report]: Woodland Hills, California, Robert Stone and Associates, Inc., Job. No. 66-136C.

Robert Stone and Associates, Inc., 1979, Dewatering test results and related analyses, Abalone Cove landslide, Rancho Palos Verdes, California [unpublished consulting report]: Canoga Park, California, Robert Stone and Associates Job No. 1372-98, 29 p., 2 boring logs, 3 plates.

Rowell, H. C., 1982, Chronostratigraphy of the Monterey Formation of the Palos Verdes Hills, in Cooper, J. D., comp., Landslides and landslide abatement, Palos Verdes Peninsula, southern California, Geological Society of America Cordilleran Section annual meeting, Anaheim, California: Association of Engineering Geologists, Southern California Section, p. 7–13.

Skempton, A. W., 1985, Residual strength of clays in landslides, folded strata and the laboratory: Géotechnique, v. 35, p. 3–18.

Skempton, A. W., and Petley, D. J., 1967, The strength along structural discontinuities in stiff clays, in Proceedings, Geotechnical Conference, Oslo: Oslo, Norwegian Geotechnical Institute, v. 1, p. 29–46.

Skempton, A. W., Schuster, R. L., and Petley, D. J., 1969, Joints and fissures in the London Clay at Wraysbury and Edgware: Géotechnique, v. 19, p. 205–217.

Stone Geological Service, Inc., 1967, Geologic and soils report, South Shores Mobile Home Park, 25th Street, Paseo Del Mar, Palos Verdes Hills [unpublished consulting report]: Project No. 66-136.

TAGA, Inc., 1985, TSLOPE computer program for limit equilibrium slope stability analyses: San Ramon, California, TAGA Engineering Software Services.

TAGA, Inc., 1991, TSLOPE3 computer program for three-dimensional slope stability analyses: TAGA Engineering Systems & Software, Lafayette, California.

Townsend, F. C., and Gilbert, P. A., 1973, Test to measure residual strengths of some clay shales: Geotechnique, v. 23, p. 267–271.

Tulinov, R., and Molokov, I., 1971, Role of joint filling material in shear strength of rocks: Nancy, France, International Society of Rock Mechanics, Symposium on Rock Fracture, v. 2, 13 p.

Vaughan, P. R., and Walbancke, H. J., 1975, The stability of cut and fill slopes in Boulder clay: Birmingham, England, Midland Society of Soil Mechanics, Symposium on Glacial Materials, 11 p.

Voight, B., 1973, Correlation between Atterberg plasticity limits and residual shear strength of natural soils: Géotechnique, v. 23, p. 265–267.

Vonder Linden, K., 1972, An analysis of the Portuguese Bend landslide, Palos Verdes Hills, California [Ph.D. thesis]: Stanford, California, Stanford

University, 260 p.

Vonder Linden, K., 1989, The Portuguese Bend landslide: Engineering Geology, v. 27, p. 301–373.

Vonder Linden, K., and Lindvall, C. E., 1982, The Portuguese Bend landslide, *in* Cooper, J. D., comp., Landslides and landslide abatement, Palos Verdes Peninsula, southern California, Geological Society of America Cordilleran Section annual meeting, Anaheim, California: Association of Engineering Geologists, Southern California Section, p. 49–56.

Watry, S. M., 1992, Shear strength of bentonite from the toe of the Portuguese Bend landslide [M.S. thesis]: Los Angeles, California State University, 263 p.

Wolle, C., 1975, Uso do microscópic electrônico de Varredura na observação da microestructura das Agrilas [in Portuguese]: São Paulo, Brazilian Association of Soil Mechanics, 5th Congress, v. 2, p. 115–129.

Woodring, W. P., Bramlette, M. N., and Kew, W. S. W., 1946, Geology and paleontology of the Palos Verdes Hills, California: U.S. Geological Survey Professional Paper 207, 145 p.

MANUSCRIPT ACCEPTED BY THE SOCIETY MAY 20, 1994

Evaluation of viscoplastic slope movement based on triaxial tests

Wylie W.-H. Wong and Carlton L. Ho
Department of Civil and Environmental Engineering, Washington State University, Pullman, Washington 99164-2910
Richard M. Iverson
U.S. Geological Survey, 5400 MacArthur Boulevard, Vancouver, Washington 98661
Cynthia L. Hovind
Sweet Edwards/Emcon, 7504 S.W. Bridgeport Road, Portland, Oregon 97224

ABSTRACT

Viscoplastic soil parameters are used in a nonlinear viscoplastic constitutive model to predict time-dependent displacement of slow-moving landslides. The viscoplastic material parameters are determined by a novel method that uses a standard triaxial apparatus. This method employs data obtained from consolidated drained triaxial tests and consolidated drained stress-controlled strain-rate tests. The methodology was applied to undisturbed samples from the Minor Creek landslide in the Franciscan Terrane of northern California. Viscoplastic parameters determined from the laboratory tests were combined with boring log data to calculate the landslide's vertical velocity profile. This profile provided a reasonable match to a measured velocity profile obtained from repetitive inclinometer surveys.

INTRODUCTION

Prediction of landslide motion is difficult. Classical slope-stability analyses consider only whether a slope will fail and do not consider post-failure rates of movement. To predict post-failure movement rates, a rate-dependent constitutive model for soil behavior is needed. Iverson (1984, 1985, 1986a, 1986b, 1986c) proposed and used a viscoplastic constitutive model to assess slow, time-dependent motion of the Minor Creek landslide in the Franciscan Terrane of northwestern California. This model represents the soil in the landslide shear zone as a nonlinearly viscous fluid if the soil yield strength is exceeded. Viscous behavior, combined with fluctuating hydrologic conditions, can explain time-varying downslope movement of the landslide (Iverson and Major, 1987). This paper describes a novel laboratory and field methodology that is used to determine parameters necessary for application of Iverson's (1985) viscoplastic constitutive model to calculate the velocity profile of Minor Creek landslide.

CONSTITUTIVE RELATIONSHIP

The viscoplastic model assumes that soil undergoes no deformation until a yield strength defined by the Drucker-Präger (1952) yield criterion is reached. If the effective deviator stresses exceed those required for yield, a shear zone is developed at the landslide base and viscous displacement ensues. Viscous behavior is evidenced by steady, downslope movement that occurs under a constant state of effective deviator stress. Soil atop the shear zone undergoes translational rigid-body motion. In the constitutive formulation, effective stresses are used to account for pore water pressure, and elastic displacement is neglected to simplify the model. If the effective deviator stresses return to levels insufficient for yield, slope movement ceases. The general constitutive equation for this viscoplastic deformation is (Iverson, 1985):

$$T = 2\mu D + \frac{\kappa + \alpha p'}{II_D^{1/2}} D \qquad (1)$$

Wong, W. W.-H., Ho, C. L., Iverson, R. M., and Hovind, C. L., 1995, Evaluation of viscoplastic slope movement based on triaxial texts, *in* Haneberg, W. C., and Anderson, S. A., eds., Clay and Shale Slope Instability: Boulder, Colorado, Geological Society of America Reviews in Engineering Geology, v. X.

where T is the effective deviator stress tensor, D is the deviatoric strain-rate tensor, II_D is the second invariant of D, and p' is the mean normal effective stress. The material properties are represented by μ, the apparent viscosity, and α and κ, the Drucker-Präger yield strength parameters. Soil within the shear zone undergoes nonlinearly viscous displacement, which can be characterized by a power law function as follows:

$$\mu = \mu_0 \left(\frac{II_D^{1/2}}{D_0} \right)^{(1-n)/n} \quad (2)$$

where μ is the apparent viscosity, μ_0 is the equivalent Newtonian viscosity, D_0 is a reference value of $II_D^{1/2}$ and n is the power-law exponent. For a nonlinearly viscous material, the apparent viscosity is a rate-dependent function instead of a constant. The power law function can describe a wide range of viscous behaviors. For $n > 1$, it describes shear thinning behavior; the apparent viscosity decreases with increasing strain rate as a result of progressive alignment of asymmetric particles. It describes Newtonian flow for $n = 1$. For $n < 1$, it describes shear thickening behavior; the apparent viscosity increases with strain rate, as dense soils dilate under shearing. Nonlinear viscous flow can be depicted by three parameters, n, D_0 and μ_0. The n parameter describes the style of viscous flow, and the D_0 and μ_0 parameters control the magnitude of the deformation rate. Therefore, five parameters are needed to characterize the material for this constitutive model: two yield strength parameters, κ and α, and three strain-rate parameters, μ_0, D_0, and n. The Drucker-Präger (Drucker and Präger, 1952) parameters can be estimated from the Mohr-Coulomb parameters in two-dimensional stress space based on triaxial test results as follows:

$$\kappa = c'_y \cos \phi'_y \quad (3)$$

$$\alpha = \sin \phi'_y \quad (4)$$

where c'_y and ϕ'_y are the drained Mohr-Coulomb cohesion and friction angle at yield.

The deviator stress tensor of Equation 1 can be broken into two parts, T_e, the excess deviator stress tensor, and T_t, the deviator stress tensor at yield as follows:

$$T = T_e + T_t \quad (5)$$

where

$$T_e = 2\mu_0 \left(\frac{II_D^{1/2}}{D_0} \right)^{(1-n)/n} D \quad (6)$$

and

$$T_t = \left[\frac{(\kappa + \alpha p')}{II_D^{1/2}} \right] D \quad (7)$$

If these equations are re-written in terms of invariants, they become

$$\sqrt{II_e} = 2\mu_0 \left(\frac{\sqrt{II_D}}{D_0} \right)^{(1-n)/n} \sqrt{II_D} \quad (8)$$

and

$$\sqrt{II_t} = \kappa + \alpha p' \quad (9)$$

where II_e and II_t are the second invariants of T_e and T_t respectively. Equation 9 is the Drucker-Präger yield criterion, which controls the onset of nonlinear viscous displacement. The nonlinear viscous displacement is directly related to the excess deviator stress as shown in Equation 8.

APPLICATION OF THE VISCOPLASTIC MODEL

The material parameters discussed above will now be used in the nonlinear viscoplastic model to predict a vertical velocity profile in a landslide. The boundary conditions in the landslide are zero velocity at the bottom margin of the shear zone and a velocity gradient of zero (maximum velocity) at the top margin of the shear zone. Therefore, the rigid body above the shear zone displaces at the maximum velocity. The velocity distribution within the shear zone is expressed as (Iverson, 1986a)

$$V_x = \left[1 - \left(\frac{y - T}{h - T} \right)^{(n+1)} \right] V_{x\,max} \quad (10)$$

where V_x is the velocity in the x direction (downslope direction), $V_{x\,max}$ is the maximum (ground-surface) velocity in the x direction, y is the depth of interest within the shear zone, h is the depth of landslide base, T is the thickness of the rigid body, and $(h - T)$ is the thickness of shear zone. The depths y, h, and T are measured in the slope-normal direction. The velocity gradient can be found by differentiating Equation 10 with respect to y as follows:

$$\frac{dV_x}{dy} = -V_{x\,max} \left[\frac{(n+1)(y-T)^n}{(h-T)^{n+1}} \right] \quad (11)$$

In simple shear gravity flow, the strain-rate, D, is equal to the absolute value of the velocity gradient,

$$D = \left| \frac{1}{2} \frac{dV_x}{dy} \right| \quad (12)$$

The reference strain-rate, D_0, occurs at a reference depth, y_{ref}, within the shear zone, where the velocity gradient is identical for the linear viscoplastic case ($n = 1$) and the nonlinear viscoplastic case ($n \neq 1$) in response to the same stress state. The reference depth can be evaluated from Equation 11 by equating the velocity gradients of both linear and nonlinear cases (Iverson, 1984),

$$y_{ref} = (h - T)\left(\frac{2}{n+1}\right)^{1/(n-1)} + T \quad (13)$$

Substituting Equation 13 into Equation 11, D_0 can be evaluated at the reference depth from Equation 14 as follows

$$D_0 = \frac{2^{1/(n-1)} V_{x\,max}}{(h - T)(n + 1)^{1/(n-1)}} \quad (14)$$

The maximum velocity is given by (Iverson, 1986a)

$$V_{x\,max} = \frac{(-A_1)^n (2D_0)^{(1-n)}}{\mu_0^n (n + 1)} (h - T)^{(n+1)} \quad (15)$$

where A_1 is a function of the landslide geometry and material properties, and T can be evaluated at the limit equilibrium condition and expressed as

$$T = \frac{-A_2}{A_1} \quad (16)$$

with

$$A_1 = \cos \theta \left[(\gamma_{sat} - \gamma_w) \tan \phi_y' - \gamma_{sat} \tan \theta \right] \quad (17)$$

and

$$A_2 = d\left[(\gamma_{moist} - \gamma_{sat} - \gamma_w) \cos \theta \tan \phi_y' + (\gamma_{sat} - \gamma_{moist}) \sin \theta \right] + c \quad (18)$$

where γ_{sat} is the saturated unit weight of soil; γ_{moist} is the moist unit weight of soil above the water table; γ_w is the unit weight of water; θ is the landslide slope angle; ϕ_y' is the effective friction angle at the yield condition; d is the depth of the water table; and c is the cohesion of soil.

To calculate a velocity profile using laboratory data, the yield friction angle and material properties obtained from the experimental tests can be substituted into Equations 17, 18, and then 16 to calculate the rigid-body thickness. Then, the parameters derived from the triaxial tests can be substituted into Equation 15 to calculate the maximum velocity and a predicted velocity profile can finally be evaluated from Equation 10.

DETERMINATION OF PARAMETERS WITH TRIAXIAL TESTS

Soil does not have a distinct yield point. However, the yield strength of landslide soil must be determined to apply the viscoplastic model. Therefore, a method must be devised to evaluate the yield point.

Secant modulus method

Soil behavior changes from elastic to plastic when the yield point is reached. Selection of the yield point must consider this idealized behavior and the actual, more complex stress-strain behavior. A 0.2% strain-offset method is commonly used in mechanics of materials to define the yield stress. The yield stress is defined at the intersection of the stress-strain curve and a line parallel to the initial tangent modulus with a horizontal offset of 0.2%. Thus, the yield point is determined by an arbitrary method and is not an inherent property of the material.

An alternative method, the secant modulus method, is proposed here for determination of yield stress. It is similar to the Taylor (1948) method for determining the time of 90% consolidation, which uses a proportionality coefficient to scale the secant slope from the tangent slope. In the secant modulus method, the yield point is defined at the intersection of the stress-strain curve and a line having a slope equal to a secant modulus, E_s, where E_s is a fraction of the initial tangent modulus, E_i (Fig. 1):

$$E_s = AE_i \quad (19)$$

in which A is a proportionality factor. Therefore, the yield stress can be expressed as

$$\sigma_y' = AE_i \varepsilon_y \quad (20)$$

where σ_y' is the effective yield stress, ε_y is the axial strain at yield. In order to evaluate the yield stress, E_i and A must be determined. The experimental stress-strain curve of soil can be represented by the hyperbolic function of Kondner (1963), from which E_i can be estimated. The hyperbolic equation is

$$\sigma' = \frac{\varepsilon}{(1/E_i) + (\varepsilon/\sigma_{ult}')} \quad (21)$$

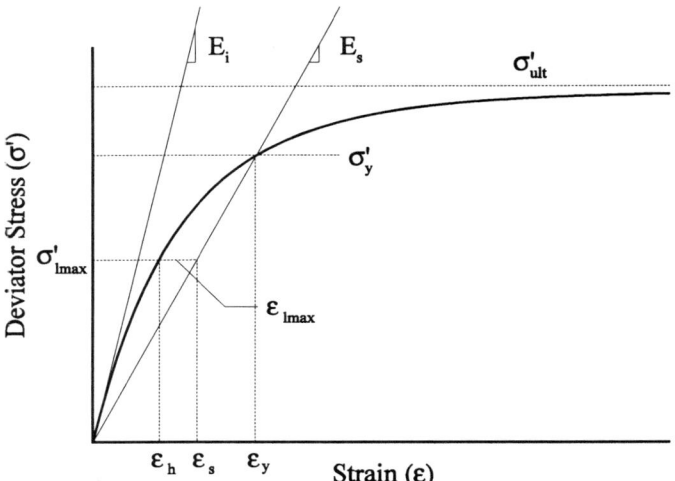

Figure 1. The secant modulus method for yield point determination. The experimental stress-strain curve is fitted by a hyperbolic function with parameters E_i and σ'_{ult}, the initial tangent modulus and hypothetical stress respectively. The yield stress, σ'_y, is found at the intersection of the secant modulus slope, E_s, and the hyperbolic curve. The maximum limiting strain, ε_{lmax}, is the strain difference between the secant modulus strain, ε_s, and the hyperbolic strain, ε_h. The maximum limiting strain controls the deviation of the secant modulus slope from the hyerbolic curve at σ'_{lmax}, the deviator stress at the maximum limiting strain. The yield strain is ε_y.

where σ' is the effective deviator stress, ε is the axial strain, and σ'_{ult} is the hypothetical value of the effective deviator stress at infinite strain. The observed deviator stress at failure occurs at a finite strain (Mesri et al., 1981). A linear least-square regression on a transformed version of Equation 21 is used to determine the values of E_i and σ'_{ult} (Duncan and Chang, 1970). Combining Equations 20 and 21, the yield point is found at the intersection of the secant modulus and hyperbolic functions. The yield stress can be expressed in terms of the hyperbolic stress as follows:

$$\sigma'_y = (1 - A)\, \sigma'_{ult} \qquad (22)$$

Next, the value of A must be determined. A limiting strain is defined as the strain difference between the secant modulus strain, ε_s, and the hyperbolic strain, ε_h, at the same deviator stress (Fig. 1). The limiting strain, which controls the deviation of secant modulus line from the hyperbolic curve, is expressed as

$$\varepsilon_l = \varepsilon_s - \varepsilon_h \qquad (23)$$

in which ε_l is the limiting strain. The secant modulus strain is expressed by rewriting Equation 20 as

$$\varepsilon_s = \frac{\sigma'}{AE_i} \qquad (24)$$

and the hyperbolic strain is given by rearranging Equation 21 as

$$\varepsilon_h = \frac{\sigma'_{ult}\, \sigma'}{(\sigma'_{ult} - \sigma')E_i} \qquad (25)$$

Substituting Equations 24 and 25 into Equation 23 produces

$$\varepsilon_l = \frac{\sigma'}{AE_i} - \frac{\sigma'_{ult}\, \sigma'}{(\sigma'_{ult} - \sigma')E_i} \qquad (26)$$

The maximum limiting strain, ε_{lmax}, between ε_s and ε_h is evaluated where $d\varepsilon_l/d\sigma' = 0$. This can be mathematically expressed as

$$\frac{d\varepsilon_l}{d\sigma'} = \frac{1}{AE_i} - \frac{\sigma'_{ult}}{(\sigma'_{ult} - \sigma')E_i}\left(1 + \frac{\sigma}{\sigma'_{ult} - \sigma'}\right) = 0 \qquad (27)$$

The effective deviator stress at the maximum limiting strain, σ'_{lmax}, can be calculated by solving Equation 27 as follows

$$\sigma'_{lmax} = (1 - \sqrt{A})\, \sigma'_{ult} \qquad (28)$$

Substituting Equation 28 into Equation 26 and evaluating the equation at $\varepsilon_l = \varepsilon_{lmax}$, the following relationship between A and ε_{lmax} can be established:

$$A = \frac{1}{\left[\sqrt{(\varepsilon_{lmax} E_i / \sigma'_{ult})} + 1\right]^2} \qquad (29)$$

Finally, the yield stress is determined by substituting Equation 29 into Equation 22.

In summary, the parameters for the secant modulus method are ε_{lmax}, A, E_i and σ'_{ult}. Once the value of ε_{lmax} is chosen, the yield stress is found. The objective of the secant modulus method is to approximate the nonlinear stress-strain behavior as closely as possible. The proposed method is able to simulate the idealized elastic behavior by means of the secant modulus line, yet preserve the nonlinear behavior by using a limiting strain criterion. The limiting strain criterion is employed in the selection of ε_{lmax} in determination of yield stress. Since the yield stress is evaluated by the secant modulus method, a novel testing method has been developed to determine the parameters of the viscoplastic model.

Two-test method

Tests are needed to evaluate the yield parameters and strain-rate parameters of the viscoplastic model. A two-test method, which consisted of a consolidated drained triaxial (CDTX) test and a consolidated drained stress-controlled strain-rate (CDSR) test, conducted on a triaxial apparatus, was developed for this purpose. The CDTX test was used to evaluate the yield parameters of landslide soil, and the CDSR test was used to determine the strain-rate parameters of shear zone soil. Although the CDTX and CDSR tests are independent tests, the determinations of yield and strain-rate parameters are interrelated because the viscous displacement depends on the deviator stress in excess of the yield stress.

For the CDTX test, the specimen was saturated and isotropically consolidated. Then, the drained strength was recorded until the specimen reached 20% strain. Three strength tests were needed to define the Drucker-Präger yield envelope, which was plotted in stress defined by the square root of the second invariant deviator stress tensor versus the effective mean normal stress space. The second invariant of effective deviator stress tensor, II, for triaxial test is found by,

$$II = \frac{1}{3}(\sigma'_1 - \sigma'_3)^2 \qquad (30)$$

where σ'_1 is the effective major principal stress, and σ'_3 is the effective minor principal stress. Similarly, the second invariant of deviatoric strain-rate tensor, II_D, is determined by,

$$II_D = \frac{1}{3}(\dot{\varepsilon}_1 - \dot{\varepsilon}_3)^2 \qquad (31)$$

where $\dot{\varepsilon}_1$ is the major principal strain-rate, and $\dot{\varepsilon}_3$ is the minor principal strain-rate. It should be noted that $II^{1/2}$ is the deviatoric stress intensity and $II_D^{1/2}$ is the strain-rate intensity, both terms represent the magnitudes of the corresponding tensors. The yield stress intensity, $II_t^{1/2}$, is calculated by taking the square-root of Equation 30 at yield condition.

For the CDSR test, the specimen was also saturated and isotropically consolidated. Then, the yield stress intensity, $II_t^{1/2}$, was evaluated according to the Drucker-Präger yield envelope from the CDTX data. The CDSR test was conducted in a stress-controlled manner, a constant stress level greater than $II_t^{1/2}$ was applied to the specimen to induce viscous deformation. Small loading increments were applied to the specimen to obtain as many constant stress levels as possible. At least three constant stress levels were needed for the test, and each constant stress level was maintained for twelve hours. To determine the strain-rate parameters, Equation 8 can be re-written in the form

$$II_e^{1/2} = 2\mu_0 D_0^{(n-1)/n}(II_D^{1/2})^{1/n} \qquad (32)$$

If the following substitution is made,

$$b = 2\mu_0 D_0^{(n-1)/n} \qquad (33)$$

and

$$m = 1/n \qquad (34)$$

the equation becomes a simple power function,

$$II_e^{1/2} = b(II_D^{1/2})^m \qquad (35)$$

in which $II_e^{1/2}$ is the excess stress intensity, $II_D^{1/2}$ is the strain-rate intensity. $II_e^{1/2}$ and $II_D^{1/2}$ were evaluated for each stress level at the end of the test. The excess stress intensity is calculated by subtracting the yield stress intensity from the deviator stress intensity. A least-square regression analysis of a log-linearized form of equation 35 can then be used to determine b and m. Although n can be obtained, μ_0 and D_0 cannot be individually determined from the log-linearized regression of excess stress intensity versus strain-rate intensity data. However, once the shear zone thickness is determined, D_0 can be evaluated from Equation 14, and then μ_0 is calculated by using Equation 33.

RESULTS AND ANALYSIS

The two-test method was employed using undisturbed specimens obtained from the shear zone of Minor Creek landslide, California. The effect of using undisturbed versus remolded specimens was also evaluated. The physical properties of shear zone soil were determined by grain-size analysis (ASTM, 1991, D422) and Atterberg limits tests (ASTM, 1991, D423 and D424) prior to the two-test program. It was found that the liquid limit was 26%; the plastic limit was 15.7%; and the plasticity index was 10.3%. Grain-size analysis showed that 85.3% of the soil passed through a No. 4 sieve and 46.8% of the soil passed through a No. 200 sieve. The shear zone soil was identified as silty, clayey sand using the Unified Soil Classification System. The saturated unit weight was 21.3 kN/m³ and the in situ moist unit weight was 19.9 kN/m³.

Yield strength parameters

To determine the yield strength parameters, CDTX tests were conducted on undisturbed and remolded samples obtained from the shear zone. The stress-strain curves of all CDTX tests were fitted using Kondner's hyperbolic function. The parameters of the hyperbolic function and the coefficients of determination for the least-square regression on CDTX tests are listed in Table 1. Then, a value of the maximum limiting strain, ε_{1max}, must be chosen for the secant modulus method to determine the yield strength.

The yield stress depends on the choice of ε_{1max}. A greater value of ε_{1max} gives a greater yield stress. The relationship of A versus ε_{1max} for an undisturbed test is shown in Figure 2.

TABLE 1. HYPERBOLIC PARAMETERS OF CDTX TESTS FOR UNDISTURBED AND REMOLDED SAMPLES OBTAINED FROM THE SHEAR ZONE

CDTX Test	Effective Confining Stress (kPa)	Initial Tangent Modulus (kPa)	Hypothetical Stress (kPa)	Coefficient of Determination for the Least-Square Regression
411 Undisturbed	69.0	5,520	102.3	0.99
421 Remolded	55.2	6,000	71.6	0.98
	103.4	12,700	130.3	0.98
	151.7	209,000	291.9	0.99

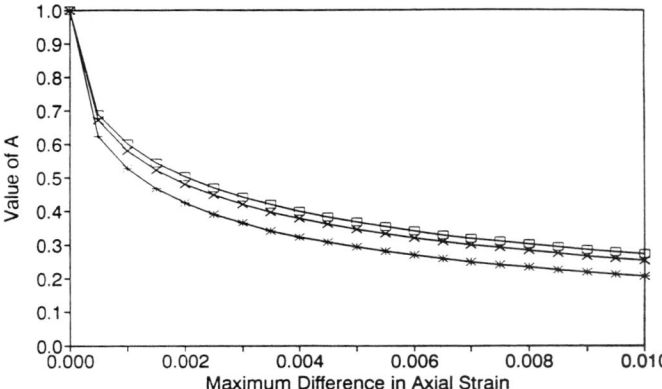

Figure 3. The relationship of the secant modulus proportionality factor, A, and the maximum limiting strain, ε_{lmax}, for a consolidated drained triaxial tests on remolded samples.

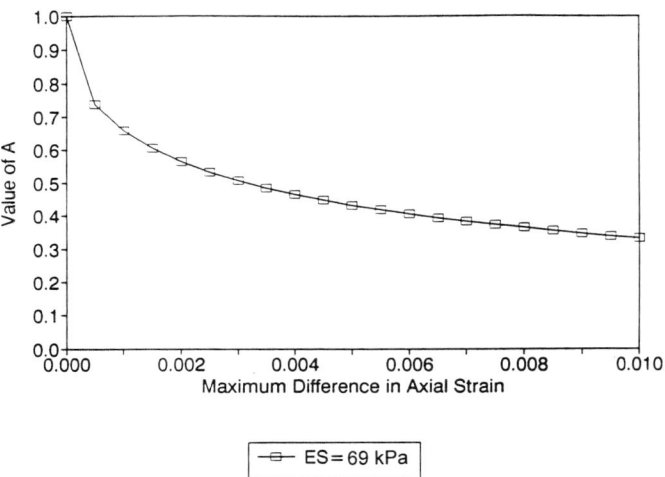

Figure 2. The relationship of the secant modulus proportionality factor, A, and the maximum limiting strain, ε_{lmax}, for a consolidated drained triaxial test on an undisturbed sample.

The value of A decreases significantly with ε_{lmax} in the range of 0% to 0.4%, and then it approaches a steady value with ε_{lmax} up to 1%. The remolded curves in Figure 3 are almost the same as that in Figure 2 with limited scatter.

With the known values of hyperbolic parameters in Table 1, the yield stresses were evaluated with ε_{lmax} equal to 0.2% and 0.4%. It was found that the actual nonlinear stress-strain behavior cannot be preserved when the maximum limiting strain is greater than 0.4%. Should the maximum limiting strain be less than 0.2%, the yield strength will be too low. By restricting the deviation of the secant modulus line to 0.2% and 0.4% maximum limiting strains, the secant modulus method closely approximates the nonlinear stress-strain behavior. Therefore, the 0.2% and 0.4% maximum limiting strains were used to define the yield stresses.

Once the yield stresses were found, the Drucker-Präger yield envelope and the effective yield friction angles, ϕ'_y, were determined for undisturbed and remolded samples. The cohesion parameter, κ, was found to be zero from the yield envelope, because the strength of shear zone soil is at residual strength and the cohesive strength was destroyed by prolonged shearing (Skempton, 1964, 1985). For each type of sample, two sets of yield parameters were evaluated with ε_{lmax} equal to 0.2% and 0.4%. The yield parameters are summarized in Table 2, which include the parameters A, α, and ϕ'_y.

The choice of the maximum limiting strain can be justified by comparing the effective yield friction angle with the effective in-situ friction angle. The effective in-situ friction angle is 17.5° determined by back calculation using an infinite-slope limit equilibrium analysis (Iverson and Major, 1987). Based on the assumptions of Iverson's viscoplastic model, time-dependent displacement occurs whenever the yield strength is ex-

TABLE 2. YIELD PARAMETERS FOR UNDISURBED AND REMOLDED SAMPLES EVALUATED AT 0.2% AND 0.4 MAXIMUM LIMITING STRAIN (ε_{lmax}) CRITERIA

CDTX Test	A	α	ϕ'_y (degrees)
411 ε_{lmax} = 0.2 % Undisurbed	0.56	0.3069	14.13
421 ε_{lmax} = 0.2 % Remolded	0.55	0.2925	13.50
411 ε_{lmax} = 0.4 % Undisurbed	0.43	0.3807	17.30
421 ε_{lmax} = 0.4 % Remolded	0.36	0.3858	17.50

ceeded. Hence, the limit-equilibrium condition of the infinite slope method, which also depicts the yielding of soil before any incipient displacement, provides a means to estimate the in-situ friction angle and residual strength of shear zone soil. The effective yield friction angles based on 0.2% maximum limiting strain are smaller than the in-situ friction angle, and the effective yield friction angles based on 0.4% maximum limiting strain are equal to the in-situ friction angle. Therefore, the 0.4% limiting strain criterion provides a justifiable means for determination of yield strength.

Strain-rate parameters

Two CDSR tests were conducted on undisturbed and remolded samples of shear zone soil to determine the strain-rate parameters. The samples were consolidated at 69 kPa corresponding to the effective in-situ confining pressure. Then, the stress-strain-time response of the samples were recorded at three constant stress levels greater than the yield stress intensity. More constant stress levels can be obtained if a greater effective confining pressure is used. However, the effective confining pressure should ideally correspond to the effective in-situ confining pressure. For each stress level, the strain rate was evaluated to determine the strain-rate intensity, and the excess stress intensity was calculated by subtracting the yield stress intensity from the deviator stress intensity of that stress level. Then, b and m were determined on the log-linearized regression of excess stress intensity versus strain-rate intensity data.

For each type of sample, b and m were evaluated at both 0.2% and 0.4% maximum limiting strains as summarized in Table 3. Student t-tests were conducted to determine if the regression function and the experimental data represent the same set. The t-test two-tailed probabilities ranged from 0.990 to 0.999 as shown in Table 3. For the undisturbed sample, Figures 4 and 5 show the nonlinear viscous deformation determined at 0.2% and 0.4% maximum limiting strains respectively; n was evaluated as 1.16 for $\varepsilon_{lmax} = 0.2\%$ and 0.51 for $\varepsilon_{lmax} = 0.4\%$. For the remolded sample, Figures 6 and 7 display the nonlinear viscous deformation based on 0.2% and 0.4% limiting strains respectively; n was evaluated as 2.97 for $\varepsilon_{lmax} = 0.2\%$ and 2.15 for $\varepsilon_{lmax} = 0.4\%$. The value of n decreases with a greater ε_{lmax} value. This indicates that the soils apparently change deformation style from $n > 1$ (shear-thinning behavior) to $n < 1$ (shear-

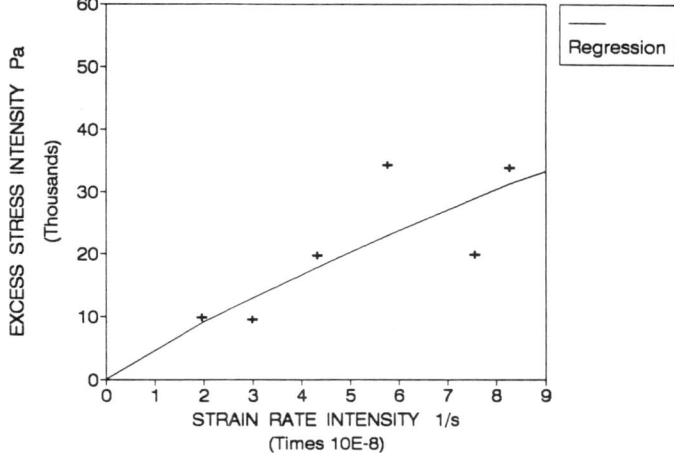

Figure 4. The nonlinear viscous deformation is represented by the regression on excess stress intensity versus strain rate intensity data for CDSR test 412 on an undisturbed sample evaluated at 0.2% maximum limiting strain. Each constant stress level was maintained for twelve hours, two strain rate intensities were evaluated at the second and the eighth hours of that stress level.

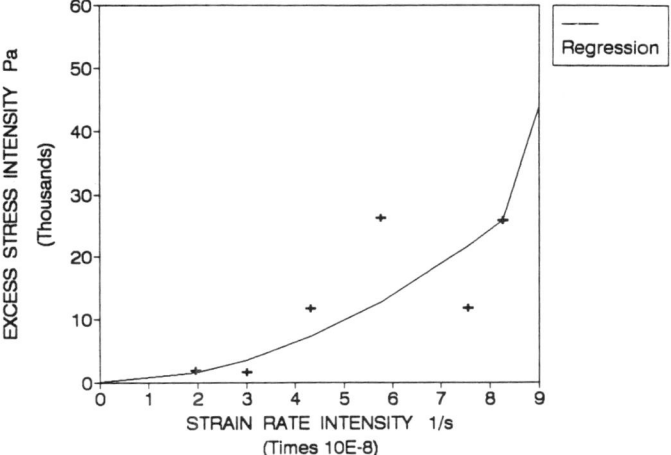

Figure 5. The nonlinear viscous deformation is depicted by the regression on excess stress intensity versus strain rate intensity data for CDSR test 412 on undisturbed sample evaluated at 0.4% maximum limiting strain. Each constant stress level was maintained for twelve hours, two strain rate intensities were evaluated at the second and the eighth hours of that stress level. The regression curve is concave upward because the power-law exponent is less than unity.

TABLE 3. LOG-LINEARIZED REGRESSION PARAMETERS OF CDSR TESTS FOR UNDISTURBED AND REMOLDED SAMPLES BASED ON 0.2% AND 0.4% MAXIMUM LIMITING STRAIN (e_{lmax})

CDSR Tests	m	log b	n	b (N-s/m²)	t-test Probability
412 $\varepsilon_{lmax} = 0.2\%$ Undisturbed	0.863	10.61	1.16	4.07×10^{10}	0.996
422 $\varepsilon_{lmax} = 0.2\%$ Remolded	0.337	6.90	2.97	7.92×10^{6}	0.990
412 $\varepsilon_{lmax} = 0.4\%$ Undisturbed	1.969	18.36	0.51	2.29×10^{18}	0.996
422 $\varepsilon_{lmax} = 0.4\%$ Remolded	0.465	7.64	2.15	4.32×10^{7}	0.999

thickening behavior), which contradicts previous inferences that long-term slope movement exhibits shear-thinning behavior (Mitchell 1976, Iverson 1984). The apparent change of deformation mode results from the selection of the maximum limiting strain (Figure 8).

Figure 8 is a plot of deviator stress intensity versus strain-rate intensity for the undisturbed sample (Test No. 412). Viscoplastic flow was observed at three constant stress levels, and the yield stress intensity was determined at 0.2% and 0.4% maximum limiting strains. Since the regression of strain-

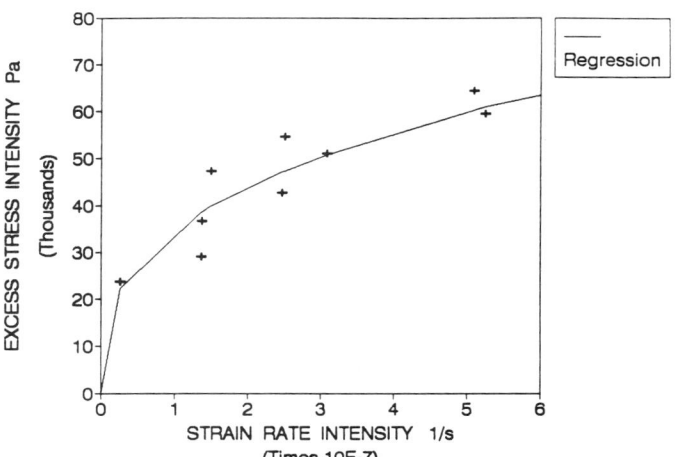

Figure 6. The nonlinear viscous deformation is displayed by the regression on excess stress intensity versus strain rate intensity data for CDSR test 422 on a remolded sample evaluated at 0.2% maximum limiting strain.

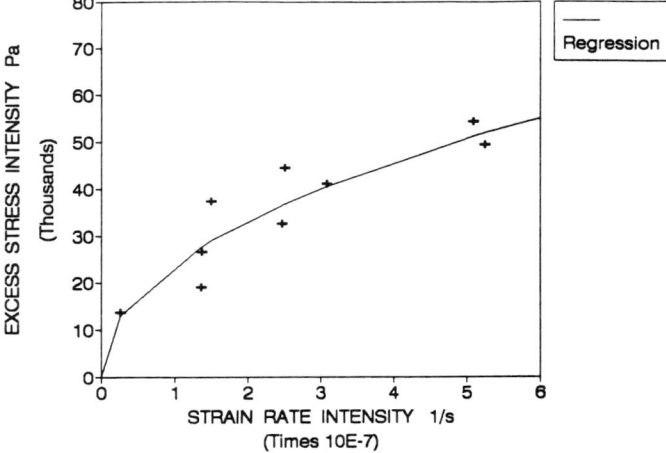

Figure 7. The nonlinear viscous deformation is exhibited by the regression on excess stress intensity versus strain rate intensity data for CDSR test 422 on a remolded sample evaluated at 0.4% maximum limiting strain.

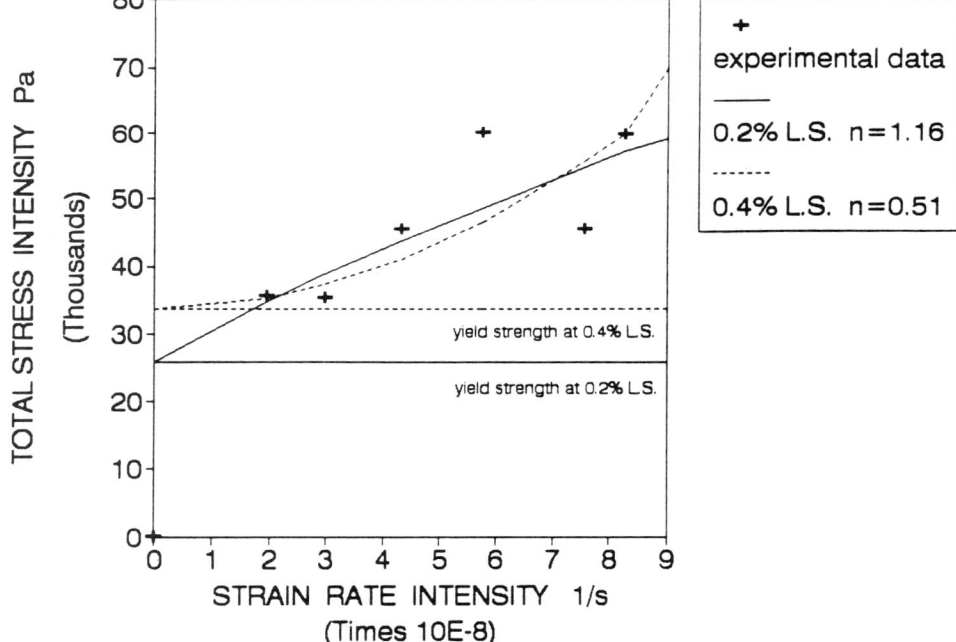

Figure 8. Comparison of 0.2% and 0.4% limiting strain criteria on the evaluation of strain-rate parameters. The solid horizontal line is the yield strength evaluated at 0.2% maximum limiting strain (L.S.), and the dotted horizontal line is the yield strength evaluated at 0.4% maximum limiting strain (L.S.). The solid curve is the regression based on 0.2% maximum limiting strain (0.2% L.S.), and the dotted curve is the regression based on 0.4% maximum limiting strain (0.4% L.S.). The 0.4% limiting strain criterion has a higher yield strength than that of the 0.2% limiting strain criterion. A higher yield strength causes the regression curve to be concave upward and the power-law exponent, n to be less than unity.

rate parameters is based on the excess stress intensity instead of the deviator stress intensity (Equation 35), the 0.4% limiting strain criterion has a yield stress intensity greater than that of the 0.2% limiting strain criterion. A higher cut-off point on the excess stress intensity (from the 0.4% criterion) made the regression curve concave upward, resulting in $n < 1$. The lower cut-off point from the 0.2% criterion results in $n > 1$. Hence, the 0.2% limiting strain criterion gives a more reasonable result.

The yield friction angles based on the 0.4% limiting strain criterion match the in-situ friction angle better than those of the 0.2% limiting strain criterion. However, the 0.2% limiting strain criterion gives a more reasonable viscous behavior than the 0.4% limiting strain criterion. This phenomenon possibly results from the existence of coarse particles in the soil at the Minor Creek landslide (Iverson and Major, 1987). Ten boreholes were drilled at the Minor Creek site for this study, and the locations of the boreholes are shown in Figure 9 (Hovind, 1990). Undisturbed samples were obtained by 76.2-mm diameter Shelby tubes. Half of the time during the drilling process, the Shelby tubes encountered boulders at various depths, which indicates that cobble- and boulder-size particles are mixed in a finer soil matrix. The percentage of these oversized particles may be as high as 50% by weight.

Su (1989) and Fragaszy et al. (1990, 1992) studied the effect of oversized particles on static strength of cohesionless soil and defined the fraction of oversized particles, p, as the ratio of the oversize-particles mass to the total soil-mixture mass. When p is less than 0.40, the oversized particles are in the floating state. The strength of the mixture soil can be modeled by removing the oversized particles, and testing the matrix material prepared at the density that exists within the

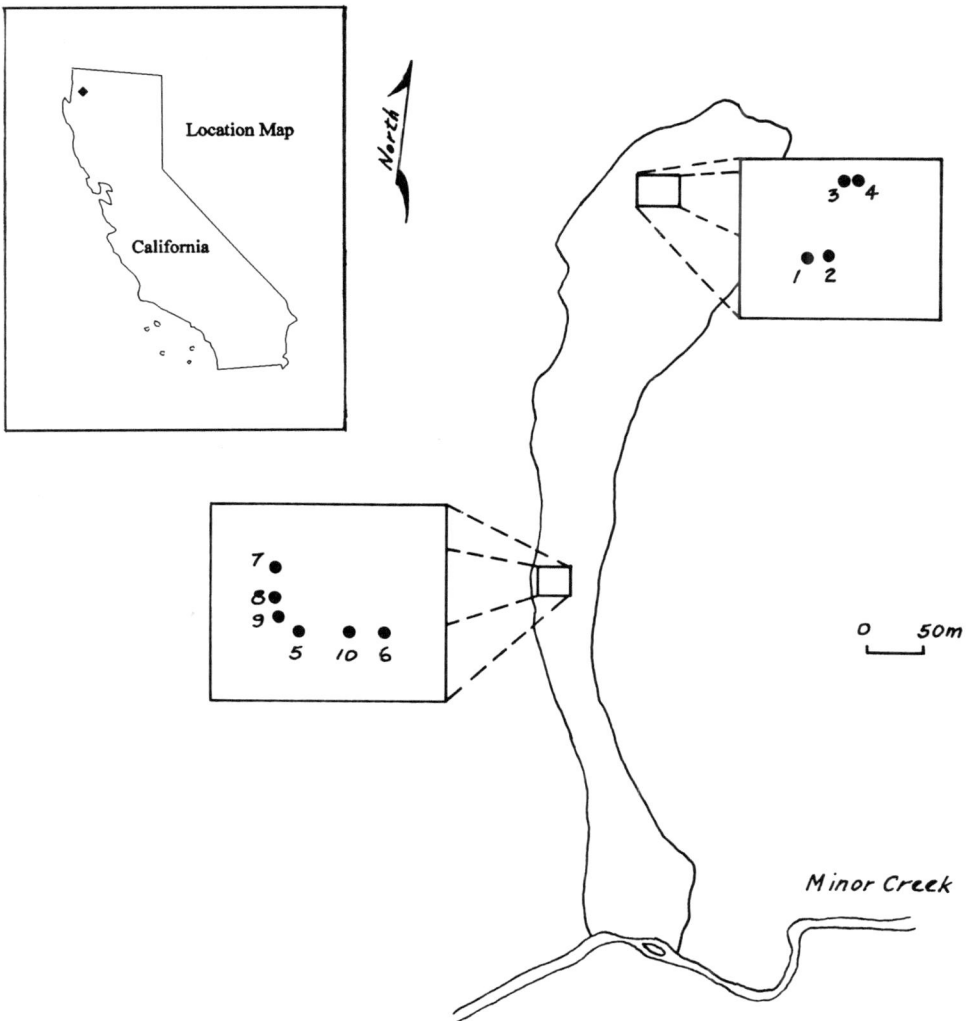

Figure 9. Locations of ten boreholes at the Minor Creek landslide. Boreholes No. 1 through No. 4 are located at the head of the slide, and boreholes No. 5 through No. 10 are located at the middle portion of the slide. Undisturbed samples were obtained at ten boreholes. The undisturbed samples used for the two-test method were taken from the shear zone of boreholes No. 5 through No. 10 (partially based on map from Iverson and Major, 1987).

mixture soil away from the oversized particles. When p is greater than 0.65, the oversized particles are in the nonfloating state. The strength of the mixture soil can be modeled by testing the oversized particles alone.

Because of the existence of the oversized particles in the landslide, tests on the soil specimens only represented the properties of the matrix material and did not represent the properties of the overall mixture material. It is reasonable to assume that the determinations of the yield strength and viscoplastic behavior based on the 0.2% limiting strain criterion only represented the characteristics of the matrix material. If the coarse material is in a nonfloating state or has a higher strength than the matrix material, the yield friction angles based on the 0.2% limiting strain criterion should be less than the in-situ friction angle (based on the infinite-slope analysis) of the landslide. On the other hand, if the coarse material is in a floating state or has a lower strength than the matrix material, the yield friction angles based on the 0.2% limiting strain criterion should be equal to the in-situ friction angles.

Once the slope starts to move, the coefficient of friction is reduced from static friction to dynamic friction and the influence of oversized particles on the yield strength no longer exists. Prior and Coleman (1978), and Vallejo (1979, 1980) observed that the coarse material loses grain-to-grain contact and is dispersed in the matrix material as the slope displaces. Therefore, the viscoplastic behavior of time-dependent slope movement is controlled by the matrix material. This may explain why the 0.4% limiting strain criterion matched the in-situ yield strength better and the 0.2% limiting strain criterion matched the viscoplastic behavior better.

VELOCITY PROFILE CALCULATIONS FOR MINOR CREEK LANDSLIDE

The Minor Creek landslide is located 20 miles northeast of Eureka, California. This landslide was selected to test the proposed methodology because of extensive data already collected there (Iverson, 1984; Iverson and Major, 1987). Geometry, movement, and hydrology are defined by data taken from borings, stake-line surveys, inclinometers, extensometers, and piezometers over the past 11 years. The landslide is approximately 760 m in length and 80 m in width, with an average inclination of 15°. Down-slope movement occurs within a basal shear zone, approximately 1 m thick, at an average depth of 5 m.

Seasonal changes in the effective-stress state in the Minor Creek landslide occur as a result of pronounced changes in rainfall and groundwater flow (Iverson and Major, 1987). During the wet season, the water table rises and reduces the ratio of resisting stress to driving stress to a value less than one. During the period when the soil is failing, the landslide movement rate is controlled by the value of this stress ratio and the viscous resistance in the shear zone (Iverson, 1986c).

A predicted velocity profile can be made from the procedures described previously. Iverson (1984) reported that the maximum field velocities of the Minor Creek landslide at four inclinometers RC9B, RC9A, RC9E, and 11A were in the range of 2.5×10^{-9} m/s to 4.9×10^{-9} m/s. All inclinometers were located at the middle portion of the landslide and the shear zone thickness varied from 0.7 m to 1.2 m. Inclinometer RC9A was used to compare the calculated velocity profiles with the field velocity profile for this study.

Four maximum velocities at inclinometer RC9A were calculated at 0.2% and 0.4% maximum limiting strains on undisturbed and remolded samples. The strain-rate parameters and the predicted maximum velocities at RC9A are listed in Table 4. For the undisturbed sample, $V_{x\,max}$ calculated with parameters evaluated at 0.2% limiting strain is close to the maximum field velocity; $V_{x\,max}$ calculated with parameters evaluated at 0.4% limiting strain is one order of magnitude larger than the field value. For the remolded sample, the calculated maximum velocities based on 0.2% and 0.4% limiting strain criteria are two orders of magnitude and one order of magnitude less than the field value, respectively. Figure 10 shows the comparison of calculated and field velocity profiles at inclinometer RC9A evaluated at 0.2% limiting strain criterion on undisturbed and remolded samples. The figure indicates the calculated profile based on an undisturbed sample most closely approximates the field velocity profile. Hence, the 0.2% limiting strain criterion on an undisturbed sample gives the best calculation of maxi-

TABLE 4. STRAIN-RATE PARAMETERS AND PREDICTED MAXIMUM VELOCITIES ON UNDISTURBED AND REMOLDED SAMPLES AT INCLINOMETER BORING RC9A

CDSR Tests	0.2 Percent Limiting Strain Criterion				0.4 Percent Limiting Strain Criterion			
	$V_{x\,max}$ m/s	n	D_0 1/s	μ_0 N-s/m²	$V_{x\,max}$ m/s	n	D_0 1/s	μ_0 N-s/m²
412 Undisturbed	4.40×10^{-9}	1.16	3.03×10^{-9}	3.0×10^{11}	2.81×10^{-8}	0.51	3.82×10^{-8}	8.56×10^{-10}
422 Remolded	2.98×10^{-11}	2.97	2.41×10^{-11}	4.4×10^{13}	3.09×10^{-10}	2.15	9.7×10^{-10}	1.43×10^{12}

Note: The maximum field velocity measured at boring RC9A is 3.9×10^{-9} m/s.

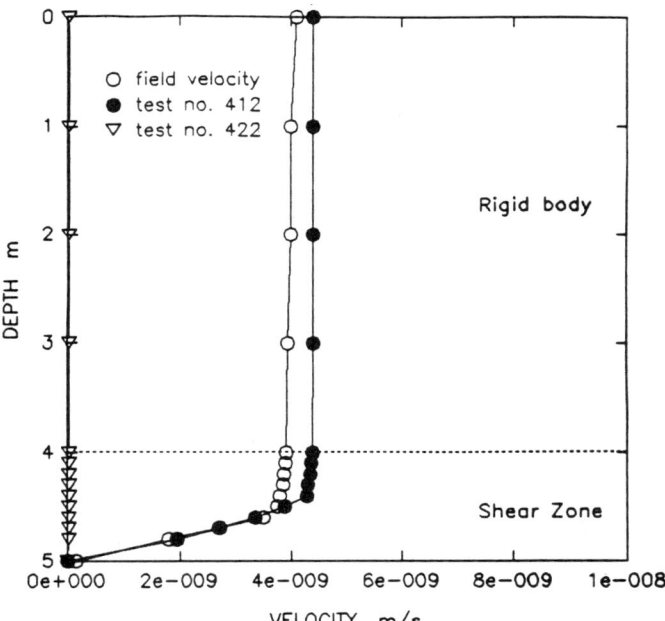

Figure 10. Comparison of predicted and field velocity profiles at inclinometer RC9A evaluated at 0.2% maximum limiting strain on both an undisturbed sample and a remolded sample. The field velocity profile was derived from repetitive inclinometer surveys. The predicted velocity profile from test No. 412 was based on an undisturbed sample, and the predicted velocity profile from test No. 422 was based on a remolded sample.

mum velocity and velocity profile. The accuracy of one order of magnitude in calculated maximum velocity is considered to be acceptable, because the maximum field velocities also have an order of magnitude variation.

By comparison of calculated maximum velocities with the maximum field velocity, it was found that the undisturbed sample performed better than the remolded sample. Undisturbed samples are representative of the in-situ soil conditions, which include soil structure, particle orientation, and mechanical properties. Undisturbed samples obtained from the shear zone have particles oriented in the shear force direction. However, remolded samples cannot maintain the same particle orientation as the undisturbed samples.

CONCLUSIONS

A laboratory method has been established to evaluate rate-dependent slope movement. This method utilizes a nonlinear viscoplastic constitutive model and new testing procedures to predict the displacement rate.

The new test procedures involved a two-test method on triaxial apparatus. The testing program, conducted on undisturbed and remolded samples from the Minor Creek landslide, included consolidated drained strength tests and stress-controlled strain-rate tests. Consolidated drained strength tests were used to evaluate the yield behavior, and consolidated drained stress-controlled strain-rate tests were used to evaluate the viscous behavior. The effect of using undisturbed versus remolded specimens was also studied. The new procedures included the secant modulus method for yield strength determination and a limiting strain criterion for nonlinear viscous flow evaluation.

The following conclusions are made from this study:

(1) The laboratory method can be used effectively to calculate the velocity profile at the Minor Creek landslide in California. The calculated maximum velocity evaluated at 0.2% maximum limiting strain on an undisturbed sample is within an order of magnitude of the field value, and the calculated profile closely approximates the field velocity profile.

(2) The secant modulus method can determine the yield strength of soils. The yield strength of soil was arbitrarily defined in the past because of the nonlinear nature of the stress-strain curve of soil; however, it can consistently be defined by four parameters of the secant modulus method. This method provides a new definition of yield strength of soils, which can be used in general soil-mechanics applications and to evaluate time-dependent deformation.

(3) Based on the results of Student t-tests, measured values of strain-rate parameters represent nonlinear viscoplastic flow very well.

(4) Calculation of the velocity profile using parameters obtained from tests on undisturbed samples is better than that using remolded samples. It is worthwhile to obtain high-quality, undisturbed specimens.

ACKNOWLEDGMENTS

The research presented in this paper is based upon work partially supported by the National Science Foundation under Grant No. BCS-8815704. The authors would also like to acknowledge Washington State University and the United States Geological Survey for their support.

APPENDIX I. SYMBOLS

Symbols	Definitions
A	proportionality factor for secant modulus method
A_1	coefficient defined in Equation 17
A_2	coefficient defined in Equation 18
b	power function regression parameter
c	cohesion of soil
c'_y	drained Mohr-Coulomb cohesion at yield
d	depth of water table
D	deviatoric strain rate tensor
D_0	reference strain rate
E_i	initial tangent modulus
E_s	secant modulus
h	thickness of landslide
I_1	first invariant of stress tensor
II	second invariant of the deviatoric stress tensor
II_D	second invariant of the deviatoric strain rate tensor

II_e	second invariant of the deviatoric stress tensor in excess of yield	ε_h	strain on the hyperbolic stress-strain curve
II_t	second invariant of the deviatoric stress tensor at yield	ε_l	limiting value of strain for secant modulus method
m	power function regression parameter	ε_{lmax}	maximum limiting strain for secant modulus method
n	power-law exponent	ε_s	strain on the secant modulus line
p	mass ratio of oversized particles to total soil mixture	ε_y	axial strain at yield point for secant modulus method
p'	effective mean normal stress	$\dot{\varepsilon}_1$	major principal strain-rate
T	deviatoric stress tensor	$\dot{\varepsilon}_3$	minor principal strain-rate
T_e	excess deviatoric stress tensor	θ	slope angle of landslide
T_t	deviatoric stress tensor at yield	κ	cohesive Drucker-Präger yield parameter
T	thickness of pseudo-rigid body of landslide	μ	the apparent viscosity of a material
v_x	velocity of landslide in the downslope direction	μ_0	equivalent Newtonian viscosity measured at $II_D^{1/2} = D_0$
y	depth within the shear zone of landslide	ϕ'_y	effective yield friction angle
y_{ref}	reference depth	σ'	effective deviator stress
α	frictional Drucker-Präger yield parameter	σ'_1	effective major principal stress
γ_{moist}	moist unit weight of soil	σ'_3	effective minor principal stress
γ_{sat}	saturated unit weight of soil	σ'_y	effective yield stress
γ_w	unit weight of water	σ'_{lmax}	effective deviator stress evaluated at the maximum limiting strain
ε	axial strain	σ'_{uh}	hypothetical deviator stress at infinite strain

REFERENCES CITED

American Society of Testing and Materials, 1991, Annual book of standards: Soil and rock, building stones; geotextiles.

Drucker, D. C., and Präger, W., 1952, Soil mechanics and plastic analysis or limit design: Quarterly Applied Mathematics, v. 10, p. 157–165.

Duncan, J. M., and Chang, C. Y., 1970, Nonlinear analysis of stress and strain in soils: Journal of Soil Mechanics and Foundation Engineering, American Society of Civil Engineers, v. 96, p. 1629–1653.

Fragaszy, R. J., Su, W., and Siddiqi, F. H., 1990, Effects of oversize particles on density of clean granular soils: Geotechnical Testing Journal, v. 13, p. 106–114.

Fragaszy, R. J., Su, J., Siddiqi, F. H., and Ho, C. L., 1992, Modeling strength of sandy gravel: Journal of Geotechnical Engineering, American Society of Civil Engineers, v. 118, p. 920–935.

Hovind, C. L., 1990, Determination of time-dependent soil parameters by triaxial tests [M.S. Civil Engineering thesis]: Pullman, Washington, Washington State University, 109 p.

Iverson, R. M., 1984, Unsteady, non-uniform landslide motion theory and measurement [Ph.D. dissert.]: Stanford, California, Stanford University, 303 p.

Iverson, R. M., 1985, A constitutive equation for mass-movement behavior: Journal of Geology, v. 93, p. 143–160.

Iverson, R. M., 1986a, Unsteady, nonuniform landslide motion: 1. Theoretical dynamics and the steady datum state: Journal of Geology, v. 94, p. 1–15.

Iverson, R. M., 1986b, Unsteady, nonuniform landslide motion: 2. linearized theory and the kinematics of transient response: Journal of Geology, v. 94, p. 349–364.

Iverson, R. M., 1986c, Dynamics of slow landslides: A theory for time-dependent behavior, *in* Hillslope processes: Boston, Allen and Unwin, p. 297–317.

Iverson, R. M., and Major, J. J., 1987, Rainfall, ground-water flow, and seasonal movement at Minor Creek landslide, northwestern California: Physical interpretation of empirical relations: Geological Society of America Bulletin, v. 99, p. 579–594.

Kondner, R. L., 1963, Hyperbolic Stress-Strain Response: Cohesive Soil: Journal of Soil Mechanics and Foundations Division, American Society of Civil Engineers, v. 89, p. 115–143.

Mesri, G., Febres-Cordero, E., Shield, D. R., and Castro, A., 1981, Shear stress-strain-time behavior of clays: Géotechnique, v. 31, p. 537–552.

Mitchell, J. K., 1976, Fundamentals of soil behavior: New York, Wiley, 422 p.

Prior, D. B., and Coleman, J. M., 1978, Disintegrating retrogressive landslides on very low angled subaqueous slope, Mississippi Delta: Marine Geotechnology, v. 3, p. 37–60.

Skempton, A. W., 1964, Long term stability of clay slopes. Géotechnique, v. 14, p. 77.

Skempton, A. W., 1985, Residual strength of clays in landslides, folded strata and the laboratory. Géotechnique, v. 35, p. 3–18.

Su, W., 1989, Static strength evaluation of cohesionless soils with oversize particles [Ph.D. dissert.]: Pullman, Washington, Washington State University, 247 p.

Taylor, D. W., 1948, Fundamentals of soil mechanics: London, England, Wiley, 700 p.

Vallejo, L. E., 1979, An explanation for mudflow: Géotechnique, v. 29, p. 351–354.

Vallejo, L. E., 1980, A new approach to the stability analysis of thawing slopes: Canadian Geotechnical Journal, v. 17, p. 607–612.

MANUSCRIPT ACCEPTED BY THE SOCIETY MAY 20, 1994

Stability of slopes: Limit analysis approach

Radoslaw L. Michalowski
Department of Civil Engineering, The Johns Hopkins University, Baltimore, Maryland 21218

ABSTRACT

Traditional analyses of earth slope stability are based on equilibrium considerations of "slices" into which the collapsing soil mass is divided. Such analyses require some static assumptions so that the system of equilibrium equations becomes determinate. It is proposed that the kinematics-based limit analysis be used to calculate the stability (factor of safety or the critical height), rather than using the limit equilibrium technique with arbitrary static assumptions (such as, for instance, inclination of forces between slices). The method described is straightforward, the analysis has a clear mechanical interpretation, and it stands on its own as a means of performing stability analyses for earth slopes. The method proposed, when used with identical failure mechanisms as in traditional slice methods, can be used to assess the static assumptions used in these conventional methods. It can be argued that all solutions using conventional slice methods can be bound by a rigorous upper bound solution and another approximate solution, both presented in this paper. Traditional solutions which fall beyond the range indicated lead to unrealistic consequences when interpreted in view of the collapse mechanism.

A convenient technique for including the influence of the pore pressure on stability of slopes is shown, and results of critical height calculations for homogeneous slopes are presented.

INTRODUCTION

There is an abundance of literature on the analysis of stability of slopes, yet the methods for evaluation of safety of earth slopes are not standardized. A wide survey of methods was presented recently by Duncan (1992) for both two- and three-dimensional failure modes. Virtually all techniques used in practice for estimations of the safety factor are based on the division of the failing soil mass into "slices" and analyzing the static equilibrium of such slices. As the problem is statically indeterminate, arbitrary assumptions are introduced as to the location or inclination of forces between the slices (see e.g., Morgenstern and Price, 1965; Spencer, 1967; Janbu, 1973). Perhaps the simplified Bishop's method (Bishop, 1955) has found the widest acceptance in practice, because of its relative simplicity. However, none of the methods can be considered accurate, or more accurate than others, as the significance of different static assumptions cannot be assessed readily. The slice technique has been used more recently in the context of variational analysis. This method, as presented by Leshchinsky (1990), seems to be a very convenient tool in slope analysis, despite an earlier controversy (De Jong, 1981).

A different approach to stability of slopes was proposed by Chen et al. (1969), who used the upper bound theorem of limit analysis to arrive at critical heights of slopes (see also Chen and Giger, 1971). Chen et al. (1969) considered plane rotational collapse mechanisms. The upper bound technique was utilized more recently in three-dimensional analysis of slopes collapsing in a translational mode (Michalowski, 1989). An attempt also was made to use limit analysis in the framework of traditional slope stability considerations where the failure pattern consists of vertical rigid blocks which coincide with the division into slices in "traditional slice methods" (Michalowski, 1994). Such an analysis was found to be more straightforward in formulation than the slice methods, as it does not require any additional static assumptions. This method is presented here.

Michalowski, R. L., 1995, Stability of slopes: Limit analysis approach, *in* Haneberg, W. C., and Anderson, S. A., eds., Clay and Shale Slope Instability: Boulder, Colorado, Geological Society of America Reviews in Engineering Geology, v. X.

Specific application of the method to clays and shales is briefly discussed in the final section of the paper.

As the theorems of limit analysis are not commonly used in engineering geology, the theoretical background is presented in the next two sections. Including the pore pressure effects is not a standard procedure in classical limit analysis, thus, the theoretical considerations relating to pore pressure effects are also presented. Next, an admissible translational failure mechanism is presented, and the formula for the safety factor of a slope is derived. The solution based on a rotational failure mode is also presented, accounting for the pore pressure. Finally, the results from the suggested analysis are discussed and compared to those using some existing methods.

LIMIT ANALYSIS

Limit analysis constitutes a powerful tool in solving stability problems, yet it is seldom used in practice. The upper bound approach is particularly well-suited for consideration of the stability of slopes. It is not used often, however, as the term "upper bound" immediately raises a red flag among conservative engineers. Also, so far, no method consistent with the kinematical approach of limit analysis exists for including the pore pressure effects. Rather than limit analysis, a "limit equilibrium method" is preferred by many. The latter also leads to upper bounds for limit loads, and it can be proved that it leads to identical results if the same translational or rotational collapse mechanisms are selected in both methods (see e.g., Leshchinsky et al., 1985; Drescher and Detournay, 1993). The three following subsections briefly give a theoretical background of limit analysis, and a technique for including pore water pressure in limit analysis is presented in the next section.

Material assumptions

We assume that the material (soil) can be modeled as an elastic/perfectly plastic solid (Fig. 1a), and its limit condition (or failure criterion) can be represented as a convex surface in the stress space. The surface in Figure 1b depicts the Mohr-Coulomb failure condition in the principal stress space. In analysis of soils the Mohr-Coulomb failure criterion constitutes the standard limit condition, and under plane-strain conditions it takes the form (compression taken as positive)

$$f(\sigma'_x, \sigma'_z, \tau_{xz}) = (\sigma'_x + \sigma'_z)\sin\phi - \sqrt{(\sigma'_x - \sigma'_z)^2 + 4\tau_{xz}^2} + 2c\cos\phi = 0, \quad (1)$$

where σ'_x, σ'_z and τ_{xz} are the components of the effective stress state, ϕ is the internal friction angle and c is the cohesion. Whenever the stress state in the soil reaches the level such that σ'_x, σ'_z and τ_{xz} satisfy Equation 1, the soil is at the limit state, and, in the absence of hardening, strain may occur without further increase in the stress state. A stress state which satisfies Equation 1, when transferred into the principal stresses, can be represented as a point on the surface in Figure 1b.

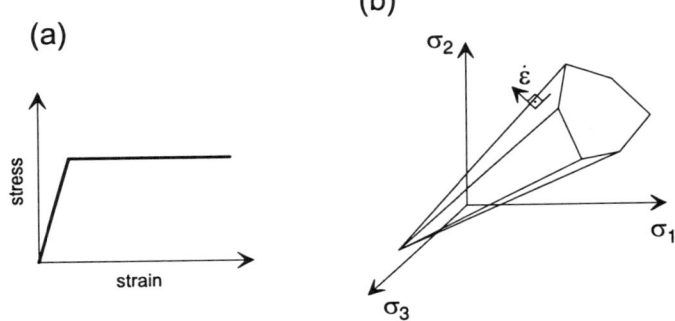

Figure 1. (a) Stress-strain behavior of an elastic/perfectly plastic solid; (b) Mohr-Coulomb failure surface in the space of principal stresses σ_1, σ_2, σ_3, with associative deformation rate $\dot{\varepsilon}$.

The stress states represented by points on the failure surface are states where the plastic deformation of the soil is possible. The soil cannot attain stress states beyond the failure surface. Plastic deformation is assumed to be governed by the normality rule, where the strain rate can be represented as a vector normal to the failure surface, i.e.,

$$\dot{\varepsilon}_{ij} = \dot{\lambda}\frac{\partial f(\sigma'_{ij})}{\partial \sigma'_{ij}}, \quad \dot{\lambda} \geq 0, \quad i, j = 1, 2, 3, \quad (2)$$

$\dot{\varepsilon}_{ij}$ is the strain rate tensor, σ'_{ij} is the tensor of effective stress, $f(\sigma'_{ij}) = 0$ is the failure criterion and $\dot{\lambda}$ is a nonnegative multiplier. The plastic deformation law in Equation 2 will be called the associative flow rule. The theorems follow from the *principle of maximum work* presented by Hill (1948), and were given by Drucker et al. (1952). The first proof of the theorems, however, was given in 1936 by Gvozdev, although his original work was not translated into English until 1960 (see Gvozdev, 1960).

Theorems

Limit analysis is based on the application of two theorems. The first one, the *lower bound theorem*, states that *the rate of work done by the actual surface tractions on the boundary where the velocity is prescribed is greater than or equal to that done by the surface tractions in any statically admissible stress field*. The statically admissible stress field must be in equilibrium, must satisfy the static boundary conditions, and must not exceed the yield condition of the material. By constructing a statically admissible stress field, it is possible to find the lower bound to the true limit load, and, theoretically, by constructing various fields, a good approximation of the limit load (the highest lower bound) can be found.

The second theorem, the *upper bound theorem*, states that *the rate of work done by the actual surface tractions and body forces is less than or equal to the rate of energy dissipation in any kinematically admissible failure mechanism*:

$$\int_v \sigma^*_{ij}\dot{\varepsilon}^*_{ij}\,dv \geq \int_S T_i V_i\,ds + \int_v X_i V^*_i\,dv \quad i, j = 1, 2, 3, \quad (3)$$

where V_i^* is the velocity vector describing the velocity field in the kinematically admissible mechanism, and $\dot{\varepsilon}_{ij}^*$ is the strain rate tensor, $\dot{\varepsilon}_{ij}^* = (V_{i,j}^* + V_{j,i}^*)/2$, and the summation convention holds in Inequality 3. Vector T_i is the true stress vector (unknown) on boundary S where the velocity is given, $V_i^* = V_i$. σ_{ij}^* is the stress tensor associated with $\dot{\varepsilon}_{ij}^*$, X_i is the vector of distributed forces (weight), and v is the volume. A failure mechanism here is the pattern of motion of the soil mass during collapse. A kinematically admissible mechanism is one which conforms to the flow rule (Equation 2) and which satisfies the kinematic boundary conditions. For a given failure mechanism, only the first integral on the right-hand side of Equation 3 is unknown. This term represents the work rate of the true collapse load in the assumed failure mechanism. If a constant velocity is specified on boundary S (kinematic boundary condition), then Inequality 3 can be used to estimate not only the work rate of the collapse (limit) load, but also the load itself (integral of the traction vector T_i on S). Reasonable upper bounds to the true limit load can be obtained by considering a collapse mechanism where the geometric parameters are varied in an optimization scheme and the least upper bound is sought.

Remarks on limit analysis

Application of limit analysis theorems does not provide the exact value of the limit load, but it does produce two bounding values. Thus, it seems natural to use the theorems to find a reasonably narrow range which bounds the true solution, so that a precise judgment of the true limit load can be made. This, however, is not always possible, as the lower bound solutions for geotechnical problems are often difficult to find.

Collapse mechanisms used in upper bound calculations have a distinct physical interpretation associated with true failure patterns. These can be known from experiments or practical experience. Stress fields used in the lower bound approach, however, are constructed without a clear relation to real stress fields other than the stress boundary conditions. Moreover, in geotechnics, where the formulation of problems often involves a half-space (semi-infinite domain), an extension of the stress field into the half-space is either cumbersome or seems to be impossible. There are only a few nontrivial solutions in geotechnical engineering for which an extension into infinity has been found. Because of the problems with constructing admissible stress fields, only the kinematic approach has been used successfully in geotechnical engineering, and this approach will be used also in this paper.

The theorems of limit analysis are most often used to find estimates of limit loads; they can also be used, as they are in this article, to find *critical heights* (failure heights) or factors of safety for slopes.

An upper bound solution to a limit load (or critical height, or a factor of safety) based on a rigid-block translational mechanism yields a solution identical to that from the limit equilibrium method based on the same geometry of failure surfaces. This was noted earlier by Mróz and Drescher (1969), Collins (1974), Michalowski (1989), and Drescher and Detournay (1993). Although the two methods yield identical solutions when used with identical rigid-block failure patterns, the kinematic approach of limit analysis is more convenient, especially when it comes to continually deforming fields and mechanisms with nonlinear failure surfaces (velocity discontinuities). The equivalency of the solutions from the upper bound approach and limit equilibrium for translational rigid-block mechanisms can be proved easily. By the principle of virtual work, a system of blocks is in equilibrium provided the virtual work done by all external forces is equal to the energy dissipated for each and every virtual displacement of the system consistent with the constraints. Therefore, equating the energy dissipation rate to work by external forces in a translational mechanism must satisfy the equilibrium of forces, as does the limit equilibrium method.

PORE PRESSURE EFFECTS

The influence of pore pressure on the stability of slopes can be incorporated in the kinematical approach of limit analysis as additional work terms in Inequality 3, as a result of buoyancy and seepage forces (Michalowski, 1994). Alternatively, a term representing the work of pore pressure on the skeleton expansion (drained process) can be added to the right-hand side of Equation 3. This term takes the form

$$\dot{E}_u = \int_v u \dot{\varepsilon}_{ii} \, dv + \int_L u n_i [V]_i \, dL =$$
$$= \int_{S+L} u n_i V_i \, ds - \int_v \frac{\partial u}{\partial x_i} V_i \, dv + \int_L u n_i [V]_i \, dL, \quad (4)$$

where u is the pore pressure, $\dot{\varepsilon}_{ii}$ is the volumetric deformation rate of the skeleton, v is the volume of the submerged soil mass, and n_i is the unit vector normal to surface S, or it is the unit vector normal to discontinuity surfaces L. The first term on the left-hand side of Equation 4 represents the work of the pore pressure on the skeleton expansion in the continually deforming field only, whereas the second term represents the work along discontinuities L; $[V]_i$ is the velocity jump vector. The Gauss theorem was used in Equation 4, index i denotes the Cartesian coordinates, and the summation convention holds. The work rate expressed in Equation 4 is positive work analogous to the work performed by a compressed fluid contained in a balloon on virtual expansion of the balloon shell. Miller and Hamilton (1989) used a similar term to reduce the energy dissipation in analysis of a slope failure, although their interpretation of this term was somewhat controversial (Miller and Hamilton, 1990).

Note that the first term on the left-hand side of Equation 4 represents the work in the continuously deforming regions only, i.e., within areas bounded by surface S and discontinuities L. The first term on the right-hand side represents integration over boundaries of all such continually deforming regions. These regions have common interfaces along L, and, after expanding the first term on the right-hand side into integrals over S and L, the integrals over L vanish

$$\dot{E}_u = \int_v u\dot{\varepsilon}_{ii}\, dv + \int_L un_i[V]_i\, dL =$$

$$= \int_S un_i V_i\, ds - \int_v \frac{\partial u}{\partial x_i} V_i\, dv. \quad (5)$$

The hydraulic head, h, (with the omission of the kinetic part) is

$$h = \frac{u}{\gamma_w} + Z, \quad (6)$$

where γ_w is the unit weight of water and Z is the elevation head. Substituting u from Equation 6 into Equation 5, one arrives at

$$\dot{E}_u = \int_v u\dot{\varepsilon}_{ii}\, dv + \int_L un_i[V]_i\, dL =$$

$$= \int_S un_i V_i\, ds - \gamma_w \int_v \frac{\partial h}{\partial x_i} V_i\, dv + \gamma_w \int_v \frac{\partial Z}{\partial x_i} V_i\, dv. \quad (7)$$

The first term on the right-hand side of Equation 7 represents the work of the pore pressure on contour S of the submerged volume v. For a slope with a phreatic surface contained within the soil, this term is equal to zero. The second term denotes the work of the seepage force, the third term is caused by the force of buoyancy; note that $(\partial Z/\partial x_i)V_i = V_z$. To account for the presence of pore water pressure in a slope, one can explicitly include the seepage and buoyancy forces in the energy balance equation or include the terms on the left-hand side of Equation 7. The latter is done in the following analysis; the procedure, however, is simple in either case.

For an incompressible (or pressure-independent) material we have $\dot{\varepsilon}_{ii} = 0$ and $n_i[V]_i = 0$ (n_i is a unit vector normal to the discontinuity), and the net work of the pore pressure from Equation 7 is zero, which indicates no influence of pore pressure on the stability analysis.

COLLAPSE OF A SLOPE

The upper bound approach requires that admissible failure mechanisms be constructed first. Two types of failures are considered here: a translational failure pattern and a rotational collapse mechanism.

Translational failure

The failure mechanism selected here resembles the traditional "slice methods," where the failing mass of the soil is divided into vertical slices. Here, the slices are considered rigid

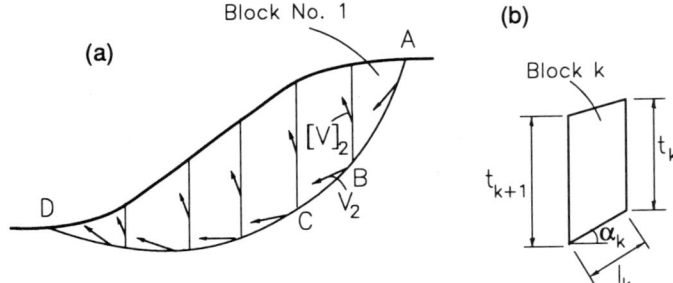

Figure 2. (a) Schematic of a translational slope failure mechanism; (b) definition of a single block geometry.

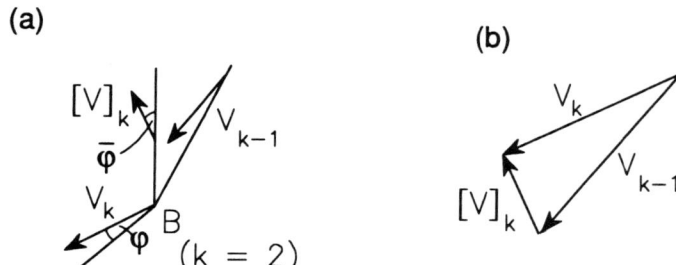

Figure 3. (a) Velocity discontinuities (solid lines) in a kinematically admissible translational collapse mechanism; (b) hodograph.

blocks (Fig. 2). These blocks move with respect to one another during failure with the relative jump in velocity inclined at the angle of internal friction on the interfaces between blocks. This requirement follows from the flow rule associated with the Mohr-Coulomb yield condition. Compatibility of the mechanism requires that the velocity jumps at rupture surfaces between the neighboring blocks (Fig. 3a) form a closed polygon as in Figure 3b. This polygon will be called a hodograph here. Assuming that the vertical component of the velocity of the first block is given (boundary condition), the velocities of all blocks and the velocity jumps between them can be calculated by repetitively constructing the hodograph in Figure 3b. The velocities of blocks, V_k, and the interfacial velocity jumps, $[V]_k$, are

$$V_k = V_{k-1} \frac{\cos(\alpha_{k-1} - \phi_{k-1} - \bar{\phi}_k)}{\cos(\phi_k + \bar{\phi}_k - \alpha_k)}, \quad [V]_k = V_k \frac{\sin(\phi_k - \phi_{k-1} - \alpha_k + \alpha_{k-1})}{\cos(\alpha_{k-1} - \phi_{k-1} - \bar{\phi}_k)}, \quad (8)$$

where index k denotes the block number, α_k is the angle of inclination of the block base to the horizontal, ϕ_k is the internal friction angle at the base of the block, and $\bar{\phi}_k$ is the internal friction angle on the right interface of the kth block (interface between blocks k and $k-1$).

Factor of safety and critical height

The *factor of safety* is defined here as

$$F = \frac{c}{c_d} = \frac{\tan \phi}{\tan \phi_d}. \quad (9)$$

Parameters c and ϕ are the cohesion and internal friction angle of the soil, and c_d and ϕ_d are their values necessary to maintain the energy balance in an admissible failure mechanism. Inequality 3 will now be used to arrive at the upper bound to the factor of safety associated with the failure mechanism as in Figure 2a. The left side in Inequality 3 represents the energy dissipation rate within the failing soil. As the blocks in the collapse mechanism are considered to move as rigid bodies, the energy is dissipated only at the rupture surface between the sliding mass and the material at rest and at the surfaces between the adjacent blocks. The rate of energy dissipation per unit area of a velocity discontinuity for the Coulomb material can be written as

$$\dot{d} = \sigma'_{ij} n_j [V]_i = c [V] \cos \phi, \quad (10)$$

where n_j is the unit vector perpendicular to the rupture surface, $[V]_i$ is the velocity jump vector and $[V]$ is its magnitude. Indices i and j in Equation 10 denote the Cartesian coordinates (the summation convention holds). Thus the left side of Inequality 3 can be written for the mechanism in Figure 2a as

$$\dot{D} = \sum_{k=1}^{n} [l_k c_k V_k \cos \phi_k + t_k \bar{c}_k [V]_k \cos \bar{\phi}_k], \quad (11)$$

where l_k is the length of the block base (Fig. 2b), t_k is the length of the interface of the kth block with block $k-1$, overbars denote material parameters on that interface, velocities V_k and $[V]_k$ are given in Equation 8, and n is the number of blocks. Any vertical load on the blocks will be lumped with the weight of the blocks, hence, the first term on the right-hand side of Inequality 3 is zero, and the last term is

$$\dot{E}_\gamma = \sum_{k=1}^{n} (W_k + Q_k) V_k \sin (\alpha_k - \phi_k), \quad (12)$$

where the weight of block k is W_k, and any additional vertical load on the block is Q_k. The rate of work resulting from the presence of pore water pressure in a slope collapsing according to the rigid-block mechanism as in Figure 2a can be calculated according to Equation 7, and it becomes

$$\dot{E}_u = \sum_{k=1}^{n} \{V_k l_k u_k \sin \phi_k + [V]_k t_k \bar{u}_k \sin \bar{\phi}_k\}, \quad (13)$$

where u_k is the average pore water pressure at the base of block k, and $\bar{u}_k$ is the average pore pressure on interface t_k. Introducing c_d, ϕ_d, $\bar{c}_d$, and $\bar{\phi}_d$ into Equations 11, 12, and 13, defined by Equation 9 as $\phi_{dk} = \arctan (\tan \phi_k / F)$, $c_{dk} = c_k/F$, $\bar{\phi}_{dk} = \arctan (\tan \bar{\phi}_k / F)$, and $\bar{c}_{dk} = \bar{c}_k/F$, and requiring that $\dot{D} = \dot{E}_\gamma + \dot{E}_u$, one obtains the upper bound to the factor of safety as

$$F = \frac{\sum_{k=1}^{n} \{l_k c_k V_k \cos \phi_{dk} + t_k \bar{c}_k [V]_k \cos \bar{\phi}_{dk}\}}{\sum_{i=k}^{n} \{V_k [(W_k + Q_k) \sin (\alpha_k - \phi_{dk}) + l_k u_k \sin \phi_{dk}] + [V]_k t_k \bar{u}_k \sin \bar{\phi}_{dk}\}} \quad (14)$$

Velocities V_k and $[V]_k$ are given in Equation 8 where angles ϕ_{dk} and $\bar{\phi}_{dk}$ need to be used in place of ϕ_k and $\bar{\phi}_k$. The factor of safety is, of course, independent of the magnitude of the velocity taken as the boundary condition (velocity of the first block). Note that Equation 14 is an implicit function with respect to the factor of safety F, because F appears on the right-hand side of Equation 14 in ϕ_{dk} and $\bar{\phi}_{dk}$.

In traditional slice techniques some arbitrary static assumptions are made about the forces on interfaces between slices (such as, for instance, their inclination). Assumptions of similar significance may be introduced in the kinematical approach of limit analysis through specifying different soil strengths at surfaces between blocks. In the particular case where this strength is assumed to be zero between all blocks ($\bar{\phi}_k = 0$ and $\bar{c}_k = 0$), velocities in Equation 14 can be eliminated easily, and the factor of safety can be transformed into

$$F = \frac{\sum_{k=1}^{n} \frac{l_k c_k}{\cos \alpha_k + \sin \alpha_k \tan \phi_k / F}}{\sum_{k=1}^{n} \left[(W_k + Q_k) \frac{\tan \alpha_k - \tan \phi_k / F}{1 + \tan \alpha_k \tan \phi_k / F} + \frac{l_k u_k}{\sin \alpha_k + F \cot \phi_k \cos \alpha_k} \right]}. \quad (15)$$

Note that the safety factor becomes indeterminate in Equations 14 and 15 when $c = \bar{c} = 0$; in such a case F must be calculated by equating the denominator of Equation 14 (or Equation 15) to zero (or $\dot{E}_\gamma + \dot{E}_u = 0$).

The critical (or failure) height is defined here as the maximum height of a stable slope. It is convenient to represent the critical height of a homogeneous slope in a dimensionless manner as $\gamma H/c$, sometimes called a *stability factor* (N_s). For a homogeneous slope ($c_k = \bar{c}_k = c$, $\phi_k = \bar{\phi}_k = \phi$, $k = 1, 2, 3, \ldots n$) the stability factor (N_s) can be obtained by specifying $F = 1$, then calculating cohesion c from Equation 14 for the slope of a given height and unit weight and using $c = \gamma H/N_s$ to find N_s.

Rotational failure

Stability analyses based on slice methods with cylindrical failure surfaces are often falsely called "rotational failure" analyses. In these analyses, the flow rule for the soil is not specified and, consequently, the admissibility of a specific collapse pattern cannot even be considered. It is fair to say, however, that, in view of plasticity theory, the static slice equilibrium analyses indirectly imply the assumption of the normality rule (see Equation 2) for the material along the failure surface between the collapsing soil mass and the material at rest. Nonassociative flow rules (although retaining coaxiality of the principal directions of the stress and strain-rate tensors) would require reducing the internal friction angle and cohesion (Drescher and Detournay, 1993). A rigid rotation of the soil mass with respect to the center of a cylindrical failure surface implies a constant velocity discontinuity ("jump") along that surface. Kinematical admissibility, however, requires that, for a frictional soil with the associative flow rule, this velocity jump propagates according to an exponential law. Thus, in many cases of analysis of deep-seated slope failures, the usual term "rotational failure" may not be appropriate, as rotational failure is not consistent with the analysis.

Kinematically admissible failure patterns for slopes were considered by Chen et al. (1969) and Chen and Giger (1971). As pointed out by Chen (1975), rotational mechanisms are most efficient in the kinematical approach and lead to lower critical heights than translational mechanisms do. Kinematical admissibility of a rigid rotation collapse pattern requires that the failure surface be a log spiral:

$$r = r_0 \, e^{(\theta - \theta_0) \tan \phi}, \tag{16}$$

where r is the radius of the spiral related to angle θ, and r_0 and θ_0 are the initial values (Fig. 4). The critical height of slopes was found by Chen and his coworkers as

$$\frac{\gamma H}{c} = \frac{H}{r_0} \frac{e^{2(\theta_h - \theta_0) \tan \phi} - 1}{2 \tan \phi \, (f_1 - f_2 - f_3 - f_4)}, \tag{17}$$

where H/r_0 and functions f_1 through f_4 depend on variables θ_0, θ_h, and β' (Fig. 4). See Appendix I for functions f_1 through f_4. An optimization scheme can be used to find the minimum of $\gamma H/c$ with θ_0, θ_h, and β' being the variables. Results of such calculations are used here as a reference (for the absence of pore water pressure these results are available, for instance, in Chen, 1975). For a slope of a known height, the rotational failure analysis can be formulated alternatively in terms of the factor of safety to yield

$$F = \frac{2(\theta_h - \theta_0) \tan \phi}{\mathrm{Ln} \left[1 + 2 \dfrac{\gamma H}{c} \dfrac{r_0}{H} \tan \phi \, (f_1 - f_2 - f_3 - f_4) \right]}, \tag{18}$$

where $\gamma H/c$ is known, and H/r_0 and functions f_1 through f_4 need to be calculated using the formulae in Appendix I, but with $\tan\phi$ replaced with $\tan\phi/F$. In the particular case where $\phi = 0$

$$F = \frac{H}{r_0} \frac{\theta_h - \theta_0}{\dfrac{\gamma H}{c} (f_1 - f_2 - f_3 - f_4)} \tag{19}$$

The influence of the pore pressure on the stability of a slope can be included now into the analysis by including the work expressed in Equation 7 into the energy balance equation. This requires that the pore pressure be specified. The pore pressure distribution can be found from constructing the groundwater flow net for given boundary conditions, or, as is often done for practical purposes, the distribution of the pore pressure may be assumed a priori, and be given by coefficient r_u as described by Bishop and Morgenstern (1960)

$$r_u = \frac{u}{\gamma z}, \tag{20}$$

where u is the pore pressure, γ is the unit weight of the soil, and z is the depth of the point considered below the soil surface. The stability factor $\gamma H/c$ can now be derived as

$$\frac{\gamma H}{c} = \frac{H}{r_0} \frac{e^{2(\theta_h - \theta_0) \tan \phi} - 1}{2 \tan \phi \, (f_1 - f_2 - f_3 - f_4 + r_u f_5)} \tag{21}$$

where f_5 is a function of geometrical parameters and the internal friction angle (Appendix II). The factor of safety now becomes

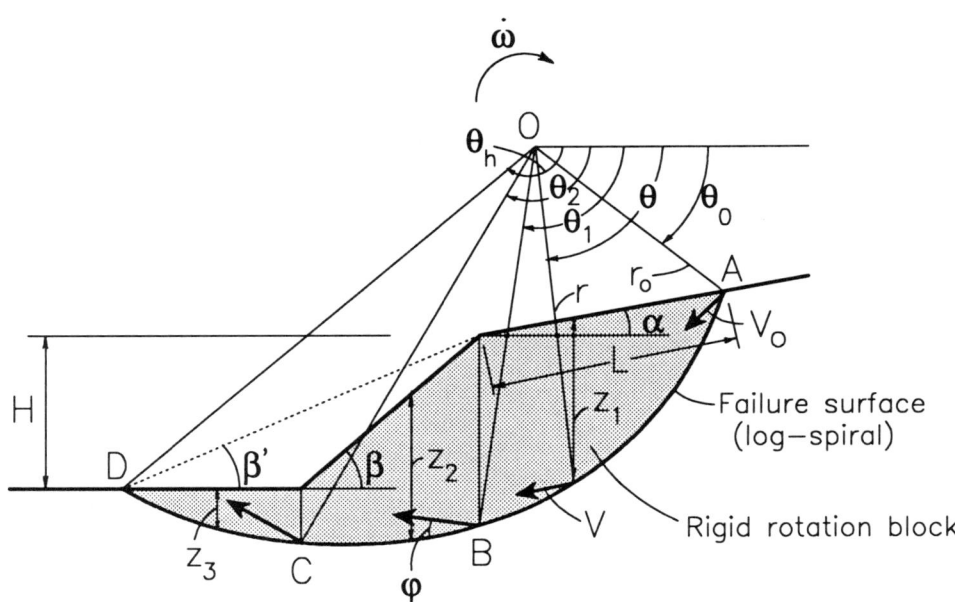

Figure 4. Kinematically admissible rigid-rotation failure mechanism.

$$F = \frac{2(\theta_h - \theta_0)\tan\phi}{\mathrm{Ln}\left[1 + 2\frac{\gamma H}{c}\frac{r_0}{H}\tan\phi(f_1 - f_2 - f_3 - f_4 + r_u f_5)\right]} \quad (22)$$

When $\phi = 0$ the value of function f_5 becomes zero, which indicates no influence of the pore pressure on the stability analysis when the internal friction of the soil is zero.

COMPUTATIONAL RESULTS

A computer program was written to perform calculations of failure heights of slopes using the method suggested here. Additional calculations using the rotational failure mode were also performed. The latter calculations were performed including the influence of the pore pressure not included in the original works of Chen and his coworkers. Results of calculations by the simplified Bishop's method (Bishop, 1955) are also presented for comparison.

Failure heights are presented here in terms of stability factor $\gamma H/c$. This allows one to compare different methods in a somewhat more efficient way than comparing safety factors for a limited number of slopes with different height and soil parameters. A cylindrical failure surface was assumed in all calculations. The number of blocks used was in the range of 20 to 40, depending on the slope inclination. The number of blocks was chosen in such a way so that a further increase in this number would produce no significant change in the stability factor (less than 1%). Calculations were performed with two limiting assumptions as to the strength of the soil between the blocks: (a) the shear strength of the soil on interfaces between blocks is equal to the strength of the soil along the base, and (b) the shear strength on interfaces between blocks is neglected ($\bar{\phi} = 0$, $\bar{c} = 0$). The first assumption leads to the upper bound for the true critical height, whereas the second assumption is expected to supply an approximate but conservative solution, which no longer can be proved an upper bound.

Figure 5 presents dimensionless critical heights ($\gamma H/c$) for uniform slopes with inclination angle β in the range from 15° to 90°, and with internal friction angle ϕ in the range from 0° to 40°. The results in Fig. 5 are all for zero pore pressure in the slope. The upper and lower limits of the shaded areas represent the solutions for assumptions of full shear strength, and zero shear strength of the soil on interfaces between blocks, respectively. It can be demonstrated that assuming interfacial shear strength between blocks anywhere between zero and its full magnitude will yield a solution between the two limiting ones. It will be argued later that the arbitrary static assumptions used in the slice methods can be interpreted in view of the collapse mechanism as prescribing a certain strength to block interfaces. The solution based on the rotational mechanism is also presented in the figure, as well as that using the simplified Bishop's method and results given by Spencer (1967). Because critical heights increase significantly with a decrease in slope inclination angle, $\gamma H/c$ in Figure 5 is pre-

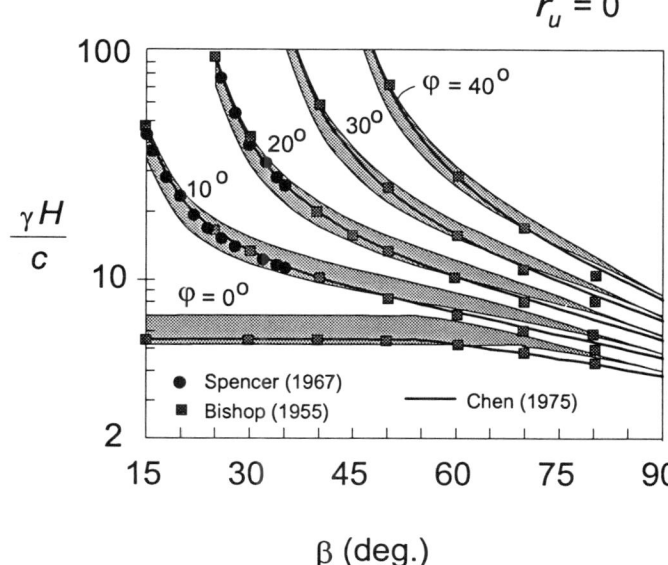

Figure 5. Dimensionless failure height of slopes as function of slope inclination angle for different internal friction angles; case for horizontal top surface of the slope ($\alpha = 0$) and no pore pressure (after Michalowski, 1994). See discussion in text.

sented in a semi-log plot. The corresponding numerical results are presented in Table 1. Three factors are given in Table 1 for each slope inclination angle β and internal friction angle ϕ. The first factor is calculated assuming full strength on interfaces between blocks, the second factor represents the result when the interfacial strength is neglected, and the third one is a factor from the analysis of the rotational collapse (these factors, only in Table 1, are from Chen and Giger, 1971).

The difference between the limits of the shaded area in Figure 5 becomes smaller with an increase in slope angle β. This is because, for steep slopes, the radius of the failure surface obtained from the procedure of minimizing factor $\gamma H/c$ increases significantly with the increase in β. This leads to "flattening" of the failure surface and to a decrease in the relative velocity of the blocks. The assumption of zero or full soil strength on interfaces between blocks then becomes less significant, as the contribution of the block interfaces to the total energy dissipation decreases. When the failure surface becomes a plane, the blocks no longer move with respect to one another, and the assumption of a specific strength on the block interfaces has no influence on the solution at all (see case for $\beta = 90°$ and $\phi = 0$ in Table 1).

The results for pore pressure expressed in terms of pore pressure coefficient r_u (Equation 20) are presented in Figure 6 ($r_u = 0.25$) and Figure 7 ($r_u = 0.5$). The numerical results are given in Tables 2 and 3. For slopes with negligible pore pressure, stability factor $\gamma H/c$ becomes infinity when the internal

TABLE 1. STABILITY FACTORS $\gamma H/c$ ($r_u = 0$)

β	$\phi = 0$	$\phi = 10°$	$\phi = 20°$	$\phi = 30°$	$\phi = 40°$
15°	6.99	47.88			
	5.22	34.03			
	5.53	45.49			
20°	6.99	25.20			
	5.21	18.27			
	5.53	23.14			
25°	6.98	18.61	97.53		
	5.22	13.86	70.26		
	5.53	16.64	94.63		
30°	6.98	15.49	43.62		
	5.21	11.81	32.18		
	5.53	13.50	41.22		
35°	6.98	13.60	28.92	147.66	
	5.21	10.57	21.91	107.40	
	5.53	11.61	26.66	144.20	
40°	6.98	12.28	22.22	61.06	
	5.21	9.70	17.27	45.69	
	5.53	10.30	19.99	58.27	
45°	6.98	11.23	18.34	38.13	189.54
	5.21	9.03	14.59	29.42	139.40
	5.53	9.31	16.16	35.54	185.49
50°	6.98	10.35	15.74	27.90	74.72
	5.21	8.46	12.79	22.17	56.93
	5.52	8.51	13.63	25.41	71.49
55°	6.94	9.54	13.79	22.10	44.88
	5.21	7.95	11.45	18.05	35.41
	5.46	7.84	11.80	19.71	41.89
60°	6.57	8.79	12.22	18.27	31.74
	5.21	7.48	10.37	15.31	25.92
	5.25	7.26	10.39	16.04	28.91
65°	6.16	8.05	10.86	15.45	24.32
	5.21	7.03	9.44	13.28	20.50
	5.03	6.73	9.25	13.44	21.72
70°	5.70	7.32	9.63	13.19	19.42
	5.12	6.58	8.61	11.66	16.90
	4.80	6.25	8.30	11.48	17.15
75°	5.21	6.59	8.50	11.26	15.79
	4.88	6.14	8.10	10.30	14.22
	4.56	5.80	7.48	9.94	13.97
80°	4.78	5.90	7.40	9.55	12.86
	4.61	5.70	7.18	9.09	12.06
	4.33	5.37	6.75	8.67	11.61
85°	4.37	5.30	6.49	8.10	10.45
	4.32	5.24	6.41	7.98	10.23
	4.08	4.97	6.10	7.61	9.77
90°	4.00	4.76	5.75	6.93	8.62
	4.00	4.77	5.72	6.94	8.61
	3.83	4.58	5.50	6.69	8.29

β = slope angle; φ = internal friction angle.

friction angle of the soil is equal to or larger than the slope inclination angle. This means that, theoretically, an infinitely high slope is stable. This is not true, however, in the presence of pore pressure. For instance, the height of a 30° slope built of homogeneous soil with an internal friction angle of 40° has a limit when the pore pressure is equivalent to that expressed by the pore pressure coefficient $r_u = 0.5$ (see Table 3).

It is interesting to notice that, for a very steep slope and large pore pressure, the dimensionless stability factor calculated for a soil with a large internal friction angle may be smaller than that for a soil with a smaller internal friction (see diagram for φ equal to 40° and 20° in Fig. 7b for slope inclination angles exceeding 70°). This occurs because the increase of the internal friction angle produces an increase in the ad-

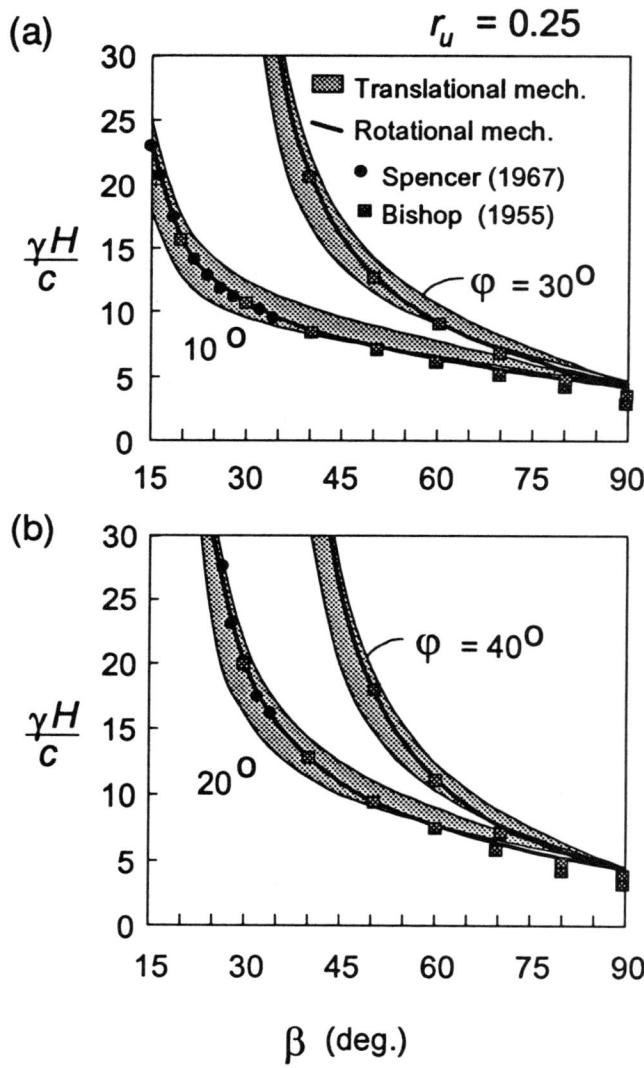

Figure 6. Dimensionless failure height of slopes for pore pressure coefficient $r_u = 0.25$, and α = 0 (after Michalowski, 1994). See discussion in text.

verse effect of the pore water pressure. For steep slopes this adverse effect increases more rapidly than the stabilizing effect due to an increase in φ.

DISCUSSION

Comparison of stability calculation results by different methods has always been an engaging issue, although it has usually been done without reference to methods other than those based on the limit equilibrium of slices (see e.g., Fredlund and Krahn, 1977, Espinoza et al., 1992). The method presented here constitutes a convenient tool for assessment of the assumptions used in existing slice methods. More specifically, it allows one to determine the level of conservatism following

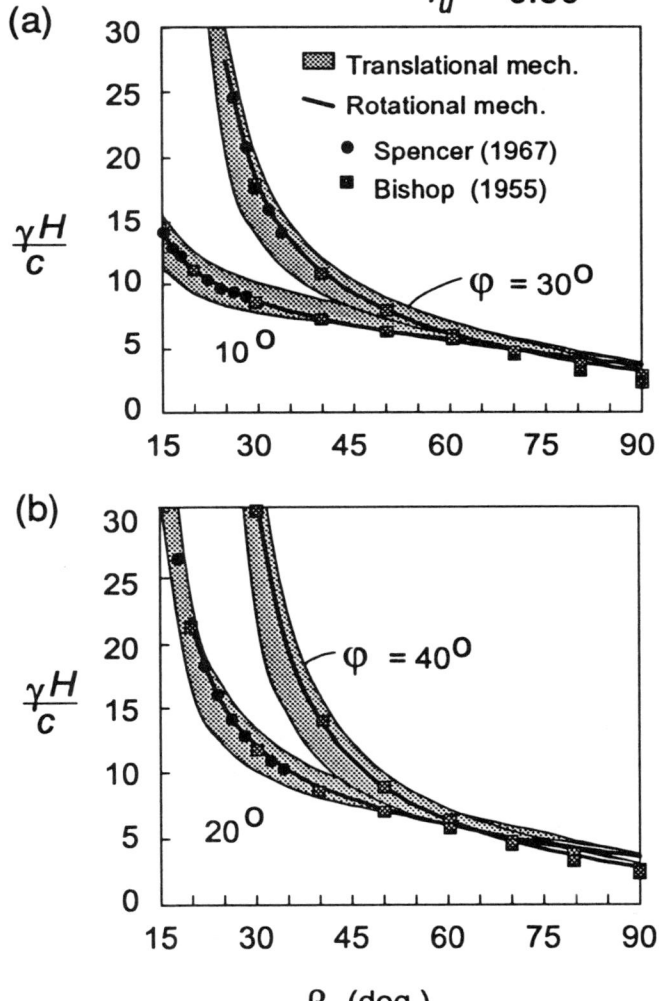

Figure 7. Dimensionless failure height of slopes for pore pressure coefficient $r_u = 0.50$, and $\alpha = 0$ (after Michalowski, 1994). See discussion in text.

TABLE 2. STABILITY FACTORS γH/c ($r_u = 0.25$)

β	φ = 10°	φ = 20°	φ = 30°	φ = 40°
15°	25.00 17.85 23.18			
20°	17.26 12.61 15.58	66.79 47.10 64.28		
25°	14.15 10.58 12.48	31.67 22.90 29.82		
30°	12.40 9.51 10.71	21.70 16.15 20.05	56.37 39.57 53.90	
35°	11.23 8.79 9.53	17.04 13.06 15.46	31.65 22.97 29.84	
40°	10.34 8.23 8.65	14.32 11.26 12.76	22.21 16.71 20.66	44.29 30.80 41.83
45°	9.60 7.77 7.95	12.48 10.00 10.94	17.29 13.48 15.84	27.43 19.45 25.70
50°	8.95 7.36 7.36	11.11 9.14 9.60	14.25 11.48 12.85	19.69 14.99 18.25
55°	8.34 6.98 6.85	10.00 8.40 8.55	12.15 10.00 10.79	15.29 12.18 13.99
60°	7.74 6.61 6.38	9.05 7.75 7.69	10.56 8.98 9.26	12.44 10.31 11.23
65°	7.15 6.25 5.96	8.19 7.17 6.97	9.28 8.08 8.06	10.43 8.94 9.27
70°	6.63 5.88 5.56	7.39 6.62 6.33	8.18 7.30 7.09	8.88 7.84 7.80
75°	5.93 5.51 5.18	6.61 6.10 5.76	7.19 6.59 6.27	7.60 6.91 6.65
80°	5.33 5.13 4.82	5.84 5.58 5.25	6.26 5.92 5.57	6.48 6.08 5.71
85°	4.80 4.73 4.47	5.14 5.07 4.77	5.38 5.28 4.95	5.44 5.30 4.92
90°	4.34 4.32 4.14	4.56 4.54 4.35	4.64 4.63 4.40	4.54 4.53 4.26

β = slope angle; φ = internal friction angle.

TABLE 3. STABILITY FACTORS $\gamma H/c$ ($r_u = 0.5$)

β	$\phi = 10°$	$\phi = 20°$	$\phi = 30°$	$\phi = 40°$
15°	15.60 11.24 14.03	45.41 31.19		
20°	12.58 9.27 11.12	23.12 16.30 21.34	65.87 43.12	
25°	11.10 8.33 9.66	16.54 12.05 15.10	29.97 20.16 27.35	86.15 51.32
30°	10.16 7.79 8.70	13.36 10.06 12.07	19.77 13.80 18.00	36.11 22.25 31.43
35°	9.43 7.43 7.97	11.45 8.88 10.23	14.98 10.91 13.63	22.26 14.41 19.59
40°	8.84 7.09 7.38	10.14 8.08 8.96	12.19 9.25 11.05	15.93 10.90 14.17
45°	8.32 6.78 6.88	9.16 7.47 7.99	10.35 8.15 9.32	12.34 8.93 11.06
50°	7.83 6.48 6.44	8.38 6.96 7.22	9.02 7.35 8.05	10.03 7.66 9.01
55°	7.36 6.19 6.05	7.70 6.52 6.57	8.01 6.71 7.06	8.43 6.75 7.54
60°	6.89 5.90 5.68	7.08 6.11 6.02	7.18 6.16 6.26	7.24 6.04 6.43
65°	6.41 5.61 5.33	6.51 5.72 5.52	6.46 5.68 5.59	6.31 5.45 5.55
70°	5.91 5.31 5.00	5.95 5.34 5.07	5.82 5.22 5.01	5.54 4.93 4.82
75°	5.39 4.99 4.68	5.39 4.97 4.66	5.21 4.79 4.50	4.87 4.46 4.21
80°	4.86 4.66 4.36	4.81 4.58 4.27	4.62 4.36 4.05	4.26 4.00 3.69
85°	4.38 4.32 4.06	4.26 4.24 3.91	4.02 3.93 3.63	3.66 3.55 3.23
90°	3.95 3.94 3.77	3.77 3.76 3.58	3.48 3.47 3.26	3.07 3.07 2.82

β = slope angle; ϕ = internal friction angle.

from static assumptions which are based on engineering intuition. Although useful for such purpose, the method presented here stands on its own and can be used in stability analyses without reference to other methods. The advantage of this method is in its straightforward formulation, where a physical interpretation can be given to every step of the analysis.

Traditional methods for slope stability analysis require static assumptions relating to forces acting on interfaces between slices (force inclination, location of the thrust line, etc.). Assessment of such assumptions is difficult, if not impossible, without resorting to more rigorous approaches. The upper bound method used here is one such approach. It should be mentioned that the upper bound technique yields solutions identical to those of the limit equilibrium method if identical translational collapse mechanisms are used in the two approaches. The collapse mechanism used here in the upper bound approach was purposely selected in the form of vertical slices, so that it corresponds to those used in slice techniques.

The interslice forces in the traditional slice methods depend on the static assumptions used. Interpreted in view of an admissible collapse mechanism corresponding to the specific division into slices, they become limit forces, because only then is the relative movement of the slices (blocks) admissible. Hence, the arbitrary static assumptions can be interpreted as assigning a certain strength to interfaces between blocks (in general, different on different interfaces). A reasonable assumption for practical purposes is to limit such strength between zero (conservative) and the full strength of the soil. Thus, the shaded regions in Figures 5–7 limit all reasonable solutions which may be expected from equilibrium equations applied to the system of slices over cylindrical failure surfaces. Only if the strength of the soil on interfaces were assumed larger than the actual strength of the soil could the minimized limit equilibrium solution give a stability factor $\gamma H/c$ higher than the upper limit of the shaded range in Figures 5–7. Stability factors below the lower limit, when interpreted in terms of the kinematics-based analysis, would be indicative of negative energy dissipation, or production of energy, on interfaces during collapse (thermodynamically inadmissible).

It is interesting to notice that the simplified Bishop's method produces results which fall below the shaded areas in Figure 5–7 for very steep slopes. Once $\gamma H/c$ is known, the interslice forces can be back calculated, and, when interpreted in terms of kinematics-based analysis, it can be shown that, indeed, an inadmissible situation arises for very steep slopes where the work dissipation rate becomes negative on some interfaces. For gentle slopes, the simplified Bishop's method yields results near rigorous upper bound solutions based on the translational mechanism, and it very closely follows the rigorous upper bound solution based on the rotational mechanism for a very wide range of slope inclinations.

Results from the analysis of the rotational mechanism also fall below the shaded areas in Figure 5–7 for steep slopes even though they are rigorous upper bounds; the rotational mechanism is simply a better mechanism (it leads to the least upper bound) than the translational mechanism with rigid blocks.

The method presented can be used for an arbitrary (not just cylindrical) shape of the failure surface, and for nonhomogeneous soils. When layers of soils with different internal friction angles (ϕ) are present, $\bar{\phi}$ may vary along the interfaces between blocks. Consequently, the associative low rule will require that the velocity jump vector between blocks vary. In such a case the blocks could no longer move as rigid bodies. It is suggested that for nonhomogeneous soils the rigorous rules of the kinematic approach be relaxed, and the dilatancy angle (angle at which the velocity jump vector is inclined to the failure surface) on the interfaces between blocks be taken conservatively as equal to the lowest internal friction angle of the soil encountered, or, even more conservatively, as zero. This has a clear interpretation of reducing the true shear strength of the soil at interfaces between blocks. Although approximate, this is still a more clear path than that where arbitrary characteristics of forces between blocks are assumed, without any concern for non-homogeneity of the soil within the failing mass.

Application of the presented method to clays is straightforward, and it requires only that, as for other soils, the strength parameters in the Mohr-Coulomb yield criterion (cohesion and internal friction) be obtained from the drained tests. The influence of the pore pressure on the slope stability is then accounted for by including the work of the pore pressure in the energy balance equation.

The method presented is also applicable to calculating stability of shale slopes. Here, however, the anisotropy of the shale strength must be included in the calculations. Shear strength along the surfaces in the direction of lamination may be considerably lower than that in the transverse direction. The energy dissipation rate in the kinematically admissible mechanism must then account for the dependence of the shear strength on the direction of the rupture surface. The process of optimization of the slope collapse mechanism must allow for the rupture surfaces to assume the most adverse directions, not an assumed a priori shape, such as a cylinder.

Extension of the method to analyzing shale slopes becomes more complicated when the surfaces of weakening in the shale are not perpendicular to the assumed plane of deformation (collapse). In such a case, a nonsymmetrical, three-dimensional analysis must be performed, which will significantly increase the degree of complexity. Such analysis, however, retains its tractability and conceptual simplicity. An attempt at such analysis was shown earlier for isotropic soils (Michalowski, 1989).

ACKNOWLEDGMENTS

The material presented in this paper is based upon work supported by the National Science Foundation under Grant No. MSS-9301494. This support is gratefully acknowledged.

APPENDIX I. ROTATIONAL FAILURE MECHANISM: FUNCTIONS $f_1 - f_4$

Functions f_1 through f_4 in Equations 17, 18, 19, 21, and 22 depend on the geometrical parameters and the internal friction angle (Chen, 1975). For completeness they are given below. Definitions of geometrical parameters can be found in Figure 4.

Geometrical relations in Figure 4 allow one to find the ratio of the slope height to radius r_0 and the ratio of length L to r_0 as

$$\frac{H}{r_0} = \frac{\sin \beta}{\sin (\beta - \alpha)} [\sin (\alpha + \theta_h) e^{(\theta_h - \theta_0) \tan \phi} - \sin (\alpha + \theta_0)] \quad (23)$$

and

$$\frac{L}{r_0} = \left[\sin (\theta_h - \theta_0) - \frac{H}{r_0} \frac{\sin (\beta + \theta_h)}{\sin \beta} \right] / \sin (\alpha + \theta_h) \quad (24)$$

The following are the expressions for functions f_1 through f_4

$$f_1 = \frac{1}{3(1 + 9 \tan^2 \phi)} [(3 \tan \phi \cos \theta_h - \sin \theta_h) e^{3(\theta_h - \theta_0) \tan \phi}$$

$$- 3 \tan \phi \cos \theta_0 - \sin \theta_0] \quad (25)$$

$$f_2 = \frac{1}{6} \frac{L}{r_0} \left(2 \cos \theta_0 - \frac{L}{r_0} \cos \alpha \right) \sin (\alpha + \theta_0) \quad (26)$$

$$f_3 = \frac{1}{6} \frac{H}{r_0} \frac{\sin (\beta + \theta_h)}{\sin \beta} \left(2 \cos \theta_h e^{(\theta_h - \theta_0) \tan \phi} + \frac{H}{r_0} \cot \beta \right) e^{(\theta_h - \theta_0) \tan \phi} \quad (27)$$

$$f_4 = \frac{1}{2} \left(\frac{H}{r_0} \right)^2 (\cot \beta' - \cot \beta)$$

$$[\cos \theta_0 - \frac{L}{r_0} \cos \alpha - \frac{1}{3} \frac{H}{r_0} (\cot \beta' + \cot \beta)] \quad (28)$$

APPENDIX II. ROTATIONAL FAILURE MECHANISM: PORE PRESSURE EFFECTS

The influence of pore pressure in the analysis based on the rotational failure mechanism is accounted for by including the work of the pore pressure on the skeleton expansion during failure, on the right side of Inequality 3. In a rigid rotation mechanism this work is performed along the log-spiral surface ABCD (Fig. 4). Pore pressure is represented here according to the proposal by Bishop and Morgenstern (1960) by coefficient r_u (Equation 20). Distance z associated with part AB of the slip surface (Fig. 4) is denoted here by z_1, and z_2 and z_3 are related to parts BC and CD, respectively. Angles θ_1 and θ_2 were found from equations

$$\cos \theta_1 e^{(\theta_1 - \theta_0) \tan \phi} = \cos \theta_0 - \frac{L}{r_0} \cos \alpha$$

and $\quad (29)$

$$\cos \theta_2 e^{(\theta_2 - \theta_0) \tan \phi} = \cos \theta_0 - \frac{L}{r_0} \cos \alpha - \frac{H}{r_0} \cot \beta,$$

where L/r_0 and H/r_0 are given in Equations 23 and 24, and all symbols are presented in Figure 4. For toe failures $\theta_2 = \theta_h$. Expressions for z_1, z_2, and z_3 can be written as

$$\frac{z_1}{r_0} = \frac{r}{r_0}\sin\theta - \sin\theta_0 - (\cos\theta_0 - \frac{r}{r_0}\cos\theta)\tan\alpha,$$

$$\frac{z_2}{r_0} = \frac{r}{r_0}\sin\theta - \sin\theta_h\, e^{(\theta_h-\theta_0)\tan\phi} +$$

$$(\frac{r}{r_0}\cos\theta - \cos\theta_2\, e^{(\theta_2-\theta_0)\tan\phi})\tan\beta, \quad (30)$$

and

$$\frac{z_3}{r_0} = \frac{r}{r_0}\sin\theta - \sin\theta_h\, e^{(\theta_h-\theta_0)\tan\phi}$$

where r is described in Equation 16. The work resulting from pore pressure along failure surface ABCD in incipient failure with rotation rate $\dot\omega$ about point O is

$$\dot E_u = \int_{\theta_0}^{\theta_h} u V_i n_i \frac{r}{\cos\phi}\, d\theta, \quad (31)$$

where u is the pore pressure, V_i is the velocity jump vector along the failure surface, n_i is the unit vector normal to this surface, and r is the radius of the spiral (Equation 16). A velocity jump vector along the log-spiral discontinuity in a kinematically admissible rigid rotation mechanism must propagate according to the following equation

$$V = V_0 e^{(\theta-\theta_0)\tan\phi} = r_0\dot\omega e^{(\theta-\theta_0)\tan\phi}. \quad (32)$$

Equation (31) can now be written as

$$\dot E_u = \gamma r_0^2 \dot\omega r_u \tan\phi \left[\int_{\theta_0}^{\theta_1} z_1 e^{2(\theta-\theta_0)\tan\phi}\, d\theta + \right.$$

$$\left.\int_{\theta_1}^{\theta_2} z_2 e^{2(\theta-\theta_0)\tan\phi}\, d\theta + \int_{\theta_2}^{\theta_h} z_3 e^{2(\theta-\theta_0)\tan\phi}\, d\theta \right], \quad (33)$$

where z_1, z_2, and z_3 are given in Equation 30. The third integral in Equation 33 is equal to zero for toe failures. Analytical solutions were found for integrals in Equation 33. Equation 33 can be rewritten as

$$\dot E_w = \gamma r_0^3 \dot\omega r_u f_5, \quad (34)$$

where f_5 is the coefficient dependent on the geometrical parameters and ϕ. Coefficient f_5 was used in Equation 21 for calculations of stability factor $\gamma H/c$.

REFERENCES CITED

Bishop, A. W., 1955, The use of slip circle in the stability analysis of slopes: Géotechnique, v. 5, p. 7–17.
Bishop, A. W., and Morgenstern, N. R., 1960, Stability coefficients for earth slopes: Géotechnique, v. 10, p. 129–150.
Chen, W. F., 1975, Limit analysis and soil plasticity: Amsterdam, Elsevier, 638 p.
Chen, W. F., and Giger, M. W., 1971, Limit analysis of stability of slopes: Journal of Soil Mechanics and Foundations Division, American Society of Civil Engineers, v. 97, p. 19–26.
Chen, W. F., Giger, M. W., and Fang, H. Y., 1969, On the limit analysis of stability of slopes: Soils and Foundations, v. 9, p. 23–32.
Collins, I. F., 1974, A note on the interpretation of Coulomb's analysis of the thrust on a rough retaining wall in terms of the limit theorems of limit plasticity: Géotechnique, v. 24, p. 106–108.
De Jong, D. J. G., 1981, A variational fallacy: Géotechnique, v. 31, p. 289–290.
Drescher, A., and Detournay, E., 1993, Limit load in translational failure mechanisms for associative and non-associative materials: Géotechnique, v. 43, p. 443–456.
Drucker, D. C., Greenberg, H. J., Prager, W., 1952, Extended limit design theorems for continuous media: Quarterly of Applied Mathematics, v. 9, p. 381–389.
Duncan, J. M., 1992, State-of-the-art: Static stability and deformation analysis, *in* Proceedings, Stability and Performance of Slopes and Embankments—II, San Francisco: New York, American Society of Civil Engineers, v. 1, p. 222–266.
Espinoza, R. D., Repetto, P. C., and Muhunthan, B., 1992, General framework for stability analysis of slopes: Géotechnique, v. 42, p. 603–615.
Fredlund, D. G., and Krahn, J., 1977, Comparison of slope stability methods of analysis: Canadian Geotechnical Journal, v. 14, p. 429–439.
Gvozdev, A. A., 1960, The determination of the value of the collapse load for statically indeterminate systems undergoing plastic deformation (English translation by R. M. Haythornthwaite): International Journal of Mechanical Sciences, v. 1, p. 322–335.
Hill, R., 1948, A variational principle of maximum plastic work in classical plasticity: Quarterly Journal of Mechanics and Applied Mathematics, v. 1, p. 18–28.
Janbu, N., 1973, Slope stability computations, *in* Hirschfeld, R. C., and Poulos, S. J., eds., Embankment-dam engineering, Casagrande volume: New York, Wiley, p. 47–86.
Leshchinsky, D., 1990, Slope stability analysis: Generalized approach: Journal of Geotechnical Engineering, American Society of Civil Engineers, v. 116, p. 851–867.
Leshchinsky, D., Baker, R., and Silver, M. L., 1985, Three dimensional analysis of slope stability: International Journal of Numerical and Analytical Methods in Geomechanics, v. 9, p. 199–223.
Michalowski, R. L., 1989, Three-dimensional analysis of locally loaded slopes: Géotechnique, v. 39, p. 27–38.
Michalowski, R. L., 1994, Slope stability analysis revisited: Géotechnique, in press.
Miller, T. W., and Hamilton, J. H., 1989, A new analysis procedure to explain a slope failure at the Martin Lake mine: Géotechnique, v. 39, p. 107–123.
Miller, T. W., and Hamilton, J. H., 1990, A new analysis procedure to explain a slope failure at the Martin Lake mine. Discussion: Géotechnique, v. 40, p. 145–147.
Morgenstern, N. R., and Price, V. E., 1965, The analysis of the stability of general slip surface: Géotechnique, v. 15, p. 79–93.
Mróz, Z., and Drescher, A., 1969, Limit plasticity approach to some cases of flow of bulk solids: Journal of Engineering for Industry, Transactions of the American Society of Mechanical Engineers, v. 51, p. 357–364.
Spencer, E., 1967, A method of analysis of the stability of embankments assuming parallel interslice forces: Géotechnique, v. 17, p. 11–26.

Manuscript Accepted by the Society May 20, 1994

Geological Society of America
Reviews in Engineering Geology, Volume X
1995

Groundwater flow and the stability of heterogeneous infinite slopes underlain by impervious substrata

William C. Haneberg
New Mexico Bureau of Mines and Mineral Resources, Socorro, New Mexico 87801

ABSTRACT

The effects of steady groundwater flow on the stability of heterogeneous infinite slopes underlain by impervious substrata were investigated using a series of computer simulations. A preliminary one-dimensional analysis shows that the magnitude of hydraulic head perturbation that can be attributed to flow across an idealized heterogeneity is controlled by both the size of the heterogeneity and the hydraulic conductivity contrast. In the limit, a perfectly impermeable heterogeneity will give rise to hydraulic gradients that are controlled by the normalized length of the heterogeneity, whereas an infinitely permeable heterogeneity will cause the hydraulic gradient to vanish across the heterogeneity. A series of two-dimensional finite-difference models shows that individual heterogeneities occupying 10% of the slope area can significantly perturb flow fields and influence local factors of safety in hypothetical slopes. Although local factors of safety are reduced enough to indicate the development of small secondary slides as a consequence of flow through or around heterogeneities, overall factors of safety remain unchanged. When heterogeneities are closely spaced, however, their effects on the stability of the model slopes are cumulative and in one case the overall factor of safety was increased by about 2%.

INTRODUCTION

The destabilizing effects of groundwater flow through both natural and engineered slopes has been understood since the pioneering work of Terzaghi (1950), and a geometric analysis of gravity-driven hillside flow systems was included in the first unified analysis groundwater flow (Hubbert, 1940). Even earlier, Terzaghi (1929) constructed flow nets to illustrate the destabilizing effect of flow through high-permeability zones beneath a dam. Nonetheless, the heterogeneity of clayey colluvium and landslide debris, which may range in particle size from clay to boulder, makes prediction of subsurface pore pressure distributions one of the most confounding aspects of slope stability analysis.

Heterogeneous hydraulic conductivity fields can in many cases be attributed to macroscopic changes in soil properties. For example, a portion of a trench log through a thin colluvium landslide along the Ohio River valley shows that sliding has occurred along the contact between clayey colluvium and weathered shale bedrock with minor limestone interbeds (Fig. 1). Detailed descriptions of the colluvium and its hydraulic properties are contained in Fleming and Johnson (1994), Baum (1994), Haneberg and Gökce (1994), and Haneberg (1991a). The colluvium is composed of upper brown and lower yellow units. Results from limited laboratory testing by Murdoch (1988) suggest that the hydraulic conductivity of the brown colluvium may be an order of magnitude greater than the hydraulic conductivity of the yellow colluvium, whereas the results of laboratory tests on different samples from the same slope show no consistent trend with depth (Haneberg and Gökce, 1994). Nearly impermeable limestone fragments, which range from sand-sized fragments to slabs a decimeter thick and a meter long, occupy 10% to 20% of the colluvial volume. Open cracks in the colluvium, however, serve as conduits for

Haneberg, W. C., 1995, Groundwater flow and the stability of heterogeneous infinite slopes underlain by impervious substrata, *in* Haneberg, W. C., and Anderson, S. A., eds., Clay and Shale Slope Instability: Boulder, Colorado, Geological Society of America Reviews in Engineering Geology, v. X.

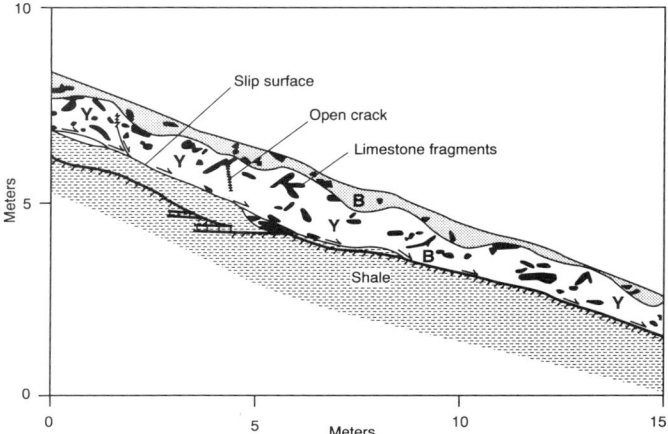

Figure 1. Portion of a trench log through a thin colluvium landslide along the Ohio River valley near Cincinnati, Ohio. B denotes brown colluvium, and Y denotes yellow colluvium.

saturated flow and barriers to unsaturated flow. In this case, the long dimension of some of the heterogeneities approaches the thickness of the colluvium. Figure 2 shows a cut in landslide debris along the valley of the Rio Costilla in northern New Mexico. Parent materials were Tertiary alluvial and volcaniclastic deposits composed of clay to boulder-sized clasts. The debris consists of cobbles to small boulders of both plutonic and volcanic origin within a fine-grained matrix. A pedogenic soil developed along the top of the exposure. Examination of outcrops and inspection of borehole logs show that the igneous clasts are heavily weathered, in some cases to granular saprolite and in others to a clayey consistency. In addition, the debris contains isolated lenses of light gray clay several centimeters to decimeters thick and several decimeters long. In this example, however, the diameters of even the largest boulders seen in outcrop are many orders of magnitude smaller than the size of the landslide mass itself.

Faced with the kinds of textural heterogeneity described above, one might logically ask what effects hydraulic conductivity heterogeneities have on groundwater flow fields and, consequently, slope stability. This paper describes the effects of large-scale hydraulic conductivity heterogeneities on pressure head distributions in hypothetical infinite slopes resting atop impervious substrata, as inferred from mathematical models of steady-state groundwater flow. Although recent advances in stochastic hydrology have provided a promising set of tools with which to rigorously analyze heterogeneity, especially with regard to scale (e.g., Gelhar, 1993), the work described in this paper was restricted to deterministic models. The influence of pressure head perturbations on slope stability is subsequently calculated using an adaptation of the well-known infinite slope factor of safety equation that can account for variable pore water pressure along the length of the slope. Complete saturation or tension saturation is assumed on the basis of observations made during several field studies of landslide hydrology

Figure 2. Outcrop of landslide debris along the Rio Costilla valley, northern New Mexico. The field notebook is for scale.

(Haneberg and Gökce, 1994; Haneberg, 1991a; Baum and Reid, this volume; Anderson and Sitar, 1993; Iverson and Major, 1987). The analysis is further restricted to steady-state flow in order to eliminate the complicating effects of transient pressure head pulses moving through the slab. Throughout this paper, pore water pressure is given in terms of equivalent pressure head, which implies that the pore water is incompressible. Pressure head, which has units of length, should be multiplied by the unit weight of water (9,810 N/m^3) to find pore water pressure with units of force/area. Variables used throughout the paper are listed in Table 1.

PREVIOUS MODELS OF GROUNDWATER FLOW IN UNSTABLE SLOPES

The hydrologic literature is rich with studies describing and analyzing hillside flow, summaries of which are contained in the works edited by Kirkby (1978) and Anderson and Burt

TABLE 1. LIST OF VARIABLES

Variable	Explanation
D	Thickness of a porous layer or fraction thereof
E	Interslice normal force
ζ	Factor of safety against sliding
h	Hydraulic head
ΔH	Hydraulic head differential of magnitude H
i	Iteration index
K	Hydraulic conductivity
L	Length of porous slab or portion thereof
N	Number of slices used in limit-equilibrium stability analysis
n_e	Effective or fillable porosity
T	Interslice shearing force
β	Slope angle
γ_{sat}	Saturated unit weight of soil
γ_w	Unit weight of water
ϕ	Angle of internal friction
ψ	Pressure head
τ	Shear strength along a potential slip surface
ξ	Distance parallel to slope measured from toe of slab
ζ	Distance normal to slope measured from base of slab
$\Delta\xi$	Slice width in limit equilibrium stability analysis

(1990). Coates (1990) reviewed primarily qualitative geomorphic aspects of groundwater and slope stability. A number of studies containing mathematical models of groundwater flow through unstable slopes, grouped according to the type of model, are summarized below.

Water balance models

The least complicated type of flow-based hillside hydrologic model involves calculation of the water balance, typically between infiltration due to rainfall and discharge due to gravity drainage. For example, Wieczorek (1987) discussed qualitative relationships among infiltration, deep vertical drainage, slope-parallel drainage for shallow soil slumps, deep soil slumps, and shallow soil slides over bedrock. Wilson and Wieczorek (1988) subsequently proposed a so-called leaky barrel water balance model that incorporated an empirical drainage coefficient and an empirical critical water level above which sliding occurred. For clayey hillside soils with thick capillary fringes, which can in many cases be approximated as fully saturated soils, Haneberg (1992) suggested that the steady-state ratio of discharge to recharge is a product of the ratio of saturated hydraulic conductivity to recharge rate, the aspect ratio of the slope (D/L), and the slope angle. For coarse-grained granular soils in which the height of the capillary fringe is negligible, the height of the water table can be substituted for soil thickness in the water balance equation. Slopes with a discharge to recharge ratio less than unity may be susceptible to sliding because they cannot dissipate rainfall-induced pore pressure increases as rapidly as slopes with ratios less than unity. For example, colluvium-covered slopes along the Ohio River valley that are prone to episodic sliding appear to have an average annual discharge to recharge ratio of approximately 1.1. This marginal ratio suggests that the slopes are barely able to drain infiltrated precipitation during average years and probably accumulate enough water to promote instability only in unusually wet years. Unsteady water balance models can also be formulated in order to model changes in water volume in response to individual storms or series of storms (Haneberg, 1992). A significant disadvantage of water balance models is that the fundamental quantity of interest is water content, not pressure head. Thus, unsaturated soil moisture characteristic curves must be used to estimate mean pressure head from calculated water content.

Two-dimensional flow models

Hodge and Freeze (1977) modeled groundwater flow through a variety of hypothetical slopes, specifically investigating the role of factors such as anisotropy, layering and folding of strata, low permeability zones, and topography. They showed that geologic details can give rise to potentially destabilizing pressure head perturbations. Rulon and Freeze (1985) modeled flow through a stratified and variably saturated slope, which allowed for the development of perched water tables, and also showed that their calculated pressure head distributions would produce higher factors of safety than those based upon simple water table configurations. Reid et al. (1988) reconstructed the sequence of events prior to a landslide in northern California. Based upon the results of finite element models of flow through a layered and variably saturated hillslope, they concluded that a subtle decrease in soil hydraulic conductivity may have impeded downslope drainage and created a groundwater mound that increased pressure heads enough to cause failure. Murdoch (1988) combined a series of one-dimensional unsteady infiltration models to calculate a quasi two-dimensional model of infiltration into a sloping and variably-saturated infinite half-space, and used the results to interpret his field observations of pressure head fluctuations in a thin colluvium landslide along the Ohio River. Baum (1994) modeled steady-state groundwater flow through the same hillside, but included flow through interbedded shales and fractured limestones beneath the colluvial landslide debris. His model showed nearly slope-parallel flow through the entire hillside with zones of elevated head where permeable fractured limestone layers met less permeable colluvium. Moreover, Baum concluded that the destabilizing effects of slightly artesian pressures in fractured limestone layers is negligible because fractures occupied so little of the bedrock.

Modified Dupuit water table models

In cases where the slope aspect ratio (D/L) is small, slope-normal variations in hydraulic head can be ignored and water levels modeled using Dupuit assumptions modified to account for flow above a sloping, rather than horizontal, substrate. Haneberg and Gökce (1994) and Jackson and Cundy (1992), respectively, used one- and two-dimensional finite difference models of unsteady unconfined groundwater flow through an inclined layer overlaying an impervious substrate to successfully reproduce patterns of storm-induced water-table fluctuations in slopes along the Ohio River valley and in the Cascade Mountains of Washington. Although neither model incorporated the effects of spatially variable hydraulic properties, either could have been modified to include vertically averaged hydraulic properties. An important result of these water table models is that low values of pre-storm effective porosity, typically in the range of 1% to 10%, are necessary in order to reproduce observed piezometric response. Haneberg and Gökce (1994) also used a dimensional analysis of the equations governing Dupuit flow to suggest that rainfall-induced piezometric level increases will occur uniformly over the slope and on a short time scale proportional to $n_e D/K$ (see Table 1 for definitions of variables), whereas piezometric level decreases to gravity drainage of the slope will vary along the slope on a long time scale proportional to $n_e L/K$.

Pore pressure diffusion models

Iverson and Major (1987), Haneberg (1991a, 1991b), Pyles et al. (1987), and Baum and Reid (this volume) all used models of vertical or slope-normal pressure head diffusion to analyze the hydrologic response of landslides to rainfall. Wu and Swanston (1980) used a model of vertical hydraulic head diffusion to model the hydrologic response of landslides to rainfall, and ultimately predict the probability of failure following different storms. The most recent development has been development of an analytical pore pressure diffusion model in which flux, rather than pressure head, is specified along the ground surface (M. E. Reid, 1994, personal commun.). With a diffusion model, one is able to account for commonly observed phenomena such as the attenuation and time lag of pressure head pulses traveling downward from the surface. Pressure head signals from high-frequency or short duration storms propagate downward rapidly, but are also attenuated very near the surface. Signals from low-frequency or long duration storms, on the other hand, travel more slowly and are attenuated less. Therefore, the potential impact of any given storm on slope stability is a function of its duration or frequency as well as its intensity or amplitude. In thick (tens of meters) landslides with low hydraulic conductivity, rainfall-induced pressure heads may take several months to propagate from the ground surface to the failure surface, which can complicate attempts to correlate rainfall and movement rates.

Integrated analyses of flow and stability

Iverson and Major (1986) analyzed steady groundwater flow in homogeneous infinite slopes and identified combinations of seepage force magnitude, seepage force direction, and slope angle conducive to Coulomb frictional failure and static liquefaction. Iverson (1990) conducted a theoretical analysis of groundwater flow in infinite slopes, allowing pressure heads to vary only normal to the ground surface. He found that knowledge of the slope-parallel water table position and the slope-normal hydraulic gradient is sufficient to define the distribution of hydraulic head in an infinite slope. Iverson and Reid (1992) formulated a two-dimensional poroelastic model, in which steady-state seepage forces were treated as body forces and incorporated into the equations of equilibrium for elastic media, and used the model to further evaluate the effects of seepage vector orientation and magnitude on Coulomb failure potential. They found that, although groundwater flow decreases the stability of most parts of a slope and shifts the locus of minimum stability toward the toe of the slope, the effects of groundwater flow were not as significant as predicted by the one-dimensional analysis of Iverson and Major (1986). Reid and Iverson (1992) used the same two-dimensional poroelastic model to evaluate the effects of slope geometry and the presence of high- and low-permeability layers oriented horizontally, vertically, and parallel to the ground surface. One of their findings was that slope-parallel flow predominated in high permeability layers parallel to the ground surface and underlain by lower permeability material, even when the permeability contrast is as low as an order of magnitude.

SIMULATION OF GROUNDWATER FLOW AND SLOPE STABILITY

Steady one-dimensional flow across a heterogeneity

Analysis of a simple one-dimensional steady flow problem can provide some insight into the influence of heterogeneities on pressure head distribution, as well as a basis for understanding more complicated two-dimensional flow models. Consider an example in which a head differential of magnitude ΔH forces flow across a zone that has higher or lower hydraulic conductivity than the two adjacent zones (Fig. 3). In this example, zone 2 is composed of three layers, so for one-dimensional layer-parallel flow the effective zone 2 hydraulic conductivity is the weighted arithmetic mean of the three layer conductivities (e.g., Domenico and Schwartz, 1990). Head is specified at each end of the slab, giving the external boundary conditions $h = 0$ at $\xi = -L_1/2$ and $h = \Delta H$ at $\xi = L_1/2$ (see Table 1). Both head and flux are required to be continuous across the two internal boundaries ($\xi \pm L_2/2$). Analytical solutions can be easily obtained for each of the three flow zones, showing that head is a linear function of distance across each zone (see Appendix).

The effect of a high- or low-conductivity zone sandwiched between two zones of equal conductivity is reflected by the nature of the zone 2 hydraulic gradient,

$$\frac{d(h_2/\Delta H)}{d(\xi/L_1)} = \frac{1}{(L_2/L_1) + (\overline{K}_2/K_1) - (L_2/L_1)(\overline{K}_2/K_1)}, \quad (1)$$

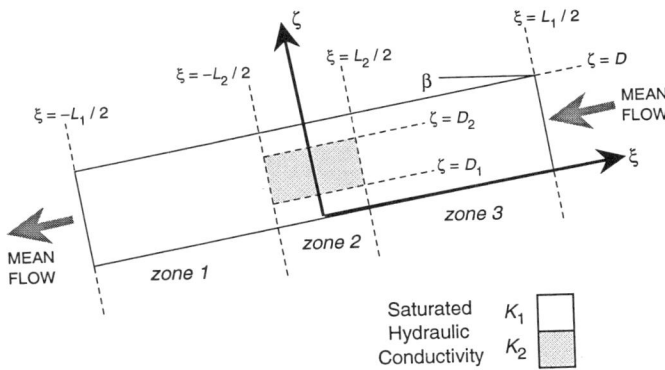

Figure 3. Illustration of variables used in the groundwater flow analyses. Zones 1, 2, and 3 refer to the one-dimensional analysis, in which the effects of layering in zone 2 are incorporated using a weighted arithmetric mean.

which is found by differentiating Equation A17 (see Appendix) with respect to ξ and rearranging to put each of the variables into dimensionless form. As illustrated in Figure 4, the nature of the zone 2 hydraulic gradient depends on whether $\bar{K}_2/K_1 > 1$ or $\bar{K}_2/K_1 < 1$. Because the inhomogeneity does not occupy the entire thickness of the slab in this example, zone 2 hydraulic conductivity must be averaged in order to apply the one-dimensional flow model. Thus, $K_2/K_1 = 10^{-2}$ is equivalent to $\bar{K}_2/K_1 = 0.50$; $K_2/K_1 = 10^{-1}$ is equivalent to $\bar{K}_2/K_1 = 0.55$; $K_2/K_1 = 10^1$ is equivalent to $\bar{K}_2/K_1 = 5.5$; and $K_2/K_1 = 10^2$ is equivalent to $\bar{K}_2/K_1 = 50$. Averaging diminishes the importance of low-conductivity heterogeneities and amplifies the importance of high-conductivity heterogeneities. As the width of zone 2 increases the gradient approaches unity regardless of the value of the hydraulic conductivity ratio. As zone 2 becomes narrower and the conductivity ratio smaller, the hydraulic gradient becomes exponentially larger. Conversely, as zone 2 becomes narrower and the conductivity ratio becomes larger, the hydraulic gradient vanishes. In the limits, the hydraulic gradient approaches zero as the conductivity ratio approaches infinity and the hydraulic gradient approaches L_1/L_2 as the conductivity ratio approaches zero. The relative importance of high- and low-conductivity zones, as measured by the magnitude of resulting head increase or decrease on either side of zone 2, will thus be controlled by a combination of geometry and hydraulic conductivity ratio.

Referring to Figure 4, consider an example in which $L_2/L_1 = 0.2$, $D_1/D = 0.25$, $D_2/D = 0.75$, and $\bar{K}_2/K_1$ is allowed to vary from 0.50 to 50. This is a one-dimensional analog of a problem that will be solved numerically for two-dimensional flow in a subsequent section of this paper. Hydraulic head profiles for each of the four cases illustrated in Figure 4 are plotted in Figure 5, along with the reference case of $\bar{K}_2/K_1 = 1$. Because of the small difference between average zone 2 hydraulic conductivities, the curves for the two low-conductivity cases are virtually indistinguishable. For these particular combinations of geometry and hydraulic conductivity, the presence of a high-conductivity zone has a greater effect than the presence of a low-conductivity zone. Along the up-gradient edge of zone 2, the low-conductivity head is 0.67 dimensionless units more than the reference case, whereas the high-conductivity heads are 0.78 and 0.98 dimensionless units less than the reference case.

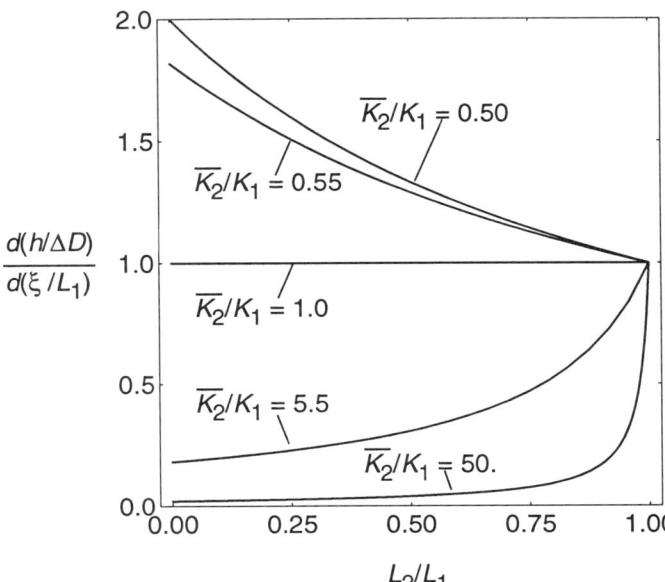

Figure 4. Dimensionless zone 2 hydraulic gradient versus heterogeneity width for various hydraulic conductivity ratios.

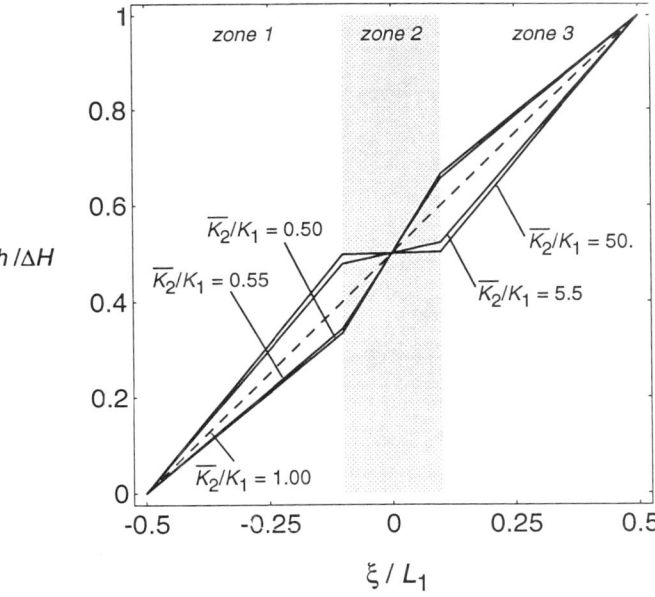

Figure 5. Normalized hydraulic head vs. normalized distance for one-dimensional flow through a low- or high-conductivity zone (solid lines) and for flow through a homogeneous medium (dashed line). The geometric significance of zones 1, 2, and 3 are illustrated in Figure 3.

Steady two-dimensional flow through a heterogeneous slab

Results obtained from the preceeding one-dimensional analysis can be expanded upon by simulating two-dimensional flow and incorporating the results into a modified version of the infinite slope stability analysis equation. Two-dimensional steady-state flow through a rectangular slab resting atop a sloping impervious substrate can be described by simplifying the transient flow equation of Haneberg (1991a, 1991b) to

$$0 = \frac{\partial}{\partial \xi}\left[K_{\xi\xi}\left(\frac{\partial \psi}{\partial \xi} + \sin \beta\right)\right] + \frac{\partial}{\partial \zeta}\left[K_{\zeta\zeta}\left(\frac{\partial \psi}{\partial \zeta} + \cos \beta\right)\right]. \quad (2)$$

For inhomogeneous, anisotropic, or variably saturated porous slabs, Equation 2 must be solved numerically. All of the solutions in this paper were obtained using a mesh-centered finite-difference approximation (e.g., Huyakorn and Pinder, 1983) solved by simple Gauss-Seidel iteration. Hydraulic conductivity values were specified for each node, and internodal values of $K_{\xi\xi}$ and $K_{\zeta\zeta}$ were approximated as the arithmetric mean of the two adjacent nodal values. The slope-parallel flow boundary condition of

$$\frac{\partial \psi}{\partial \zeta} = -\cos \beta \quad (3)$$

was used at each end of the slab ($\xi = \pm L_1/2$), and along the base of the slab ($\zeta = 0$). For the fourth boundary condition, pressure head was

$$\psi = 0 \quad (4)$$

along the upper surface of the slab ($z = D$). All of the simulations presented in this paper used a slope angle of $\beta = 10°$ and, except for one simulation, a dimensionless aspect ratio of $D/L = 1/5$. Twenty rows and one-hundred columns of finite difference nodes, including imaginary nodes along no-flow boundaries, were used to calculate the results shown in this paper. A tolerance of 10^{-6} was used for each solution, but repeated calculations showed that head distributions were relatively insensitive to changes in grid spacing and tolerance.

Sliding potential

This paper uses a conventional factor-of-safety approach, which provides estimates of stability along specific slip surfaces, to evaluate the effects of steady groundwater flow on slope stability. Others have used more general quantities such as Coulomb failure potential to evaluate the potential for sliding, which can yield fundamental insights into the differences between the buoyant effects of pressure head and the frictional effects of seepage forces or pressure head gradients but cannot provide the factors of safety widely used to evaluate many engineering slope stability problems (Iverson and Major, 1986; Iverson and Reid, 1992; Reid and Iverson, 1992).

Results from field and laboratory studies of colluvial slopes suggest that clayey colluvium is commonly composed of clasts ranging from clay- to boulder-sized, and that material properties may be more akin to frictional silty sands or fine sands than true cohesive clays. Frictional behavior would be advantageous, because it would both simplify the limit-equilibrium stability equation and reduce by one the degrees of freedom inherent in slope stability analyses. For example, samples of colluvium derived from Ordovician illitic shales and limestones near Cincinnati, Ohio (Fleming and Johnson, 1994) exhibit saturated hydraulic conductivities on the order of 10^{-5} to 10^{-6} m/s, angles of internal friction of 12° to 24°, and cohesive strength ranging from 0 to 8 kPa, with very little difference between peak and residual shear strength. In analyzing the stability of the same hillside, Baum (1994) used residual friction angles ranging from 15° to 24° and residual cohesion values of 1 to 4 kPa, increasing the cohesive strength as he decreased the angle of internal friction. During the investigation of a construction-induced slide of clayey colluvium developed above fine-grained Pennsylvanian sedimentary rocks near Wierton, West Virginia, D'Appolonia et al. (1967) obtained peak cohesive strength values of 8 to 10 kPa and peak internal friction angles of 19° to 20°, as well as residual internal friction angles of about 16°. Anderson and Sitar (1993) conducted an extensive program of shear strength testing on clayey to silty colluvium derived from sandstone and siltstone bedrock in the San Francisco Bay region. Their direct shear tests yielded a peak cohesive strength of 16 kPa, a peak friction angle of 27°, a residual cohesive strength of 9 kPa, and a residual friction angle of 24°. Anderson and Sitar concluded, however, that their direct shear parameters over-estimated shear strength at low effective normal stresses. The results of these few studies suggest that peak cohesive strength values of 1 to 10 kPa, which is a small percentage of the effective normal stress acting along slip surfaces more than a meter or so deep, may be typical of clayey colluvium in landslide-prone areas. For reactivated slip surfaces along which residual shear strength is mobilized, cohesion is commonly neglected (e.g., Skempton, 1964). Thus, there is some empirical justification for analyzing clayey colluvium more than a meter or so deep as a frictional rather than a cohesive soil.

The potential for frictional sliding along slip surfaces parallel to the ground surface, which is a reasonable approximation of many landslides developed in colluvium, can be evaluated using a modification of the well-known infinite slope method in which pressure head is allowed to vary along the length of the slip surface. Potential slip surfaces must be parallel to the ground surface so that lithostatic interslice shear and normal force derivatives vanish. Pore pressure acting along the base of a slice does not affect interslice shear and normal forces (e.g., Janbu, 1973). Therefore, the assumption of variable pressure head along the slip surface does not introduce any error as long as both soil thickness and slope angle remain constant. A slope angle of $\beta = 10°$ was chosen for use in this paper to insure an average factor of safety against sliding of $F > 1$ for all the examples given herein, assuming a typical value of $\phi = 20°$ for colluvial debris. Therefore, both the upper and lower ends of the stable slab are subjected to at-rest lateral earth pressure of equal magnitude. For a sliding block with an average value of $F < 1$, the toe would approach a state of passive earth pressure whereas the head would approach a state of active earth pressure (e.g., Baum and Fleming, 1991).

The average factor of safety against sliding for an infinite slope with variable pressure head along the slip surface is given by

$$F_s = \frac{1}{N} \sum_{i=1}^{N} \left[1 - \frac{\psi_i}{(D-\zeta)} \frac{\gamma_w}{\gamma_{sat}} \frac{1}{\cos\beta} \right] \frac{\tan\phi}{\tan\beta}. \quad (5)$$

Variables used in the stability analysis are illustrated in Figure 6. Slice width cancels in all terms if the slices are of equal width, and therefore does not appear in the final expression for the factor of safety. By varying the quantity $(D - \zeta)$, which is depth below the ground surface, one can estimate the stability of an arbitrary number of potential slip surfaces as long as pressure head is known along each potential slip surface. Moreover, one can calculate local factors of safety along each potential slip surface by considering the driving and resisting forces for each slice rather than values summed over the entire slope (Baum, 1994). The end result is a grid of F values calculated throughout the saturated block, which can be contoured to isolate zones in which groundwater flow contributes to increased or decreased stability.

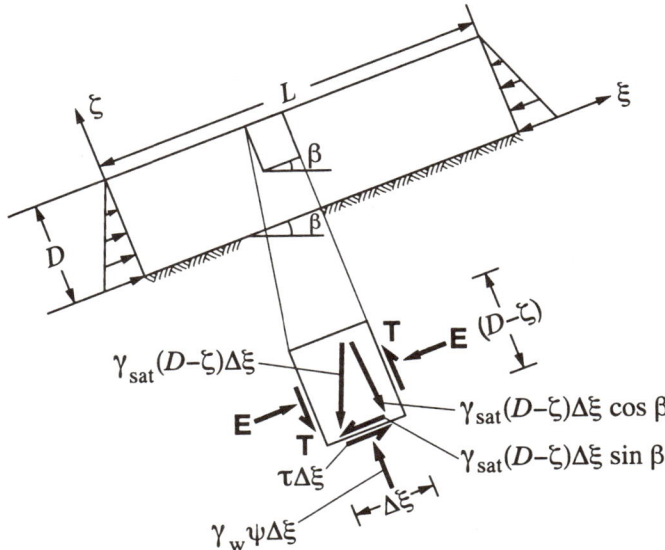

Figure 6. Illustration of variables used to formulate the equation for the limit equilibrium factor of safety against sliding.

Numerical results

The destabilizing effects of hydraulic heterogeneities can be illustrated by conducting a series of numerical experiments. Although results shown in this paper are for flow through inclined slabs (see Figs. 3 and 6), in subsequent figures the slabs are depicted in a horizontal position. The first group of simulation results is presented as a series of plots illustrating head contours and seepage force vectors, a series of plots illustrating pressure head contours, and a series of plots illustrating local factors of safety against sliding for different hydraulic conductivity ratios. Departing from the convention used throughout the preceeding one-dimensional analysis, the two-dimensional results do not make use of averaged hydraulic conductivity values. Thus, a value of $K_2/K_1 = 0.01$ in the two-dimensional analysis corresponds to a value of $\bar{K}_2/K_1 = 0.5$ used in the one-dimensional analysis, and so forth.

The two-dimensional models generally confirm the one-dimensional prediction that, for this particular problem, the high-conductivity heterogeneities should have a more significant effect on head than the low-conductivity heterogeneities (Fig. 7). Each of the head contours represents a change of 1/10 of the total head differential across the slab, or $\Delta H/10$. As might be expected, hydraulic gradients, and thus seepage forces, increase within the low-conductivity zones but decrease within the high-conductivity zones. For the cases of $K_2/K_1 = 10^{-2}$, $K_2/K_1 = 10^{-1}$, and $K_2/K_1 = 10^1$, the changes in head up-gradient and down-gradient from the heterogeneity appear to be equal in magnitude. For the case of $K_2/K_1 = 10^2$, however, the up-gradient decrease in head appears to be greater than the down-gradient increase in head. The unperturbed seepage force per unit volume due solely to slope-parallel flow is $\gamma_w \sin\beta$, or about 1.7 kN/m^3, which is less than 1/5 of the buoyant force exerted by the pore water. In both the high- and low-conductivity cases, the maximum seepage force per unit volume is approximately 4.5 kN/m^3, which is less than half of the buoyant force exerted by the pore water. Away from the heterogeneities, the seepage force vectors are parallel to the slope and groundwater has no influence beyond its buoyancy on the stability of the hypothetical slopes. Near the heterogeneities, however, seepage vectors both increase in magnitude and act either into or out of the slope.

Heads calculated along the uphill and downhill edges of the heterogeneities by the one-dimensional and two-dimensional models compare favorably, although the down-gradient errors are slightly larger than the up-gradient errors (Table 2). This asymmetry of errors exists in part because the magnitudes of the down-gradient heads are smaller than those of the up-gradient heads, and in part because the absolute differences between the two models are greater for down-gradient heads than for up-gradient heads.

Contour plots of pressure head distributions also show that the high-conductivity zones in this case have a greater effect than the low-conductivity zones (Fig. 8). As with total

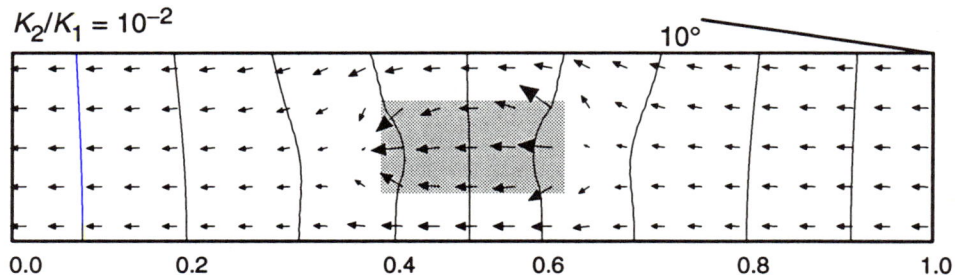
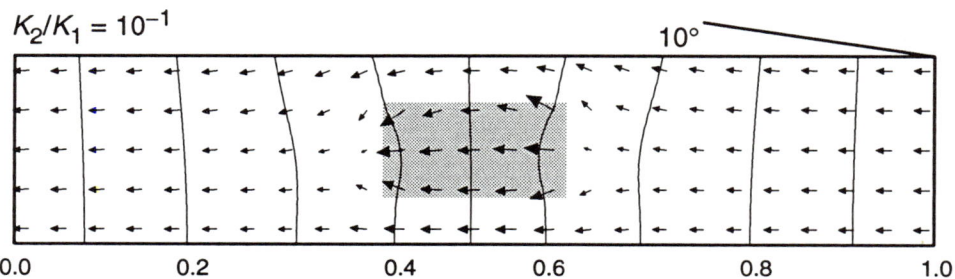
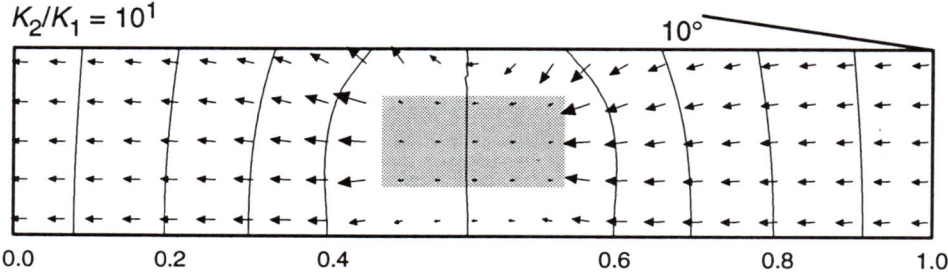
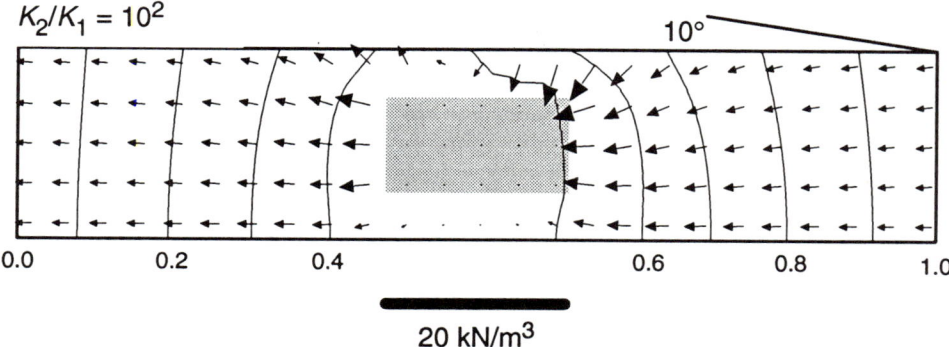

Figure 7. Hydraulic head contours and seepage force vectors for flow through sloping layers containing either low- or high-conductivity heterogeneities. The gray zones represent heterogeneities with hydraulic conductivity contrasts as indicated above each layer. The hydraulic head contour interval is $\Delta H/10$, and arrow lengths are proportional to seepage force magnitude. For comparison, the static buoyant force exerted by water is 9.8 kN/m^3.

TABLE 2. COMPARISON OF ONE-DIMENSIONAL AND TWO-DIMENSIONAL MODEL CALCULATIONS OF HEAD ALONG UP-GRADIENT AND DOWN-GRADIENT EDGES OF HETEROGENEITIES

K_2/K_1	Up-Gradient $\Delta h/H$			Down-Gradient $\Delta h/H$		
	One Dimensional	Two Dimensional	Error (%)	One Dimensional	Two Dimensional	Error (%)
10^{-2}	0.66	0.63	+5	0.34	0.38	-11
10^{-1}	0.66	0.63	+5	0.34	0.38	-11
10^{1}	0.52	0.56	-7	0.48	0.41	+17
10^{2}	0.50	0.49	+2	0.50	0.43	+16

head, pressure head is presented in dimensionless form. Each contour represents a change of 1/10 of the unperturbed hydrostatic pressure head at the base of the slab, or $D/10$. Contoured local F values reflect variations in seepage force direction and magnitude. Low-conductivity zones produce a downhill region of increased stability and an uphill region of slightly decreased stability, both of which are approximately the same size in each case (Fig. 9). The terms *decreased stability* and *increased stability* are relative to the overall factor of safety $F = 1.05$. In contrast, high-conductivity zones produce a downhill zone of decreased stability and an uphill zone of increased stability. In these two cases, the zones of increased stability are noticeably larger than the zones of decreased stability. In all four cases the overall factor of safety remained constant at $F = 1.05$, although the zones of $F < 1$ do suggest that perturbations in the flow field induced by heterogeneities can be significant enough to promote small-scale slides within the larger slab.

The difference between uphill and downhill head perturbations in the two-dimensional $K_2/K_1 = 10^2$ model raises questions about the influence of boundary effects on head distribution. In order to investigate whether this difference might be an artifact of the fixed head along the upper boundary of the porous slab, the flow field and stability were calculated for a thick slab with an aspect ratio of $D/L_1 = 0.5$ while keeping the size of the high-conductivity zone constant. As illustrated in Figure 10, moving the fixed-head boundary away from the heterogeneity does reduce some of the asymmetry. Therefore, the nature of the upper boundary appears to have a significant effect on the calculated flow fields for the $K_2/K_1 = 10^2$ case. Saturation to the ground surface, however, seems to be a reasonable and natural boundary condition. Alternative upper boundary conditions, for example, specified flux or even spatially-variable flux, might be considered in future analyses.

A series of simulations was also run to investigate the effects of closely spaced heterogeneities. The high- and low-conductivity zones were individually smaller than in the previous set of simulations, but the overall area occupied by the heterogeneities was increased from 10% to 17% of the layer. Additionally, results were calculated only for the two end-member cases of $K_2/K_1 = 10^{-2}$ and $K_2/K_1 = 10^2$. As shown in Figures 11 and 12, the general patterns of flow around and through the heterogeneities are similar to those in the single-heterogeneity models. Notice, however, that both seepage force magnitude and pressure head increase up-gradient, suggesting that closely-spaced heterogeneities have a cumulative effect. This gradual change can be seen most easily in Figure 12, in which the lowermost pressure head contours gradually increase or decrease in elevation. As a consequence, the zones of increased stability decrease in size from the toe to the head of the slope whereas the zones of decreased stability increase from the toe to the head of the slope for the $K_2/K_1 = 10^{-2}$ case (Fig. 13). The opposite is true for the $K_2/K_1 = 10^2$ case, in which the overall F was increased from 1.05 to 1.07.

DISCUSSION AND CONCLUSIONS

The results of numerical models of steady-state flow through a fully-saturated porous slab show that large-scale heterogeneities can have a significant influence on flow field details and a lesser influence on local slope stability. It is not possible to examine every conceivable combination of slope geometry and heterogeneity. Nonetheless, several conclusions can be drawn.

First, analysis of one-dimensional flow perpendicular to an idealized heterogeneity suggests that pressure head perturbations are controlled by size of the heterogeneity relative to that of the flow domain as well as the hydraulic conductivity contrast. The relationship between the magnitude of head perturbation and the size of the heterogeneity is also predicted by analysis of flow through stochastic porous media (e.g., Gelhar, 1993). Depending on the specific problem, high conductivity zones may have a larger or smaller effect on head distribution than low permeability zones even if the conductivity contrast is of equal magnitude.

Second, isolated large heterogeneities appear to increase or decrease the stability of portions of a porous slab, but do not appear to influence the resistance to sliding of the entire slab. Closely spaced heterogeneities in the hypothetical slopes, however, appear to have a cumulative effect that can change the overall factor of safety by several percent. Local factors of safety show that pressure head perturbations associated with hydraulic heterogeneities are capable of inducing localized sliding in the hypothetical slopes.

Third, the magnitude of seepage forces directed into or out of the slope as consequence of flow through or around heterogeneities is at most one-half of the buoyant force exerted by the pore water. Nonetheless, seepage force perturbations of this magnitude can have a significant effect on local slope stability.

Fourth, calculated head distributions are sensitive to the nature of the saturated upper boundary, especially when strong heterogeneities are located near the boundary. This is important because saturation to the ground surface is a fundamental assumption in many analyses of groundwater flow through unstable slopes. In most cases, saturation to the ground surface

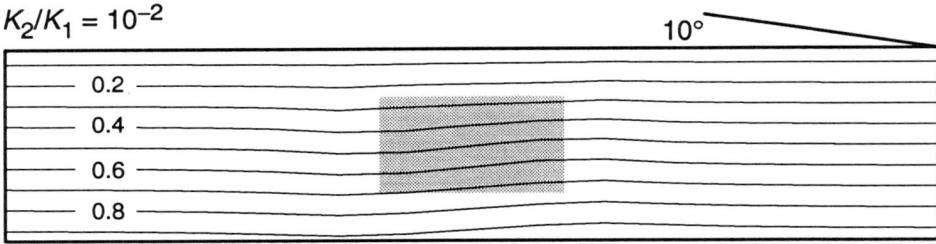
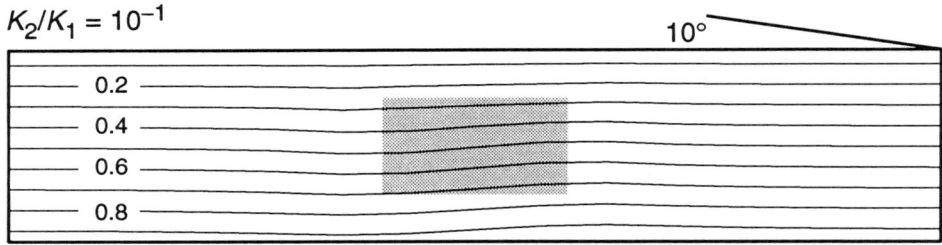
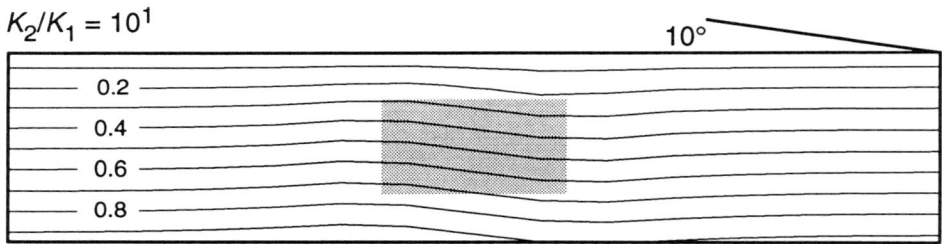
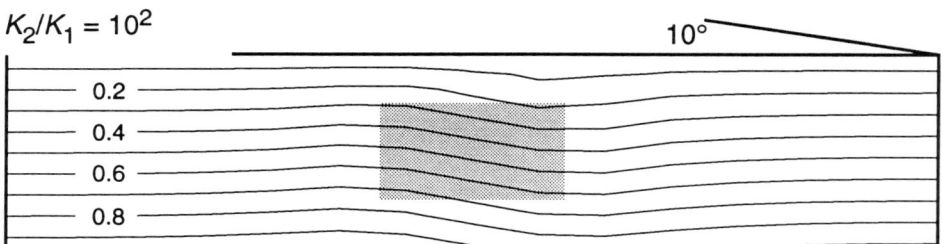

Figure 8. Pressure head contours for flow through sloping layers containing either low- or high-conductivity heterogeneities. The gray zones represent heterogeneities with hydraulic conductivity contrasts as indicated above each layer. The pressure head contour interval is $D/10$, or one-tenth of the unperturbed pressure head acting along the base of the layer under conditions of slope-parallel flow.

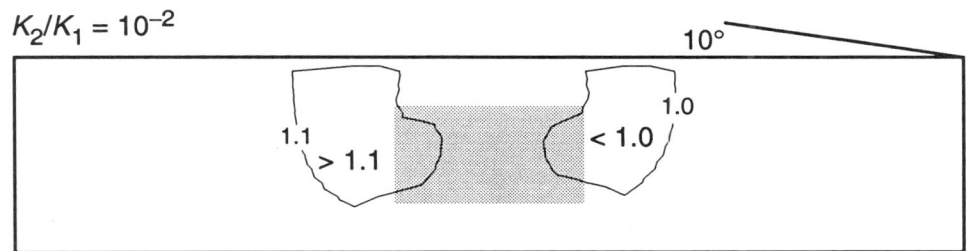
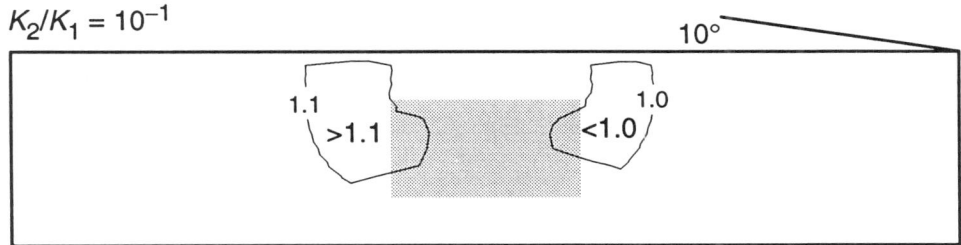
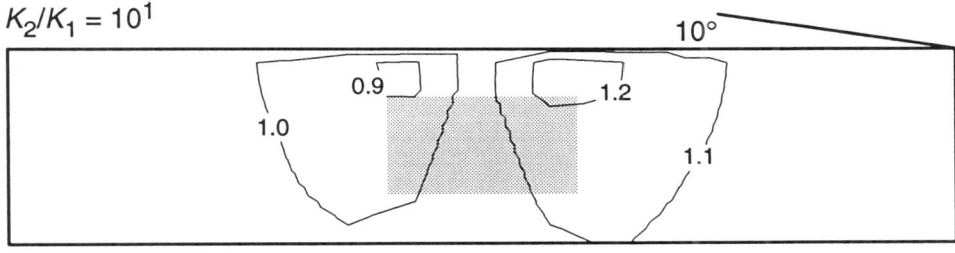
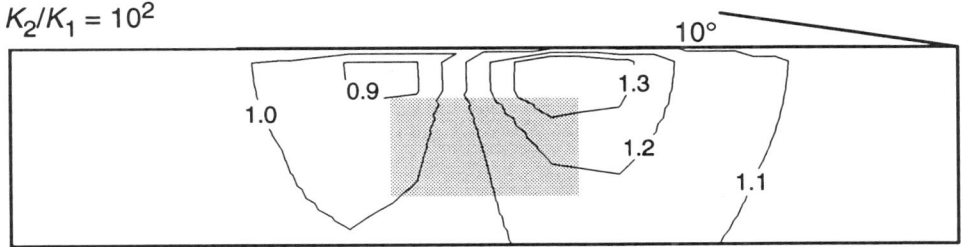

Figure 9. Local factor of safety against sliding contours for flow through sloping layers containing either low- or high-conductivity heterogeneities. The gray zones represent heterogeneities with hydraulic conductivity contrasts as indicated above each layer. The overall factor of safety for all four slabs is 1.05.

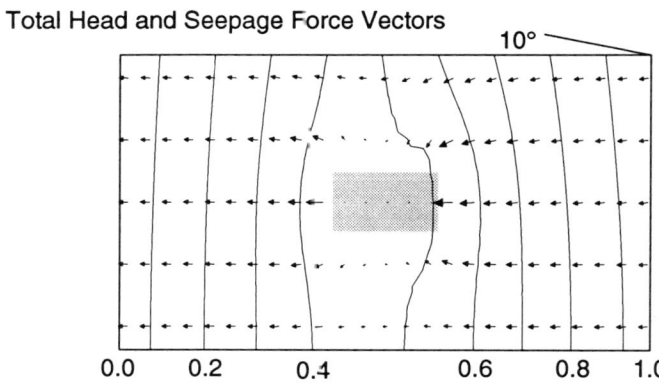

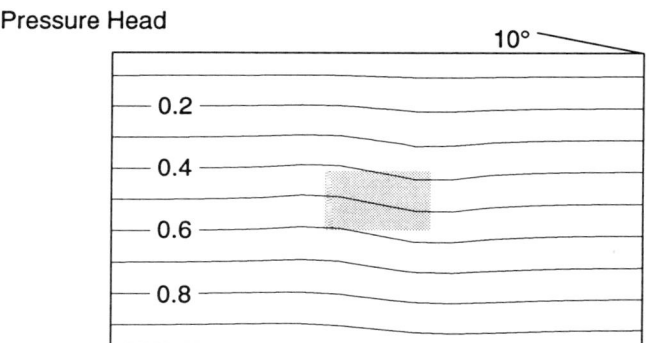

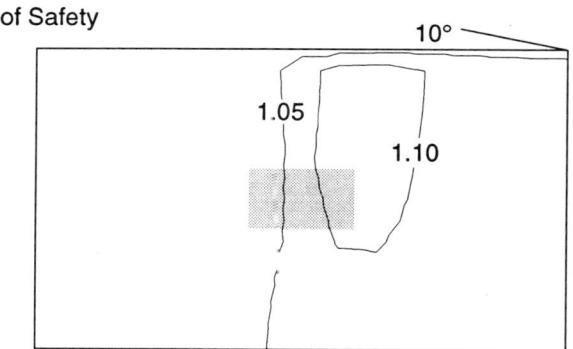

Figure 10. Hydraulic head contours, seepage force vectors, pressure head contours, and local factor of safety contours for flow through a thick heterogeneous layer with $K_2/K_1 = 100$ and an aspect ratio of $D/L - 0.5$. Overall factor of safety for this layer is 1.05.

seems to be a reasonable upper boundary condition in light of field observations. It is not clear, moreover, that alternatives such as a specified flux boundary would be an improvement except in cases where the water table is well below the ground surface.

ACKNOWLEDGMENTS

Preparation of this paper was supported by the New Mexico Bureau of Mines and Mineral Resources. Bob Fleming of the U.S. Geological Survey provided trench logs and geotechnical data from the Ohio River valley landslide. Comments by Jeff Keaton, Scott Anderson, and an anonymous reviewer helped to clarify many issues and are greatly appreciated.

APPENDIX: ONE-DIMENSIONAL STEADY FLOW SOLUTION

Steady one-dimensional flow of groundwater across the hydraulic conductivity zones illustrated in Figure 3 is governed by the three equations

$$\frac{d^2h_i}{d\xi^2} = 0, \quad i = 1, 2, 3 \quad (A1)$$

that can be twice integrated to find general solutions of the form

$$h_i = c_{i,1} + c_{i,2}\xi . \quad (A2)$$

Thus, head will be a linear function of ξ in each of the three zones. The ξ direction is taken to be perpendicular to the heterogeneity, and the i subscripts denote the zone to which each equation applies. Flow is driven by a head differential denoted by ΔH, which is specified by the boundary conditions

$$[h_1]_{-L_1/2} = 0 \quad (A3)$$

and

$$[h_3]_{L_1/2} = \Delta H . \quad (A4)$$

In this appendix, the subscripted square brackets indicate the coordinate at which an expression is evaluated. Both head and flux must be continuous across each of the two internal boundaries, giving rise to the two sets of internal boundary conditions (e.g., Strack, 1989, p. 411; Carslaw and Jaeger, 1959, p. 87):

$$[h_1]_{-L_2/2} = [h_2]_{-L_2/2} , \quad (A5)$$

$$K_1 \left[\frac{dh_1}{dx}\right]_{-L_2/2} = \overline{K}_2 \left[\frac{dh_2}{dx}\right]_{-L_2/2} ; \quad (A6)$$

and

$$[h_2]_{L_2/2} = [h_3]_{L_2/2} , \quad (A7)$$

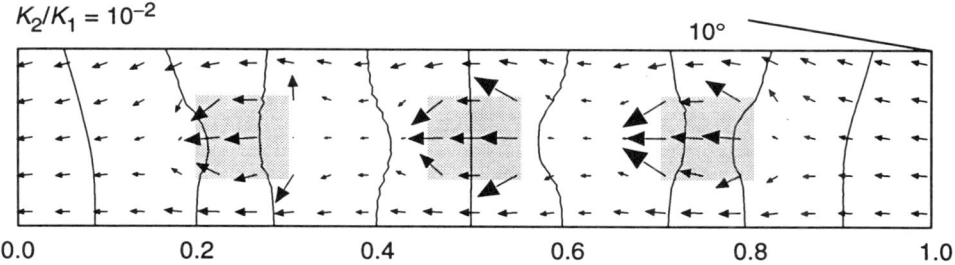

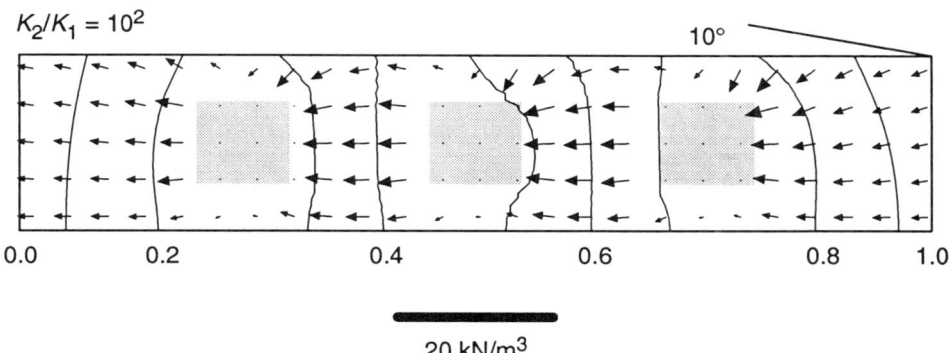

Figure 11. Hydraulic head contours and seepage force vectors for flow through layers containing three closely spaced heterogeneities with hydraulic conductivity contrasts as indicated. The hydraulic head contour interval is $\Delta H/10$, and arrow lengths are proportional to seepage force magnitude.

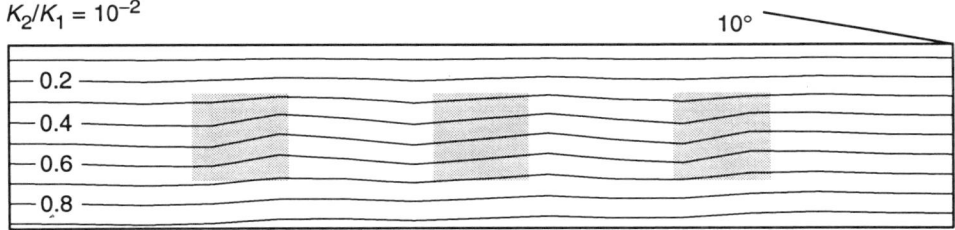

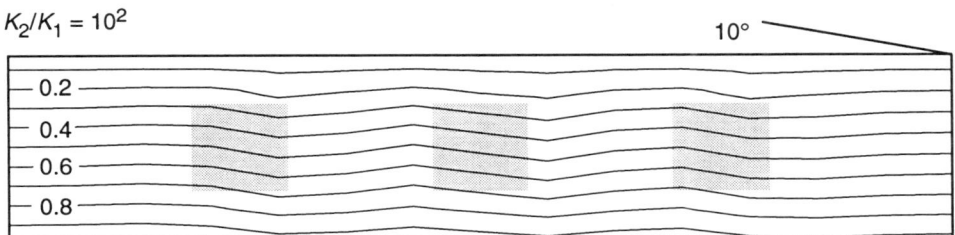

Figure 12. Pressure head contours for flow through layers containing three closely spaced heterogeneities with hydraulic conductivity contrasts as indicated. The pressure head contour interval is $D/10$.

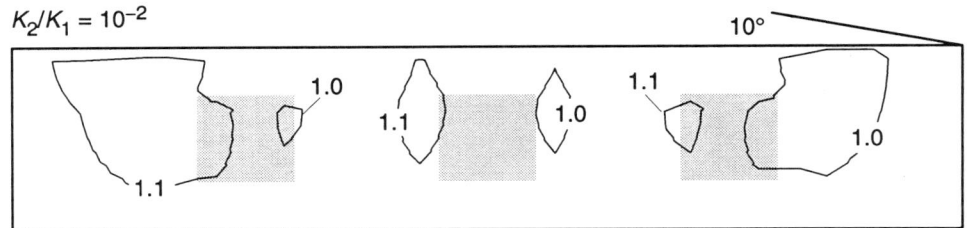

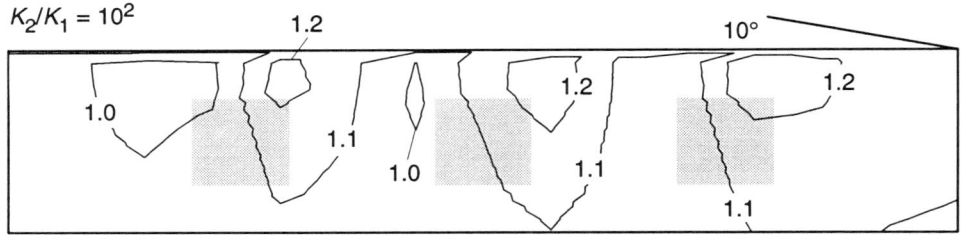

Figure 13. Local factor of safety contours for flow through layers containing three closely spaced heterogeneities with hydraulic conductivity contrasts as indicated. The overall factor of safety for each slab is as indicated.

$$\overline{K}_2 \left[\frac{dh_2}{dx}\right]_{L_2/2} = K_1 \left[\frac{dh_3}{dx}\right]_{L_2/2} . \quad (A8)$$

The overbars denote that for the problem under consideration, the zone 2 hydraulic conductivity is the weighted arithmetic mean of the values for each of the three layers illustrated in Figure 4. That is to say,

$$\overline{K}_2 = [(D_2 - D_1)K_2 + D_1 K_1 + (D - D_2)K_1]/D. \quad (A9)$$

If zone 2 does not consist of layers, then $\overline{K}_2 = K_2$. Likewise, flow through more complicated systems can be analyzed by allowing K_1, K_2, and K_3 each to be mean values. The six boundary conditions give rise to a set of six linear equations that can be solved for the constants of integration in Equation A2, yielding

$$c_{1,1} = \frac{\Delta H L_1 \overline{K}_2}{2(L_2 K_1 + L_1 \overline{K}_2 - L_2 \overline{K}_2)} \quad (A10)$$

$$c_{1,2} = \frac{\Delta H \overline{K}_2}{L_2 K_1 + L_1 \overline{K}_2 - L_2 \overline{K}_2} \quad (A11)$$

$$c_{2,1} = \frac{\Delta H}{2} \quad (A12)$$

$$c_{2,2} = \frac{\Delta H K_1}{L_2 K_1 + L_1 \overline{K}_2 - L_2 \overline{K}_2} \quad (A13)$$

$$c_{3,1} = \frac{\Delta H (2L_2 K_1 + L_1 \overline{K}_2 - 2L_2 \overline{K}_2)}{2(L_2 K_1 + L_1 \overline{K}_2 - L_2 \overline{K}_2)} \quad (A14)$$

$$c_{3,2} = c_{1,2} . \quad (A15)$$

Finally, substitution of Equation A10 through Equation A15 into the general solutions leads to particular solutions for each of the three zones, here rearranged into dimensionless form:

$$\frac{h_1}{\Delta H} = \frac{(\overline{K}_2/K_1)[1+2(\xi/L_1)]}{2[(L_2/L_1) + (\overline{K}_2/K_1) - (L_2/L_1)(\overline{K}_2/K_1)]}, \quad (A16)$$

$$\frac{h_2}{\Delta H} = \frac{1}{2} + \frac{x/L_1}{(L_2/L_1) + (\overline{K}_2/K_1) - (L_2/L_1)(\overline{K}_2/K_1)}, \quad (A17)$$

$$\frac{h_3}{\Delta H} = \frac{2(L_2/L_1) + (\overline{K}_2/K_1) - 2(L_2/L_1)(\overline{K}_2/K_1) + 2(\overline{K}_2/K_1)(x/L_1)}{2[(L_2/L_1) + (\overline{K}_2/K_1) - (L_2/L_1)(\overline{K}_2/K_1)]} . \quad (A18)$$

REFERENCES CITED

Anderson, M. G., and Burt, T. P., editors, 1990, Studies in hillslope hydrology: Chichester, United Kingdom, John Wiley and Sons, 539 p.

Anderson, S. A., and Sitar, N., 1993, Debris flow initiation—The role of hydrologic response and soil behavior: Berkeley, University of California, Department of Civil Engineering, Geotechnical Engineering Report

UCB/GT/93-02, 220 p.

Baum, R. L., 1994, Contribution of artesian water to progressive failure of the upper part of the Delhi Pike landslide complex, Cincinnati, Ohio: U.S. Geological Survey Bulletin 2059-D, 14 p.

Baum, R. L., and Fleming, R. W., 1991, Use of longitudinal strain in identifying driving and resisting elements of landslides: Geological Society of America Bulletin, v. 103, p. 1121–1132.

Carslaw, H. S., and Jaeger, J. C., 1959, Conduction of heat in solids (second edition): Oxford, Clarendon Press, 510 p.

Coates, D. R., 1990, The relation of subsurface water to downslope movement and failure, in Higgins, C. G., and Coates, D. R., eds., Groundwater geomorphology: Boulder, Colorado, Geological Society of America Special Paper 252, p. 51–76.

D'Appolonia, E., Alperstein, R., and D'Appolonia, D. J., 1967, Behavior of a colluvial slope: Journal of Soil Mechanics and Foundation Engineering Division, Proceedings of the American Society of Civil Engineers, v. 93, p. 489–515.

Domenico, P. A., and Schwartz, F. W., 1990, Physical and Chemical Hydrogeology: New York, John Wiley and Sons, 824 p.

Fleming, R. W., and Johnson, A. M., 1994, Landslides in colluvium: U.S. Geological Survey Bulletin 2059-B, 24 p.

Gelhar, L. W., 1993, Stochastic subsurface hydrology: Englewood Cliffs, New Jersey, Prentice Hall, 390 p.

Haneberg, W. C., 1991a, Observation and analysis of pore pressure fluctuations in a thin colluvium landslide complex near Cincinnati, Ohio: Engineering Geology, v. 31, p. 159–184.

Haneberg, W. C., 1991b, Pore pressure diffusion and the hydrologic response of nearly saturated, thin landslide deposits to rainfall: Journal of Geology, v. 99, p. 886–892.

Haneberg, W. C., 1992, A mass balance model for the hydrologic response of fine-grained hillside soils to rainfall [abs.]: Geological Society of America Abstracts with Programs, Annual Meeting, v. 24, p. A203.

Haneberg, W. C., and Gökce, A. Ö., 1994, Rapid water level fluctuations in a thin colluvium landslide west of Cincinnati, Ohio: U.S. Geological Survey Bulletin 2059-C, 16 p.

Hodge, R. A. L., and Freeze, R. A., 1977, Groundwater flow systems and slope stability: Canadian Geotechnical Journal, v. 14, p. 466–476.

Hubbert, M. K., 1940, The theory of groundwater motion: Journal of Geology, v. 48, p. 785–944.

Huyakorn, P. S., and Pinder, G. F., 1983, Computational methods in subsurface flow: San Diego, Academic Press, 473 p.

Iverson, R. M., 1990, Groundwater flow fields in infinite slopes: Géotechnique, v. 40, p. 139–143.

Iverson, R. M., and Major, J. J., 1986, Groundwater seepage vectors and the potential for hillslope failure and debris flow mobilization: Water Resources Research, v. 22, p. 1543–1548.

Iverson, R. M., and Major, J. J., 1987, Rainfall, ground-water flow, and seasonal movement at Minor Creek landslide, northwestern California—Physical interpretation of empirical relations: Geological Society of America Bulletin, v. 99, p. 579–594.

Iverson, R. M., and Reid, M. E., 1992, Gravity-driven groundwater flow and slope failure potential: 2. Effects of slope morphology, material properties, and hydraulic heterogeneity: Water Resources Research, v. 28, p. 939–950.

Jackson, C. R., and Cundy, T. W., 1992, A model of transient, topographically driven, saturated subsurface flow: Water Resources Research, v. 28, p. 1417–1427.

Janbu, N., 1973, Slope stability computations, in Hirschfield, R. C., and Poulos, S. J., eds., Embankment dam engineering (Casagrande volume): New York, John Wiley and Sons, p. 47–86.

Kirkby, M. J., editor, 1978, Hillslope hydrology: Chichester, United Kingdom, John Wiley and Sons, 389 p.

Murdoch, L. C., III, 1988, Pore-water pressures and unsaturated flow during infiltration in colluvial soils at the Delhi Pike landslide, Cincinnati, Ohio [M.S. thesis]: Cincinnati, Ohio, University of Cincinnati, 133 p.

Pyles, M. R., Mills, K., and Saunders, G., 1987, Mechanics and stability of the Lookout Creek earthflow: Bulletin of the Association of Engineering Geologists, v. 24, p. 267–280.

Reid, M. E., and Iverson, R. M., 1992, Gravity-driven groundwater flow and slope failure potential: 1. Elastic effective stress model: Water Resources Research, v. 28, p. 925–938.

Reid, M. E., Nielsen, H. P., and Dreiss, S. J., 1988, Hydrologic factors triggering a shallow hillslope failure: Bulletin of the Association of Engineering Geologists, v. 25, p. 349–361.

Rulon, J. J., and Freeze, R. A., 1985, Multiple seepage faces on layered slopes and their implications for slope-stability analysis: Canadian Geotechnical Journal, v. 22, p. 347–356.

Skempton, A. W., 1964, Long-term stability of clay slopes: Géotechnique, v. 14, p. 77–101.

Strack, O. D. L., 1989, Groundwater mechanics: Englewood Cliffs, New Jersey, Prentice Hall, 732 p.

Terzaghi, K., 1950, Mechanics of landslides, in Paige, S., ed., Application of geology to engineering practice (Berkey volume): New York, Geological Society of America, p. 83–123.

Terzaghi, K., 1929, The effects of minor geologic details on the safety of dams: American Institute of Mining and Metallurgical Engineers Technical Publication 215, p. 31–44.

Wieczorek, G. F., 1987, Effect of rainfall intensity and duration on debris flows in central Santa Cruz Mountains, California, in Costa, J. E., and Wieczorek, G. F., eds., Debris flows avalanches: Geological Society of America Reviews in Engineering Geology v. VII, p. 93–104.

Wilson, R. C., and Wieczorek, G. F., 1988, Rainfall thresholds for the initiation of debris flow at La Honda, California [abs.] Eos (Transactions of the American Geophysical Union), v. 69, no. 16, p. 348.

Wu, T. H., and Swanston, D. N., 1980, Risk of landslides in shallow soils and its relation to clearcutting in southeastern Alaska: Forest Science, v. 26, p. 495–510.

Manuscript Accepted by the Society May 20, 1994

Geology, hydrology, and mechanics of a slow-moving, clay-rich landslide, Honolulu, Hawaii

Rex L. Baum
U.S. Geological Survey, Box 25046, MS 966, Denver, Colorado 80225
Mark. E. Reid
U.S. Geological Survey, 345 Middlefield Road, MS 998, Menlo Park, California 94025

ABSTRACT

The Alani-Paty landslide has damaged streets, utilities, and homes built on a debris apron in Honolulu, Oahu, Hawaii. Failure of weathered, crudely stratified, highly plastic, debris-apron deposits has created several similar landslides in southeastern Oahu. The Alani-Paty landslide affects about 60 residential lots. It is about 300 m long, 160 m wide, 7–10 m thick, and consists of two main kinematic elements that are separated by a right-lateral shear zone. One element has moved about 4 m, mainly by translation, down a slope of about 12°, and the adjacent element has moved about 3 m down a slope of 9°. Longitudinal stretching in the upslope third and shortening in the downslope two-thirds characterize deformation in each element; landslides in Ohio, Utah, and Colorado have deformed similarly. Smectite-rich clay layers within the deposits are medium to stiff, and measured angles of residual friction range from 6° to 11° with cohesion intercepts less than 12.5 kPa. Saturated hydraulic conductivity within the landslide decreases with depth; below the slip surface, the hydraulic conductivity increases. Rainfall infiltrates at the ground surface, percolates downward and perches on the zone of low hydraulic conductivity near the slip surface, keeping the slide mostly saturated year round. The main body of the landslide moves during rainy periods, when the ten-day average rainfall exceeds 25 mm/day and the pore-water pressures in the upslope quarter of the landslide increase 10–30 kPa. Pore pressure increases within the landslide occur 1–2 days following the onset of rainfall and result from infiltration of rainfall and runoff; after materials above the perennial water table become saturated, downward propagating pressure waves triggered by bursts of intense rainfall produce further, short-lived increases in pore pressure. This elevated pore pressure at the slip surface triggers movement. The ground-water response in the upslope quarter of the landslide is relatively rapid compared to responses in other landslides described in the literature.

INTRODUCTION

Slow-moving landslides on the humid, subtropical island of Oahu, Hawaii, have damaged residential areas and defied stabilization efforts since the 1950s, when the Waiomao landslide (Peck, 1959) destroyed many homes and damaged public roads and utilities in Palolo Valley (Fig. 1). Since then, other landslides have destroyed homes and property in Manoa Valley (Geolabs-Hawaii, 1984; Baum et al., 1989), Aina Haina (Peck and Wilson, 1968), Kuliouou Valley, and other valleys on the leeward side of the Koolau Range of Oahu. Many of these landslides were moving during the late 1980s,

Baum, R. L., and Reid, M. E., 1995, Geology, hydrology, and mechanics of a slow-moving, clay-rich landslide, Honolulu, Hawaii, in Haneberg, W. C., and Anderson, S. A., eds., Clay and Shale Slope Instability: Boulder, Geological Society of America Reviews in Engineering Geology, v. X.

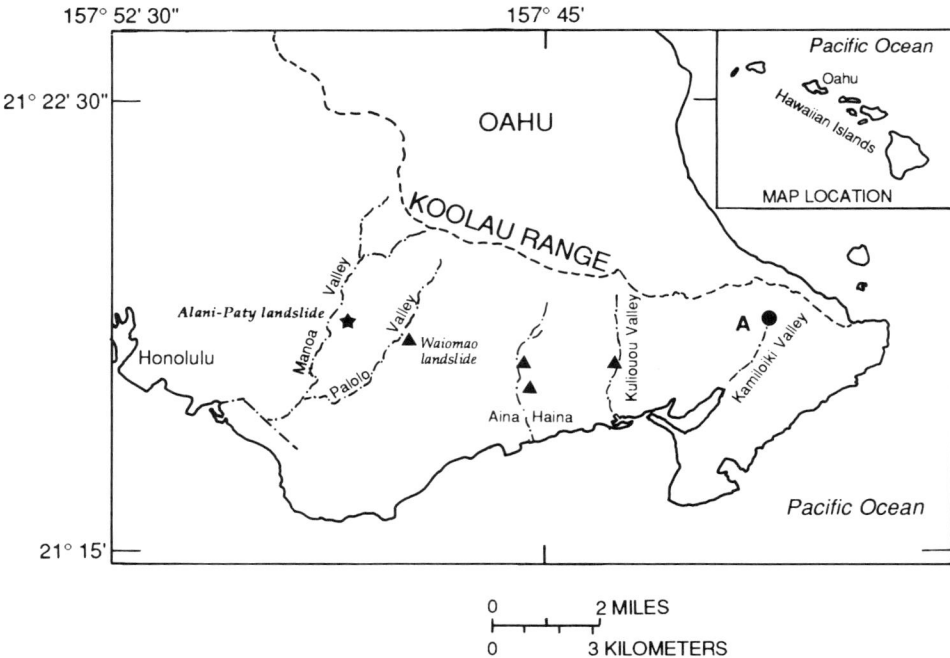

Figure 1. Location of the study area in the southeastern part of Oahu, Hawaii. The Alani-Paty landslide (star) is on the east wall of Manoa Valley. Approximate locations of other, similar landslides indicated by triangles. Solid circle A indicates approximate location of outcrop depicted in Figure 6.

and most were investigated by private consultants under contract to the City and County of Honolulu or the State of Hawaii. Despite previous investigations of landslides on Oahu, significant questions remained about these landslides, including how they form, what materials they form in and how those materials are distributed, what causes them to move, and how future landslide problems can be avoided. Consequently in 1989, the U.S. Geological Survey, in cooperation with the Department of Public Works of the City and County of Honolulu, began detailed investigations of landslides and debris flows in the Honolulu District of Oahu.

The Alani-Paty landslide, in Manoa Valley (Figs. 1 and 2), was chosen for detailed study as an example of a slow-moving landslide because it was known to be moving in January 1989, and it appeared to have many of the characteristics typical of known slow-moving landslides on Oahu. These characteristics include occurrence of the landslide in gently sloping, clay-rich deposits at the base of a valley wall and episodic movement related to rainfall. The landslide affects more than 60 lots and parts of several streets in a hillside residential area that has been plagued by water-main breaks and ground movement since the 1970s. Maximum total downslope movement through January 1989 was about 3–4 m, but annual movement was episodic and ranged from a few millimeters to about a meter.

The landslide was studied primarily by field methods, supplemented by laboratory and analytical methods. At the beginning of the study Baum et al. (1989) mapped the Alani-Paty and neighboring Hulu-Woolsey landslides to determine their size, shape, and pattern of internal strain. Subsequently, borings were made for collecting samples and installing instruments (Fig. 2; also Baum et al., 1990). Selected samples were tested by laboratories of the U.S. Geological Survey for determination of physical properties (Baum et al., 1990, 1991). Instrumentation was monitored for approximately two years to measure rainfall, movement of the landslide, and subsurface water pressures and to determine the depth of the basal slip surface (Table 1). Instrumentation included standard geotechnical instruments (inclinometers, extensometers, open-tube piezometers, and vibrating-wire pore-pressure transducers), as well as specialized instruments such as pressure-transducer soil tensiometers. Data from the bore-hole instrumentation of Geolabs-Hawaii and STV-Lyon Associates supplemented our own (Fig. 2). Analysis of aerial photographs made in various years since 1941 enabled us to locate landslide boundaries that had been masked by repairs to streets and homes and to compute displacement from coordinates of photo-identifiable points measured on successive sets of photographs using an analytical stereoplotter. We used simple, physically based theories for infiltration, ground-water flow, propagation of pressure waves, lateral earth pressure, and slope stability to analyze and interpret the hydrological and mechanical responses of the landslides.

This report briefly summarizes findings from a three-year investigation of slow-moving landslides on Oahu by describ-

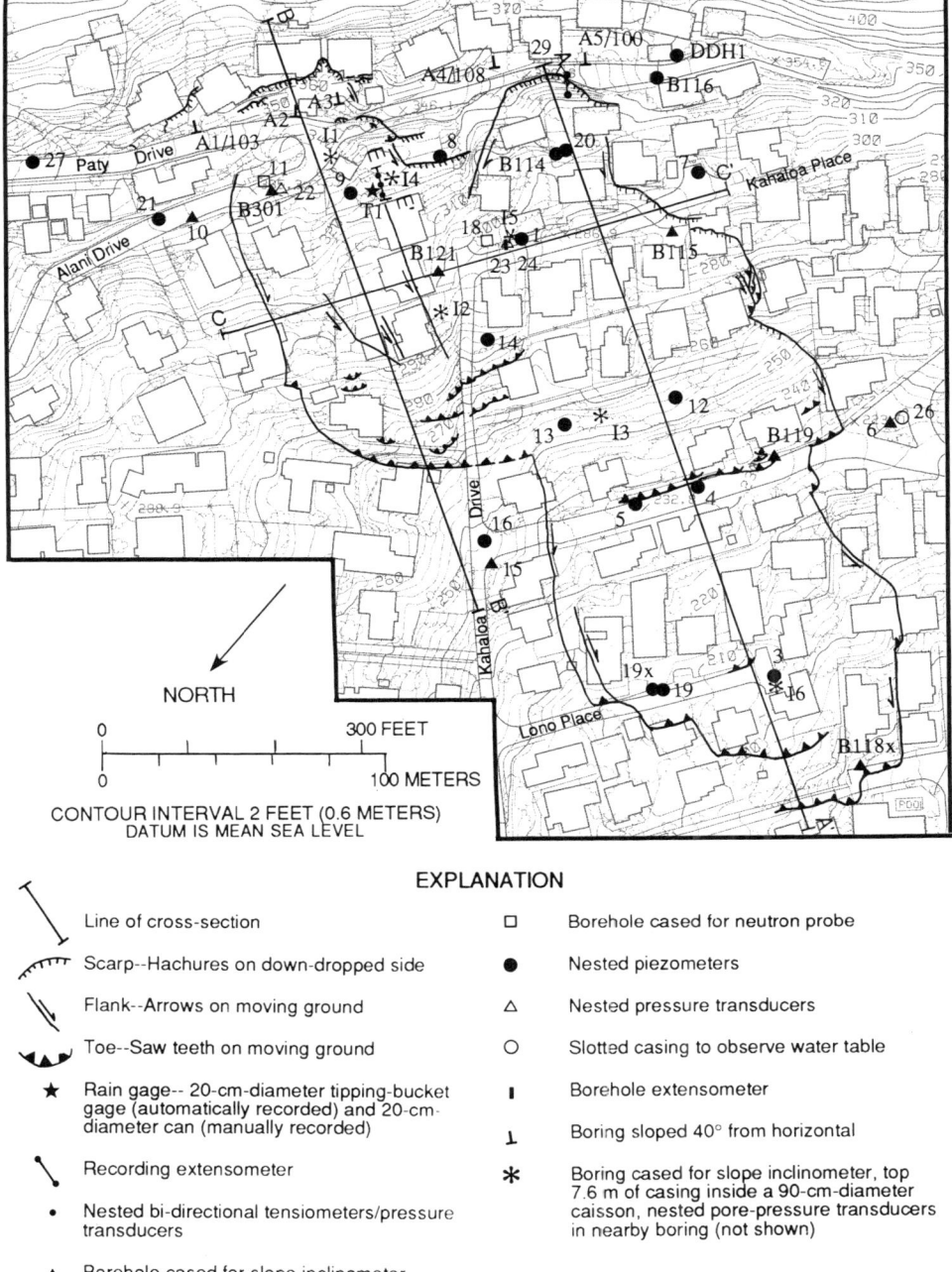

Figure 2. Map showing structural features, lines of cross-section, and locations of borings in the Alani-Paty landslide in Manoa Valley, Oahu, Hawaii. Numbers having "B" prefix designate borings for Geolabs-Hawaii (only selected borings shown), those having "I," "A," or "DDH" prefixes designate borings for STV/Lyon Associates, numbers without letter prefixes designate borings for the U.S. Geological Survey. Three inclined borings, A1, A4, and A5 contain U.S. Geological Survey piezometers identified by a three-digit number separated from the boring number by a virgule. Borehole extensometers (steel cables anchored below the slip surface, with a free end at the ground surface) were installed in holes 1, 3, 4, 5, 8, 9, 12, 13, 19, and 20 to measure downslope displacement in addition to the one in boring 23 shown on the map. Base map from R. M. Towill Corporation, 1989. Origin of coordinates: Hawaiian State Plane, Zone 1.

TABLE 1. INSTRUMENTATION USED IN INVESTIGATION OF THE ALANI-PATY LANDSLIDE

Instrument	Quantity	Frequency of Measurement	Purpose
Open-tube piezometer	41	1–4 per week	Pore-water pressure head, hydraulic conductivity*
Pore-pressure transducer	29[†]	1 per 30 minutes[§]	Pore-water pressure variations through time
Tensiometer	11	1–7 per week	Spatial and temporal distribution of soil-water tension
Tensiometer with differential, bidirectional pressure transducer	6	1 every 15 minutes[§]	Spatial and temporal changes of soil-water tension and positive pressure
Borehole cased for neutron probe	3	.......**	Variation of soil moisture with depth
Tipping bucket rain gauge	1	1 every 15 minutes[§]	Temporal distribution of rainfall
Rain gauge	3[‡]	2–4 per month	Cumulative rainfall
Recording extensometer	1	1 every 30 minutes (later 1 every 15 minutes)[§]	Time and amount of displacement of head of landslide
Borehole extensometer	8	1–4 per month	Amount of displacement in the body of the landslide[§§]
Recording borehole extensometer	3	1 every 60 minutes[§]	Time and amount of displacement in the body of the landslide[§§]
Quadrilateral	9	6 per year	Style of deformation, enlargement of the landslide
Borehole cased for slope inclinometer***	14	1–3 per month (wet periods); 6 per year (dry periods)	Depth of sliding

Note: Detailed information is available in Baum et al., 1990, 1991; Baum and Reid, 1992; and Torikai, 1992. Manual measurements were made more frequently during rainy periods than at other times.
*Hydraulic conductivity determined using slug tests near end of monitoring period.
[†]Includes 22 by STV/Lyon Associates.
[§]Data measured and recorded automatically.
**Casing poorly installed, probe yielded no useful data.
[‡]One installed on the landslide, others on Waahila ridge, upslope from the landslide.
[§§]No displacement observed, either too little displacement to pull cables down the hole or anchors below slip surface failed.
***Includes five installed by Geolabs-Hawaii for previous investigations and six by STV/Lyon Associates.

ing one example, the Alani-Paty landslide, in detail; the description includes the geomorphology, topography, and geology at the site of the landslide and the geometry, structure, hydrology, and mechanics of the landslide. Additional information is available in four open-file reports (Baum et al., 1989, 1990, 1991; Baum and Reid, 1992). The investigation was intended to determine how the landslides form, to identify the sources and understand the movement of hillside water, and to analyze the processes that trigger movement of the landslides.

PREVIOUS WORK

Several previous investigations of landslides on Oahu point out that heavy rainfall and the low strength and low hydraulic conductivity of hillside materials contribute to instability of slopes in the leeward valleys of the Koolau Range on Oahu. The slow-moving landslides form in a weak, heterogeneous material consisting of highly plastic silty clay and clayey silt containing boulder-, cobble-, and gravel-sized clasts of weathered basalt (Peck and Wilson, 1968; Walter

Lum Associates, 1979; Geolabs-Hawaii, 1984, 1985, 1988). Peck (1959, 1967) found a strong correlation between the rate of movement of the Waiomao landslide and rainfall during the previous ten days in Palolo Valley. Peck observed that the Hind Iuka landslide in Aina Haina and the Waiomao landslide, while in their early stages of development, accelerated dramatically following intense rainfall (Peck, 1959; Peck and Wilson, 1968, p. 5). He also determined that the Waiomao landslide had relatively low average permeability (Peck, 1968). Tests conducted during drilling of the Waiomao landslide indicated that the hydraulic conductivity of the landslide material ranged from approximately 5×10^{-5} to 5×10^{-9} m/s (or less). Most of the material was slow draining; two thirds of the tests indicated the hydraulic conductivity was less than 5×10^{-7} m/s (Peck, 1968).

PHYSICAL DESCRIPTION OF THE ALANI-PATY LANDSLIDE

The Alani-Paty landslide has an area of about 3.4 ha; it averages 7.6 m thick; and its volume is about 250,000 m³. The length is about 300 m, and the maximum width is about 170 m. The average slope, from head to toe, is 9° in section A–A′ and 12° in section B–B′ (Figs. 2, 3A, and 3B). The slope is significantly steeper in the upslope part, and so the ground surface profile is concave upward.

The head of the landslide is scalloped or cuspate in plan view (Fig. 4). The uppermost parts of the cusps coincide with the axes of ephemeral drainage channels that extend up the steep slopes above the landslide. A left-lateral shear zone that passes beneath several homes divides the landslide into two kinematic elements (Figs. 3C and 4). A kinematic element is a part of a landslide that moves at a different rate than and more or less independently of neighboring parts (Fleming and Johnson, 1989). The shear zone separates the Alani element, northeast of the shear, from the Kahaloa element to the southwest (Fig. 4).

The basal slip surface of the active Alani-Paty landslide was located by measuring the depths where piezometer tubes and inclinometer casing became deformed (Baum et al., 1990, Appendix C; Geolabs-Hawaii, 1990; STV/Lyon Associates, 1990, 1991). Although shear surfaces deeper than the basal slip surface have been observed in large-diameter borings (William Cotton and Associates, 1992), inclinometer measurements in deep borings yielded no evidence that the Alani-Paty landslide has moved recently on any of these deeper surfaces (STV/Lyon Associates, 1990, 1991). In a longitudinal section, the basal slip surface is concave upward (Figs. 3A and 3B); in a transverse section, it is shaped like two shallow troughs separated by a broad, low ridge (Fig. 3C). The troughs correspond in position to the Alani and Kahaloa elements. Depth of the basal slip surface ranges from 7 to perhaps 10 m in the main body (Figs. 3A and 3B).

Displacement from 1969 to 1989 varies over the surface of the landslide (Fig. 4). The displacements, computed from x-y-z coordinates of photo-identifiable points, have an average uncertainty of about ±0.15 m in the horizontal and ±0.20 m in the vertical. Displacement of the Kahaloa element ranges from about 1.5 m at the toe to 3 m near Kahaloa Place. Displacement of the Alani element ranges from 3.4 m at the intersection of Alani and Paty Drives to 4.3 m near the middle of the element (Fig. 4). The pattern of displacement is consistent with the pattern of deformation determined by mapping damage to structures (Baum et al., 1989, Plate 2); the downslope two-thirds of the Kahaloa element is shortening and the upslope third is stretching. Similarly, the downslope half of the Alani element is shortening and the upslope half is stretching (Baum et al., 1989, fig. 5).

Two areas of incipient movement are present on the northeast side of the Alani-Paty landslide. The scarp at the top of area A (Fig. 4) formed before January 1989 and constitutes the primary evidence for movement northeast of the head; no movement was detected by an inclinometer in boring 10 between November 1989 and March 1991. The second area of incipient movement (B, Fig. 4) was actively deforming during the period of observation. Evidence for this movement includes distress to houses and right-lateral offset in a driveway on the northeast side of area B, and shearing at a depth of 7.9 m in boring 15 (Baum et al., 1990). Deformation of the houses started before January 1989. The right-lateral offset in the driveway occurred between February and April 1989 and boring 15 was sheared in January 1990.

GEOLOGIC SETTING OF THE ALANI-PATY LANDSLIDE

Topography and general geology

The Alani-Paty landslide is on the southeast side of the elongate Manoa Valley, which has a relatively flat bottom and steep sides. The valley bottom in the vicinity of the landslide is underlain by alluvial and lacustrine fill deposited after basalt flows from a volcanic vent dammed the valley about 67,000 years ago (MacDonald et al., 1983). The landslide has formed in a gently sloping debris apron composed of coalescing debris fans that extend between the flat valley floor and the steep slope of exposed bedrock (Fig. 5). The surface of exposed bedrock slopes approximately 40° and is incised by several small channels, producing a scalloped bedrock-apron contact. Some of these incisions coincide approximately with scallop-shaped accumulations of weathered debris-flow deposits in the head of the Alani-Paty landslide. The surface of the debris apron slopes 8°–12° on average. The debris apron has a concave-upward profile; near the valley floor it slopes 2°–6°, whereas near its upslope edge, it slopes 10°–15°.

Overall, the debris apron consists of several adjacent fan-shaped deposits with small, intervening hollows (Baum and Reid, 1992, plate 6). A large fan, with irregular surface topog-

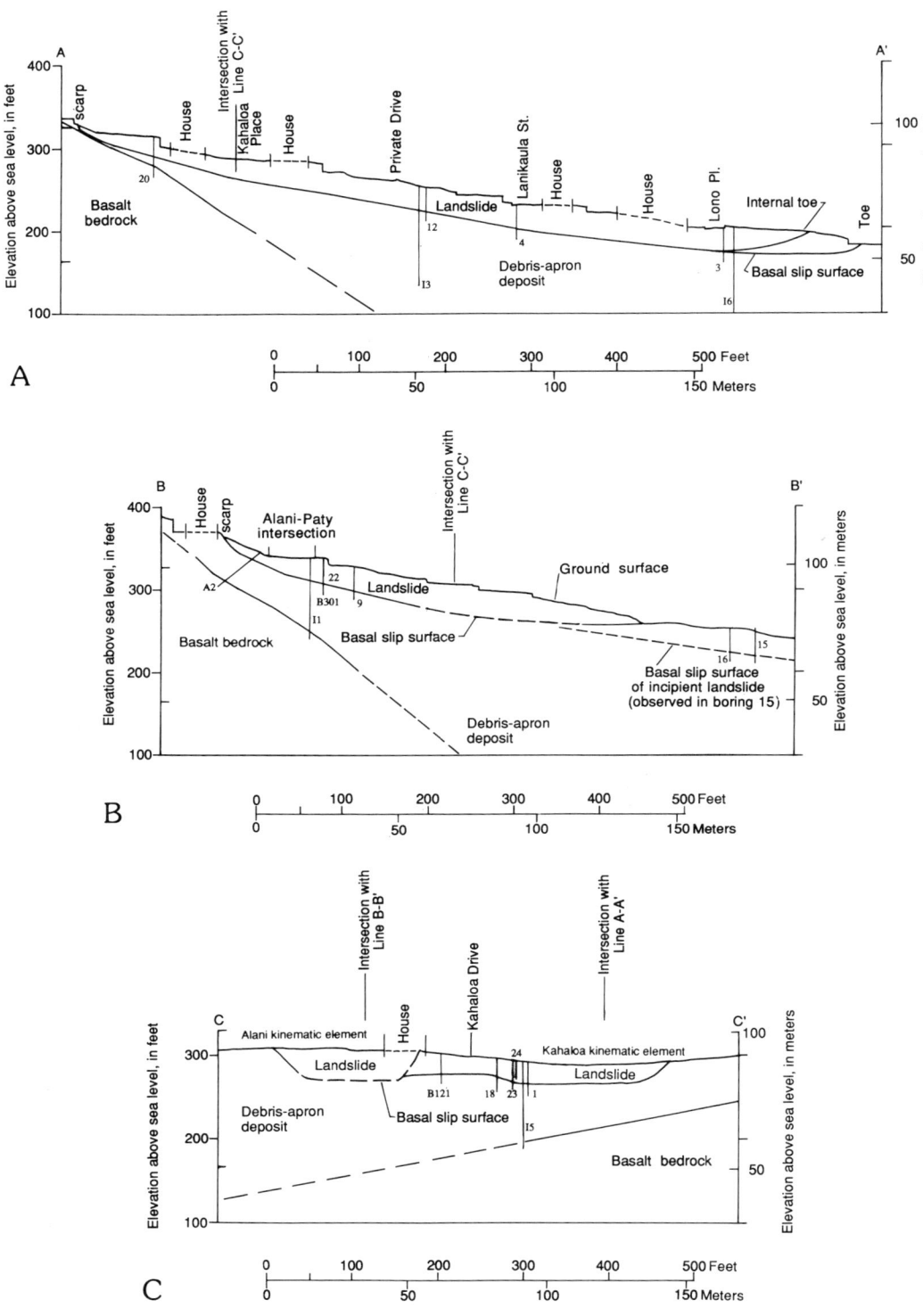

Figure 3. Cross sections through the Alani-Paty landslide. Projected locations of nearby borings indicated by vertical lines. The slip surface profiles are based on a contour map of the basal slip surface that is constrained by depth determinations at 21 borings as well as the mapped boundaries of the landslide (Baum and Reid, 1992, plate 4). Slip surface shown by dashed line where uncertain. (A) Longitudinal cross section A–A' depicts the Kahaloa kinematic element. (B) Longitudinal cross section B–B' depicts the Alani kinematic element and an area of incipient movement (downslope from the toe). (C) The Alani element is shown on the left and the Kahaloa element on the right in transverse cross section C–C'.

raphy and poorly developed surface drainage, borders the Alani-Paty landslide to the northeast. The Alani-Paty landslide occupies part of two smaller fans that are within the debris apron and southwest of the large fan. A hollow between the two smaller fans coincides with the head of the Kahaloa element of the landslide (Fig. 4). The fan apices, hollows, and lobes are visible on aerial photographs taken in 1941 and are still discernible, with difficulty, in the altered topography that has been produced by residential construction (Baum and Reid, 1992, plate 1). Shallow cuts and adjacent wedges of ar-

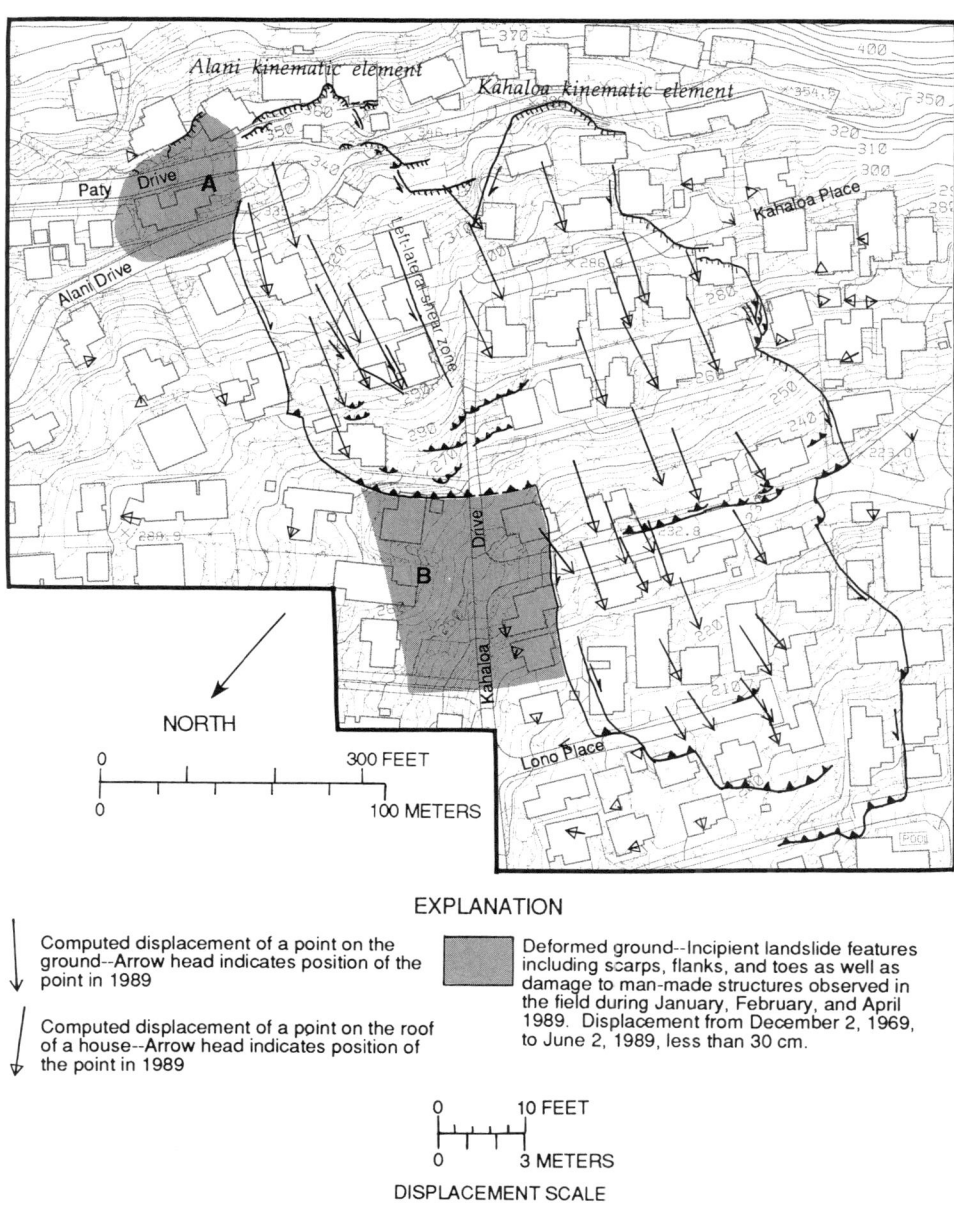

Figure 4. Map showing total displacements, structural features, kinematic elements, and deformation of the Alani-Paty landslide. Structural features mapped January, February, and April 1989 by Robert Fleming and Rex Baum. Displacements (vector sum of horizontal and vertical components) were computed from x-y-z coordinates of photo-identifiable points visible in vertical aerial photography of December 2, 1969 (approximate scale 1:12,000) and June 2, 1989 (approximate scale 1:8,500). The coordinates were measured in terms of eastings, northings, and elevation above sea level using a Kern DSR-11 analytical stereoplotter (measurements by James Messerich, U.S. Geological Survey, Denver, Colorado). The 1989 stereomodel was constrained using the same surveyed ground control as used to produce the topographic base map. The 1969 model was constrained using coordinates of photo-identifiable points in the valley bottom and on Waahila Ridge from the 1989 stereomodel. Boundaries of deformed ground (shaded) are indefinite.

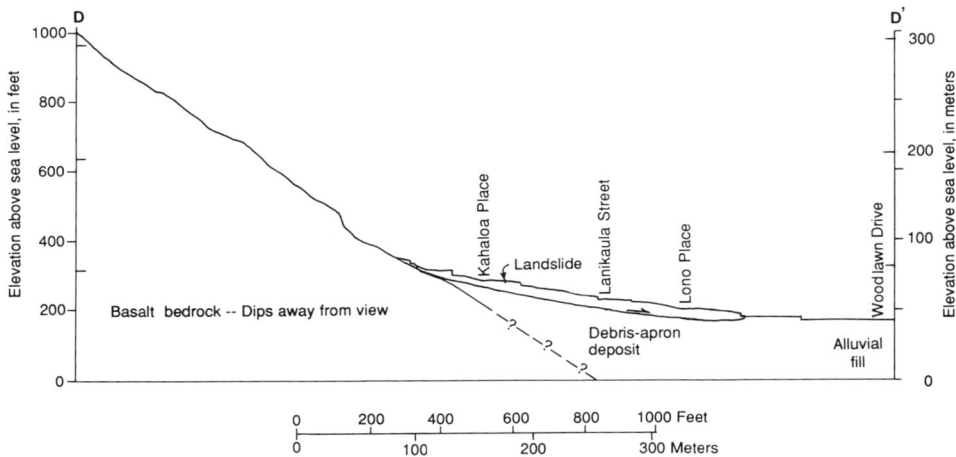

Figure 5. Cross section through the northwest side of Waahila Ridge, the debris apron, and part of the valley bottom showing the Koolau Basalt bedrock, weathered debris-flow deposits forming the apron, and alluvial fill in relation to the Alani-Paty landslide. Line of section is an extension of section A–A' (Figs. 2 and 3A).

tificial fill, consisting mainly of materials from the site, create a series of steps on the surface of the apron (Fig. 3A).

Debris apron

The debris apron is largely derived locally from the Koolau Basalt forming a ridge upslope of the landslide; some of the deposits may include materials derived from pyroclastics of the Honolulu Volcanics mapped nearby (Stearns, 1939). The bedrock consists of interbedded layers of aa and pahoehoe that dip about 8° toward the southwest (Langenheim and Clague, 1987). Weathered residuum, including clasts and soil, is transported downslope from the sides of Waahila Ridge to the fans by intermittent debris flows and rock falls, which through time have accumulated to form the debris apron (Peterson et al., 1993). The apron, wedge-shaped in cross section, overlaps the steeply sloping bedrock surface (Figs. 3A and 3B).

Once deposited on the fans, the materials begin to weather and form vertisols. Vertisols are pedologic soils containing at least 30% clay (by weight), and characterized by the presence of deep, wide surface fractures during the dry season (Bates and Jackson, 1987). A vertisol exposed near the head of Kamiloiki Valley is typical of soils in parts of Manoa Valley and other valleys on southeast Oahu. The soil is dark brown, clay rich, and highly expansive; it rests on its parent material, a friable gray clayey silt (Fig. 6). As is typical of vertisols, the soil shown in Figure 6 appears homogeneous; the internal stratification typical of many other pedologic soils is absent. In most places, the basal contact is distinct; the transition from the dark-brown vertisol to the gray parent material occurs over a few centimeters (Fig. 6). Boulders and cobbles commonly extend across the contact.

The debris apron consists of crudely stratified clayey silt and silty clay containing boulder-, cobble-, and gravel-sized fragments of weathered basalt. Samples from borings indicate that the silty clays and clayey silts are distinct, interstratified units; distinct contacts between the two materials were observed in a few core samples. The crude stratification formed as successive debris-flow deposits buried vertisols that had formed in the upper parts of older debris-flow deposits. The stratification of the debris-apron is complex and discontinuous, making it difficult to correlate layers laterally even over short distances (Baum and Reid, 1992). Where the deposits have been exposed at construction sites in neighboring areas, the stratification is subparallel with the ground surface, and the units commonly range in thickness from 0.3 to 3 m, or more. The units interfinger, have irregular shapes when viewed in cross-sections parallel to the contours and are 3–60 m wide. Simple field classification tests indicate that the clay-rich strata are highly plastic; many of the clayey silts are also plastic but have low dry strength, typical of silty materials. The silty units may be 10–1,000 times more permeable than the clayey ones. At the U.S. Geological Survey laboratory in Golden, Colorado, we performed falling head tests on materials from a cut on the west side of Palolo Valley. Two silty samples had hydraulic conductivities ranging from 10^{-6} to 10^{-7} m/s, and two clayey samples had conductivities ranging from 10^{-8} to 10^{-10} m/s.

The origin of slickensided shear surfaces, found at various depths in the deposits, including some deeper than the basal slip surface of the Alani-Paty landslide, remains undetermined. Through-going shear surfaces, subparallel to the ground surface, are commonly associated with smaller, randomly oriented, slickensided shear surfaces that are several centimeters long and with shiny, randomly oriented, fissure surfaces that are commonly <20 mm long (William Cotton and Associates, 1992). The shear and fissure surfaces may be remnant features of the buried vertisols (Robert Fleming, oral commun., 1990) or remnant slip surfaces from past episodes of landsliding during the formation of the debris apron (William Cotton, oral commun., 1992).

The crude stratification of deposits that form the debris apron contributes in several ways to the formation of translational landslides. Locally extensive layers of expansive, slickensided clay (buried vertisols) dipping approximately parallel to the slope of the ground surface provide weak zones in which failure can occur readily. Ground water moving down through the deposits may be impeded also by these clay layers, causing perched water tables above the layers and increased pore pressures within the layers.

Figure 6. Stratigraphic relations between silty and clayey units in the debris-apron deposits. Photograph of a cut in Kamiloiki Valley (location A, Fig. 1), hammer is 0.30 m long. Dark brown, blocky, expansive silty clay soil (vertisol?) overlies light gray clayey silt (parent material?). The contact is distinct, but transitional over several centimeters; clasts of weathered basalt straddle the contact.

GEOTECHNICAL PROPERTIES

The landslide material, like the rest of the debris-apron deposits, consists of gravel-, cobble-, and boulder-sized clasts of weathered basalt in a matrix of interstratified clayey silt and silty clay. The silty clay is expansive, weak, highly plastic, and barely permeable to water. The clayey silt is also plastic, but it is somewhat stronger and more permeable to water than the silty clay. Consistency of individual layers within the landslide ranges from soft to stiff; stiffness generally increases with depth. X-ray diffraction (XRD) analysis indicates that the clays in both the clayey silt and the silty clay are mainly smectite with small amounts of halloysite and kaolinite (Baum and Reid, 1992).

Engineering properties

Laboratory testing of specimens of clayey silt and silty clay, chosen to represent the range of materials sampled, indicates that the matrix material has high plasticity (Table 2; also Baum et al., 1990, 1991). Cobble- and boulder-sized clasts make up 25%–45% locally and average 38% of the material, by volume. The interstitial matrix contains about 35%–80% clay, 10%–25% silt, 5%–25% sand, and 0%–50% gravel by weight (Fig. 7). Sand coarser than about 0.5 mm is absent from most samples, and sand finer than 0.5 mm makes up only a small fraction (<10%) of many samples, particularly the silty clays (Fig. 7). Materials from the upslope half of the landslide generally are sandier and probably stronger than materials in the downslope half of the landslide (Baum and Reid, 1992, table 4). Plastic limits range from 29 to 61, and liquid limits range from 59 to 137. Despite their high clay contents, visual descriptions, and high liquid limits, three-fourths of the samples are classified as high-plasticity silts (MH) on the basis of their Atterberg limits; the remaining one-fourth were classified as high-plasticity clays (CH) (Fig. 8). Most samples classified CH have higher clay content than the MH samples (Fig. 7). One exception contained 45% gravel-sized particles; other CH samples contained no gravel.

Most of the landslide material is perennially saturated and appears to be either normally consolidated or lightly overconsolidated. Natural moisture contents typically are within ±10% of the plastic limit (Baum and Reid, 1992), and natural moisture content gradually decreases with depth at depths greater than 2–3 m (Baum et al., 1990, appendix A). Moisture content increases with depth in the upper 2–3 m of material where seasonal drying has reduced the void ratio. The average unit weight is about 21.7 kN/m^3 for saturated landslide material, based on the average moisture content of the matrix (47%), the grain density of 2.74 g/cm^3 for fines (average of five measurements), an assumed average density of 2.9 g/cm^3 for basalt clasts, and an average clast content of 38% by volume (Baum and Fleming, 1991).

Residual shear strengths. Low strength characterizes ma-

TABLE 2. RESIDUAL SHEAR STRENGTH PARAMETERS AND ATTERBERG LIMITS OF MATERIAL FROM THE ALANI-PATY LANDSLIDE

Boring	Depth Interval* (m)	Liquid Limit	Plastic Limit	Unified Soil Classification	Residual Cohesion (kPa)	Angle of Residual Friction (°)	Description
Samples from the basal slip surface							
1	7.3-8.0	94	40	MH	4.1	10.9	Brown slickensided clay
3	8.1-8.7	107	44	MH	3.9	9.3	Brown plastic clay
5	8.1-9.6	129	54	MH	6.7	6.0	Gray slickensided clay
14	5.9-6.6	84	55	MH	1.9	8.2	Gray slickensided clay
18	6.4-7.0	93	43	MH	2.2	8.9	Slightly sandy clay
20	7.0-7.3	74	49	MH	5.0	10.1	Brown sandy clayey silt
Samples from the main body of the landslide							
....	Surface	59	29	CH	11.1	25.0	Brown sandy clayey silt
....	Surface	96	48	MH	10.4	11.0	Brown sandy silty clay
3	1.8-2.3	116	48	MH	2.2	9.0	Brown slightly sandy clay
4	3.3-3.5	137	46	CH	0	8.0	Brown slightly sandy clay
12	7.9-8.4	76	42	MH	4.1	8.5	Brown sandy clay
Samples from beneath the basal slip surface or outside the landslide							
3	11.1-12.6	95	44	MH	6.1	8.0	Brown silty clay
4	11.9-12.5	76	61	MH	9.0	20.3	Brown sandy clayey silt
8	2.7-2.9	73	49	MH	8.1	7.3	Brown clayey silt
8	7.9-8.6	66	52	MH	12.5	6.1	Brown sandy clay
11	8.8-9.0	84	38	MH	2.1	7.5	Brown slickensided clay
11	9.0-9.2	65	43	MH	1.2	12.4	Sandy clayey silt
19	9.1-9.7	84	44	MH	4.3	7.0	Slightly sandy slickensided clay

Note: Residual strength parameters determined using drained direct-shear tests on remolded samples having precut failure planes (Baum et al., 1990, 1991). Samples tested under normal stresses of 34, 69, and 172 kPa. Samples were preconsolidated under twice the normal stress applied during shearing. Unified soil classification is based on Atterberg limits (Fig. 8). Direct shear tests by George Erickson and William McArthur; Atterberg limits by George Erickson and Coyn Criley.
*Includes range of uncertainty in location caused by incomplete recovery of sample.

terials from in and around the landslide; angles of residual friction range from 6° to 25°, with cohesion intercepts less than 12.5 kPa (Table 2).

Materials from the basal slip surface were identified in cores from several borings; the materials are generally silty clays and sandy silty clays (based on visual/manual examination, Table 2) of high plasticity and low to moderate strength. All samples from the basal failure surface are classified MH despite their low strength and high plasticity (Table 2). Several samples from the depth of the slip surface were slickensided; however, slickensides were also present in several other samples above and below the slip surface. Trenches and large diameter borings that penetrated the landslide exposed the active basal shear zone (as well as a few other shear zones that did not appear to be active) in plastic silty clay containing many randomly oriented, discontinuous, shiny surfaces surrounding one or a few continuous shear surfaces subparallel to the ground surface (William Cotton and Associates, 1992). The failure surface samples had strengths characterized by angles of residual friction ranging from 6° to 10.9° and residual cohesion of 1.9 to 6.7 kPa (Table 2). Liquid limits of samples from the slip surface ranged from 74 to 129.

Hydraulic properties

The saturated hydraulic conductivity, K_s, of the debris-apron materials in and near the landslide was estimated using single well (slug and recovery) test methods on 13 open-tube piezometers with measurable water levels; the estimated K_s ranged from 3.8×10^{-11} to 2.5×10^{-7} m/s (Baum et al., 1991). These values are typical of materials ranging in composition from clays to silty sands (Freeze and Cherry, 1979). In addition to the single well tests, ten permeability tests using a Guelph Permeameter were conducted on near-surface soils, at depths between 0.2 and 0.4 m (Torikai, 1992). These values ranged from 1.8×10^{-8} to 2.0×10^{-5} m/s (Fig. 9).

Materials near the basal slip surface have lower hydraulic conductivity than materials above or below, indicating that

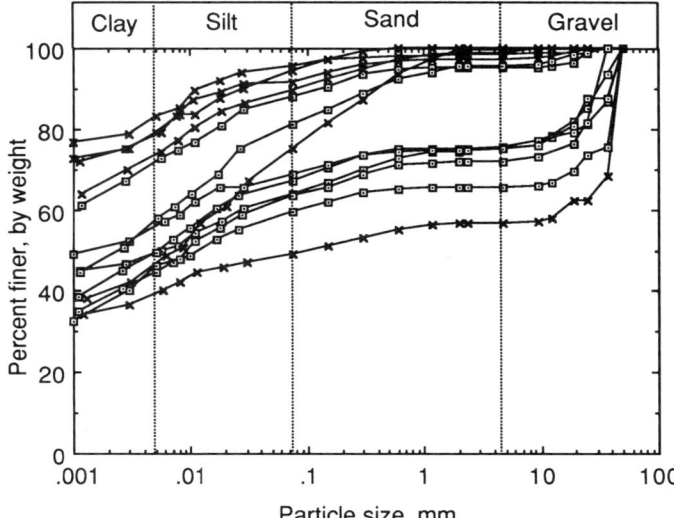

Figure 7. Particle-size analysis of samples from the Alani-Paty landslide. Cobble- and boulder-sized clasts, which make up 25%–45% (by volume) of the landslide material, were removed. Samples were air dried and analyzed according to ASTM D422-63 (ASTM, 1990). The particle classification is based on the following size ranges: *gravel* > 4.76 mm, 4.76 mm > *sand* > 0.075 mm, 0.075 mm > *silt* > 0.005 mm, *clay* < 0.005 mm. The particles are dominantly clay, silt, and fine sand. A few samples also contained abundant coarse gravel. Coarse sand and fine gravel, particles from about 0.5–10 mm in diameter, were rare or absent. Samples classified as CH (clays of high plasticity) shown by "x"; MH (silts of high plasticity) shown by squares. Analyses by George Erickson and Sue Cannon.

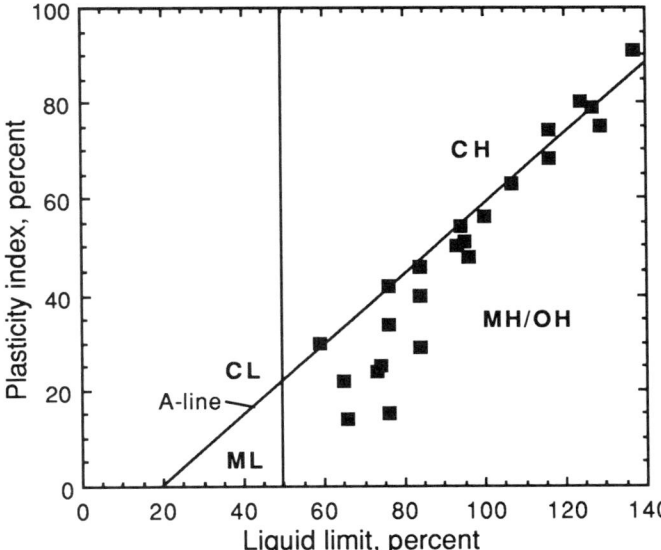

Figure 8. Plasticity chart for materials from the Alani-Paty landslide. Physical properties and locations of the samples are shown in Baum et al. (1990, 1991). The A-line is an empirical relationship that separates generally clay-like soils from silt-like soils. Abbreviations: CL, clay of low plasticity; ML, silt of low plasticity; CH, clay of high plasticity; MH/OH silt and organic clay of high plasticity. Analyses by George Erickson and Coyn Criley.

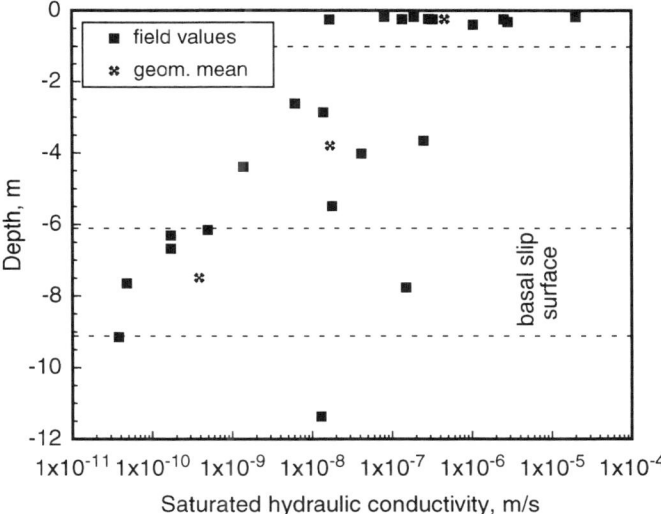

Figure 9. Saturated hydraulic conductivity determined at different depths in and near the Alani-Paty landslide. Geometric means are computed for different depth intervals identified by dashed lines. Single well and permeameter tests are described in Baum et al. (1991) and Torikai (1992).

materials near the basal slip surface can perch ground water. The geometric means of the hydraulic conductivities for each depth interval shown in Figure 9 decrease from the ground surface to the basal slip surface. Beneath the basal slip surface, the one piezometer tested showed a conductivity about the same as materials above the basal slip surface, in the interval from 2 to 6 m deep. Many piezometers beneath the basal slip surface were dry or mostly dry during the observational period. Water added to these dry open-tube piezometers drained rapidly, indicating that the piezometers were functioning properly and that the tips were in unsaturated materials. These observations indicate that materials beneath the basal slip surface have larger hydraulic conductivities than materials at the basal slip surface.

Bulk hydraulic conductivity of the landslide material may be greater than indicated by the single well and permeameter tests. Both tests only affect a small region near the piezometer or permeameter tip; therefore the hydraulic conductivity computed is only valid for this small region. Interconnected cracks and macropores may increase the bulk conductivity of the materials, particularly in the upper 2–3 m. During the drilling of several borings near the head and toe of the landslide, drilling foam was observed flowing out of cracks in the ground surface 2–8 m from the boring. Foam appeared when the drill bit reached depths of 3.9 to 6.5 m, about 1–6 hours after drilling began and foam flowed from different cracks at different times. Assuming flow in straight, nearly horizontal paths from the edge of the boring to the crack openings at the ground surface 2–8 m away, foam could reach the openings in 1–6 hours traveling at rates of 9×10^{-5} to 2×10^{-3} m/s (including the vertical distance results in faster rates). Rates calculated from near-surface hydraulic conductivity values (Fig. 9), the average porosity (26%), and assuming a unit hydraulic gradient between the edge of the boring and the crack opening range from 7×10^{-8} to 8×10^{-5} m/s and are less than indicated by the movement of drilling foam. Similarly, muddy water was observed flowing from open cracks in a private driveway on the upper flank of the landslide while drilling occurred about 50 m upslope. Based on these observations, the crack network in some parts of the landslide appears extensive enough to permit rapid flow over distances of many meters.

The unsaturated hydraulic characteristics of the near-surface soils in the upslope part of the landslide are similar to those of other clays and show that soil tension increases and hydraulic conductivity decreases with decreasing soil moisture content (Fig. 10). Because about 35% (by volume) of the near-surface material is composed of cobbles and boulders, the bulk field-saturated moisture content is only about 26% (Torikai, 1992). The moisture retention curve indicates that even when the near-surface soils experience soil water tension equivalent to a negative pressure head of 1 m, after drainage or drying, their moisture content remains near saturation (Torikai, 1992). During dry periods, near-surface soils remain about 75% saturated (Fig. 10).

HYDROLOGY OF THE LANDSLIDE

Some slow-moving landslides in the leeward valleys near Honolulu accelerate downslope during periods with large amounts of rainfall (Peck, 1959; Peck and Wilson, 1968). Residents on and near the Alani-Paty landslide have reported that damage to roads and houses on the landslide occurred during and immediately following rainy periods, especially during the winters of 1987 and 1988. To understand the effects of rainfall on the landslide, we describe the observed relations between rainfall, ground water, and movement of the landslide; the inferred flow of ground water in the landslide; and the hydrologic responses during the rainstorms.

Rainfall and landslide movement

Rainfall is the dominant source of water in the hydrologic system of the landslide. Rainfall varies substantially in Honolulu; mean annual rainfall in Waikiki is about 600 mm while near the crest of the Koolau Mountains, 8 km away, mean annual rainfall is about 4,000 mm. Mean annual rainfall on the landslide, estimated from isohyetal maps of Oahu, is about 1,800 to 2,000 mm. (Giambelluca et al., 1986). During calendar year 1990, the U.S. Geological Survey tipping bucket rain gage located on the landslide, recorded 2,053 mm (Baum et al., 1990, 1991). Rain occurs year round during orographically modified trade-wind showers, and more frequently in the winter months during Kona storms (subtropical cyclones), cold fronts, low pressure systems, and rare tropical cyclones (Giambelluca et al., 1984). Intense bursts of rain are common during prolonged storms.

The Alani-Paty landslide was stationary for most of the observation period between September 1989 and April 1991. Although no major downslope movements occurred during this period, the head of the Kahaloa kinematic element moved small amounts (3–30 mm) during several rainy periods (Fig. 11, Table 3). Between rainy periods, an extensometer recorded creeping movements at the head (Fig. 11), but inclinometers and extensometers recorded no creep of the main body of the landslide (Baum et al., 1991; STV/Lyon Associates, 1990, 1991). The largest single recorded episode of movement occurred in January 1990 (Table 3). The head of the landslide

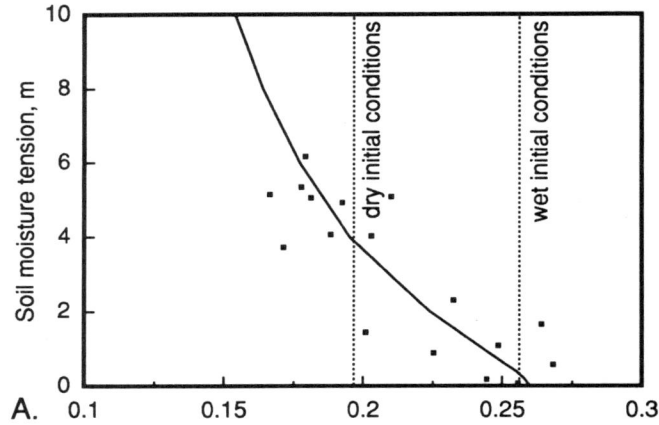

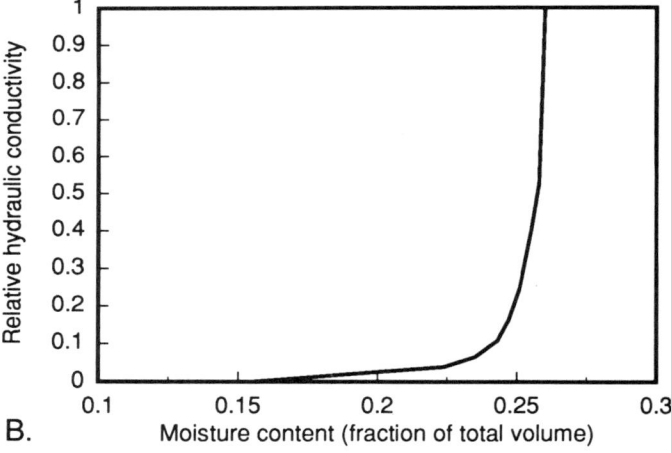

Figure 10. Moisture characteristics for unsaturated, near-surface soils in the upslope part of the landslide (after Torikai, 1992). Wet and dry values are typical of winter and summer conditions at the site, respectively. (A) Moisture retention curve. Field moisture content (as fraction of total volume) was corrected to account for 35% (by volume) cobbles and boulders using the methods of Bouwer and Rice (1984). Dots indicate corrected values, curve fitted to data using non-linear least squares method of El-Kadi (1987). (B) Relative hydraulic conductivity, $K(\psi)/K_s$, computed from parameters of the moisture retention curve using the method of van Genuchten (1980). $K(\psi)$ is the unsaturated hydraulic conductivity, and ψ is the pressure head.

moved during the storm of January 15–19, 1990 (Fig. 11) and the main body of the landslide moved downslope during the second half of January 1990 (Table 3) according to inclinometer data of STV/Lyon Associates (1990, 1991) and surveys by City and County of Honolulu (1990). Slight surface deformation near the toe of the landslide, evidenced by cracked walls and curbs, became noticeable at the end of January, about two weeks after movement started at the head scarp (S. R. Spengler, oral commun., 1990). The landslide also moved downslope about 10–30 mm sometime in the winter of 1990–1991, probably in March 1991, as detected by borehole inclinometers (STV/Lyon Associates, 1991).

Movement episodes occurred during rainy periods. However, movement was not detected during all rainy periods (Table 3 and Fig. 11). The head of the landslide moved during extended rainy periods containing high intensity rainfall bursts. Intense bursts of rain commonly occurred several times during a rainy period, and 24-hour cumulative rainfall ranged up to about 100 mm. These rainy periods resulted in high peak pressure heads within the upslope part of the landslide (Table 3). Movement in the main body of the landslide in January 1990 and March 1991 occurred during periods when the average rainfall for the previous 10 days exceeded 25 mm (Fig. 11).

Ground water

Ground water occurs in the Koolau Basalt beneath the landslide and in deposits forming the debris apron both within and outside the landslide. The response of ground water to rainfall varies significantly between different areas. Figure 12 shows schematically the main pathways of water entering and leaving perched ground water in the landslide. Ground water within the landslide is recharged from the direct infiltration of rainfall, infiltration of runoff coming from areas upslope of the landslide and from urban pavement and structures, and urban irrigation from activities such as lawn and garden watering. At times broken water pipes have contributed water to the landslide (Baum and Reid, 1992). Ground water discharges from the landslide by downslope flow toward the valley, downward drainage into the debris-apron and bedrock, seep and spring discharge, and evapotranspiration. The perennial basal freshwater table occurs at least 45 m below the landslide basal slip surface (Wentworth, 1951).

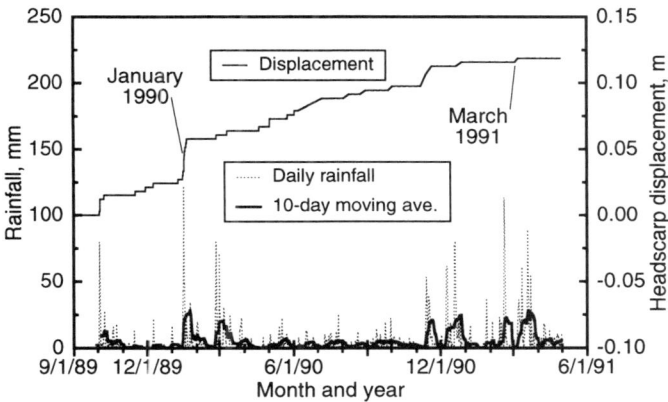

Figure 11. Measured cumulative downslope displacement at an extensometer (at the head scarp of the landslide, Fig. 2), daily rainfall, and previous 10-day average rainfall at the landslide. Most of the noise (spikes in the graph resulting from temporary deflection of the extensometer wire by wind or tree branches) has been removed from the displacement data. Creeping movements between rainy periods were limited to the head (Baum et al., 1991). Rainfall data are estimated for February 1–20, 1991 (Baum et al., 1991).

TABLE 3. RAINY PERIODS, MOVEMENT, AND PRESSURE HEADS RECORDED BETWEEN SEPTEMBER 26, 1989, AND APRIL 30, 1991, AT THE ALANI-PATY LANDSLIDE

Time Period	Cumulative Rainfall (mm)	Peak Intensity (mm per 15 minutes)	Movement at Head Scarp (mm)	Movement in Main Body (range, mm)	Peak Pressure Heads I4 at 6.1 m* (m)	P24 at 3.4 m (m)
Oct. 3, 1989	80	16	15	0	....	....
Jan. 15-17, 1990	214	14	30	50-60	4.7 (1/17)	....
Feb. 24-25, 1990	110	5	0	0	4.9 (2/25)	....
Mar. 1, 1990	71	6	0	0	....	....
Nov. 12-17, 1990	184	9	15	0	4.9 (11/18)	2.6 (11/18)
Dec. 8-9, 1990	121	4	0	0	4.9 (12/9)	3.2 (12/9)
Dec. 18-25, 1990	241	24	3	0	4.9 (12/21)	3.3 (12/19)
Feb. 18-19, 1991	170 est.	....	0	0	5.0 (2/21)	3.0 (2/20)
Mar. 6-23, 1991	452	10	3	10-30	5.0 (3/19)	3.6 (3/19)

Note: Rainfall from U.S. Geological Survey tipping bucket gauge; movement at head scarp from U.S. Geological Survey extensometer; movement in main body of landslide estimated from inclinometers (STV/Lyon Associates, 1991) and surface surveys (City and County of Honolulu, 1990); peak pressure heads are approximate, date of occurrence in parentheses.
*STV/Lyon Associates, unpublished data.

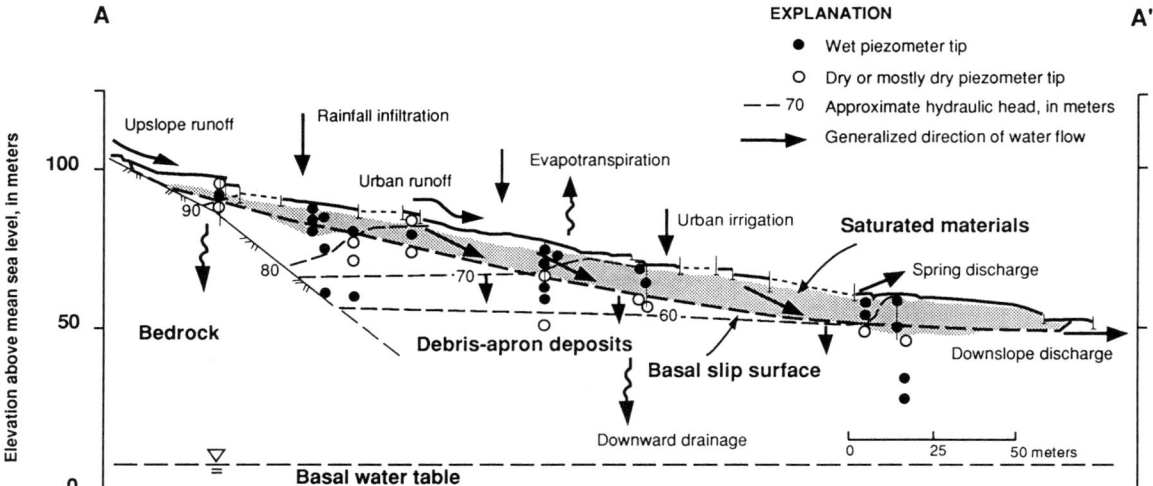

Figure 12. Cross section through the landslide showing generalized ground-water flow directions, approximate limits of the saturated materials within the landslide, and possible pathways of water entering and leaving landslide. Approximate location of the basal water table from Wentworth (1951). Dashed lines along the ground surface profile indicate its approximate position beneath houses. Piezometers are projected from borings near the line of section. Shaded area within the landslide indicates mostly saturated materials.

Occurrence of ground water. During our observation period, responses in 70 open-tube and pressure transducer piezometers varied from place to place within and near the landslide; however, there are distinct regions in which most piezometers showed similar response patterns (Baum and Reid, 1992). In general, materials within the landslide are water-saturated and have some positive pressure head (Fig. 12). The water appears to be perched near the basal slip surface by materials having lower hydraulic conductivity (Fig. 9). Materials underlying most of the landslide are largely unsaturated or have low pressure heads; this low pressure is consistent with the presence of perched water within the landslide. Deep piezometers indicate that saturated conditions exist locally beneath the unsaturated materials, within the debris-apron deposits and above the basal freshwater table. Ground water occurs intermittently in bedrock near the head of the landslide.

Most piezometers with tips located within the landslide boundaries and above the basal slip surface showed two patterns of response: (1) relatively steady pressure heads or (2) increases in pressure head during and following rainfall (Baum and Reid, 1992). In the upslope part of the slide, most piezometers responded during and following rainfall (Fig. 13). Pressure heads in this part declined during several weeks following rainy periods, indicating natural drainage from this region (Fig. 14). In the middle part of the slide, some piezometers showed small responses to rainfall while others maintained relatively steady pressure heads. In the downslope part, most shallow piezometers showed relatively steady pressure heads and many had water levels within a meter of the ground surface (Fig. 14).

In all three regions, response patterns in the pressure transducer piezometers were similar to those in the open-tube piezometers. Thus, the water table in the upslope quarter of the landslide fluctuates between the ground surface during rainy periods and a position 2–3 m below the ground surface during dry periods, but the water table in the downslope three-fourths stays near the ground surface year round.

Movement of ground water. Ground-water flow paths within the landslide are controlled by the hydraulic gradients in the materials and by the layering or anisotropy of the material hydraulic properties. Because of the heterogeneity in hydraulic properties, local ground-water flow is complex and varies in both direction and velocity. Some shallow ground-water flow emerges in springs and seeps on the surface of the landslide. In general, however, ground water within the landslide moves downslope along steeply descending flow paths in predominantly saturated conditions (Fig. 12). Ground water moves vertically downward from the predominantly saturated landslide to unsaturated materials or materials with low pressure heads beneath the basal slip surface.

All vertical components of the hydraulic head gradient between saturated shallow and deep piezometers in the same boring indicate downward ground-water flow, except for one nest, 19 and 19×, in the toe of the landslide where the component indicates upward flow (Baum and Reid, 1992). Downward gradients in the upslope one quarter of the landslide increased following rainy periods indicating recharge from the ground surface. Vertical hydraulic gradients between saturated materials above and unsaturated materials below were large year

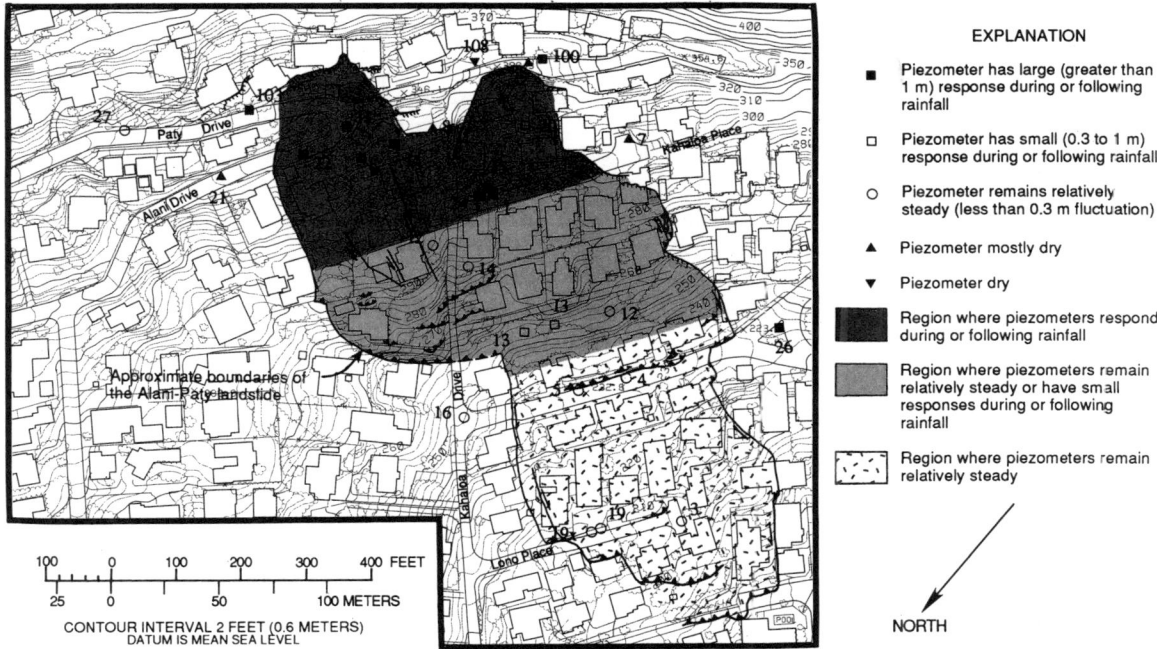

Figure 13. Map showing the distribution and response patterns during rainstorms of piezometers with tips located above the basal slip surface. Base map from R. M. Towill Corporation, 1989.

round (>–1). The perennial downslope component of the hydraulic gradient between saturated piezometers located within the landslide indicates a persistent downslope component of ground-water flow. In the upslope part of the landslide, vertically downward drainage of ground water may be reflected in the delayed responses of deeper piezometers. For example, this response lag occurs in pressure sensor nest I4 (Fig. 2), where the piezometer at 17.7 m responds 5–12 days after the piezometer at 6.1 m (STV/Lyon Associates, 1990, 1991).

Although ground water occurs intermittently in the bedrock upslope of the landslide, several observations indicate that ground-water flow from bedrock into the landslide materials is minor and probably insignificant compared to infiltration from the ground surface: (1) piezometers near the bedrock contact showed no or minor response during rainstorms; (2) hydraulic gradients within the debris-apron deposits are downward into bedrock or the deep perennial basal water table; and (3) only about 3% of the slip surface is closer to the bedrock contact than to the ground surface (Baum and Reid, 1992, fig. 3).

Discharge and recharge of ground water. We estimated the average annual vertical and downslope ground-water discharge from the landslide by assuming steady Darcian flow (Freeze and Cherry, 1979). The cross-sectional areas, saturated hydraulic conductivities, and hydraulic gradients differ between downward ground-water flow and downslope ground-water flow (Table 4). The equivalent vertical saturated hydraulic conductivity and downslope conductivity for a layered material were estimated using the methods outlined in Freeze and Cherry (1979). Any other anisotropic effects were ignored.

Using these methods and ignoring surface spring discharge and evapotranspiration, total ground-water discharge from the landslide is about 5.3×10^{-5} m^3/s (or 4.6 m^3/d, see Table 4). Although the specific discharge within the landslide is larger in the downslope direction than vertically downward (Fig. 12 and Table 4), the net ground-water discharge downward is larger because the horizontal area of the landslide is much greater than the cross-sectional area. About 80% of the natural drainage responsible for the gradual decrease in pore pressures in the upslope quarter of the landslide during dry periods appears to occur by ground water flowing vertically downward through the basal slip surface.

Average annual ground-water recharge for a medium-density urban area on Oahu, such as the landslide, receiving about 2,000 mm/yr of rain is about 50% of the annual rainfall (Giambelluca, 1986). This average recharge includes some gain from urban irrigation and some loss to runoff and evapotranspiration. Using this value over the entire landslide area, average ground-water recharge is about 90 m^3/d. This value may be an overestimate because of additional runoff from steep slopes and diversions into storm drains and is about 20 times greater than the estimated total average ground-water discharge from the landslide, shown in Table 4. This discrepancy suggests two likely possible sources of error: (1) that the average large-scale hydraulic conductivity of the landslide is greater than the equivalent conductivities computed from single well tests or (2) that the average ground-water recharge estimated using methods of Giambelluca (1986) is too large.

Hydrologic response during rainstorms

Although ground water is present in the landslide year round, the landslide moved downslope only during rainy periods. During large rainstorms water infiltrates downward into the landslide materials from direct rainfall or surface runoff

TABLE 4. ESTIMATED VERTICAL AND DOWNSLOPE GROUND-WATER DISCHARGE FROM THE ALANI-PATY LANDSLIDE

Direction	Area (m²)	Equivalent Hydraulic Conductivity* (m/s)	Hydraulic Gradient (dimensionless)	Discharge (m³/s)
Vertically Downward	33,400	1.3×10^{-9}	-1.0	4.3×10^{-5}
Downslope	1,260	6.4×10^{-8}	-0.13	1.0×10^{-5}
Total				5.3×10^{-5}

*Equivalent hydraulic conductivity calculated using three layers parallel to the average slope of the ground surface (9°) and equations 2.31 and 2.32 of Freeze and Cherry (1979). The layers, starting at the top, have the following properties: layer 1, 0.9 m thick, $K_s = 4.6 \times 10^{-7}$ m/s; layer 2, 4.5 m thick, $K_s = 1.7 \times 10^{-8}$ m/s; layer 3, 2.1 m thick, $K_s = 3.9 \times 10^{-10}$ m/s, where K_s is the saturated hydraulic conductivity. Values of K_s are geometric means of K_s measured at piezometers in each layer (Fig. 9).

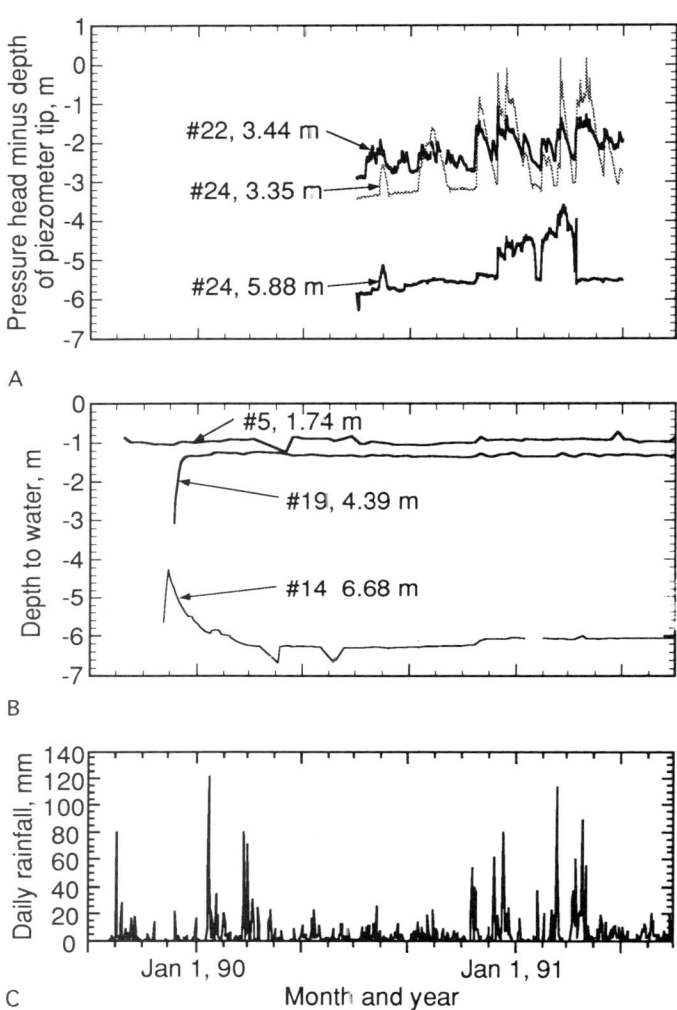

Figure 14. Selected piezometer responses. (A) Response of pressure-transducer piezometers within the upslope one-quarter of the landslide that show increased pressure during rainfall. (B) Response of selected open-tube piezometers with relatively steady water levels in the downslope three-quarters of the landslide. The large initial responses of some open-tube piezometers such as 14 and 19 indicate the gradual equilibration following drilling of water pressure in the borings with pore-water pressure in the surrounding materials. Piezometer locations shown in Figures 2 and 13. Additional responses can be found in Baum et al. (1991). (C) Daily rainfall recorded at the tipping-bucket rain gauge (Fig. 2).

onto soil-covered regions. Pore pressure responses within the landslide vary considerably with different rainstorms; however, most of the hydrologic responses during these events can be simplified into five stages (Fig. 15).

Initially during rainfall, nearly vertical infiltration into unsaturated materials occurs rapidly (Fig. 15A). However, if the application rate (from rainfall or runoff) is larger than the saturated hydraulic conductivity, the materials near the ground surface will eventually become saturated and water will start ponding on the ground surface (Fig. 15B). On a sloping surface, this water from incipient ponding will create surface runoff. With extended lower application rates, a partially saturated wetting front advances downward to the water table and the materials may become nearly saturated or tension saturated (Fig. 15C). Both vertical and downslope flow in the layered landslide materials increase after infiltrating water reaches the water table. Thus, a region within the landslide receives water from both vertical infiltration and downslope flow from the adjacent upslope materials. During periods of intense rainfall, pore pressures rise rapidly and the water table approaches the ground surface (Fig. 15D). Without continued application, the near-surface materials will become unsaturated again because of drainage and evapotranspiration (Fig. 15E). The following sections describe the surface runoff, infiltration, and ground-water response occurring on and within the landslide during rainstorms.

Surface runoff. Upslope of the landslide, the 6.8 ha drainage basin is primarily underlain by basaltic bedrock and contains two incised ephemeral channels. Field observations during the large rainstorms indicate that surface flow in these channels occurs during and immediately following long periods of heavy rain. Most of the runoff generated from the

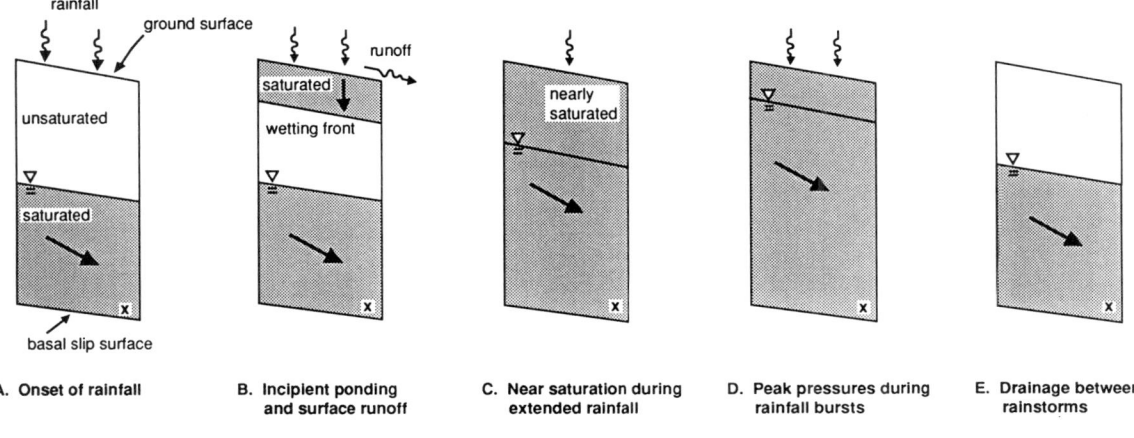

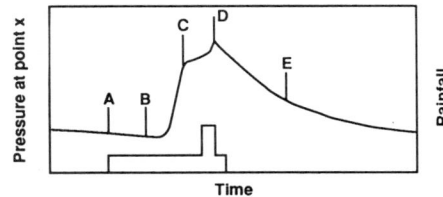

Figure 15. Schematic diagram showing the rainfall infiltration process during extended rainy periods. Large arrows indicate generalized directions of water flow.

drainage basin upslope of the slide appears to be collected by storm drains near the head of the landslide. However, some water from the channel above the Alani kinematic element is not fully captured by storm drains uphill from the landslide and may flow downslope and infiltrate into the landslide. Urban runoff from rooftops, streets, and driveways, which together cover 49% of the landslide area, flows both onto adjacent soil, where it infiltrates, and into city streets, where it flows into storm drains that carry at least part of the water off the landslide.

Infiltration. Torikai (1992) presented data on pressure responses from a vacant lot on the upslope part of the landslide during the March 1991 rains that illustrate the infiltration process. Data were collected using nests of tensiometers that measured near-surface unsaturated soil tensions and deeper nests of tensiometers with differential, bi-directional pressure transducers that measured both soil tensions and positive pressure heads.

Prior to rainfall, contours of equal hydraulic head showed predominantly downslope ground-water flow with a water table about 1–3 m beneath the ground surface (Fig. 16A). Pressure heads in the unsaturated materials were about −0.20 to −0.25 m relative to atmospheric pressure (Fig. 17). About 1–2 days after rainfall began, the tensiometers and deeper piezometer showed a rapid increase in pressure with the negative pressures becoming positive and the water table rising. The shallow tensiometer responded first, and there was a downward hydraulic gradient in the nest. Ground-water flow was predominantly downslope (with a small downward component) in the saturated zone and vertically downward in the unsaturated zone (Fig. 16B). During the rainy period, the tensiometers measured positive pressures (indicating saturation of the materials) and small pressure increases within a few hours of intense bursts of rainfall (Fig. 17). Also during this rainy period, the shallow near-surface tensiometers indicated saturated conditions, and some runoff on the soil surface was observed.

Ground-water response. During and following rainfall, piezometers in the upslope part of the landslide showed increases in pressure head that correspond well with increases in the daily rainfall amounts (Fig. 14). Periods with larger amounts of rainfall generally resulted in larger pressure heads; some piezometers showed increases of as much as 3 m over pre-rainstorm levels. Many of the piezometers in the upslope part of the landslide had peak hydraulic heads from 0 to 1 m below the ground surface; none had peaks significantly above the ground surface. However, not all piezometers had peak pressures simultaneously. Observed peak pressure heads during rainy periods for selected piezometers are shown in Table 3. During several of these periods, the landslide moved downslope. In the downslope part of the slide, pressure heads remained relatively steady and in general water levels were within a meter of the ground surface. Water levels of piezometers approaching the ground surface together with positive pore pressures in shallow tensiometers indicate that the landslide became saturated during these rainy periods. Thus, the landslide moved only when it was saturated, and pore pressures in the upslope part were relatively high.

Several patterns were recurrent in the upslope U.S. Geological Survey transducer pressure head responses recorded during rainy periods; these are well illustrated by the response during large rainstorms in March 1991 (Fig. 18). Initially, the pressure heads responded 0.5–2.5 days following the onset of rainfall. This initial response occurred earlier in the more upslope boring 22 than in boring 24. After response began, there was a relatively rapid increase in pressure head ranging from 1–2 m during the next day. After reaching these higher pressure head levels, however, even larger peaks in the pressure head occurred within 4–6 hours following intense bursts of rainfall. These rapid increases in pressure head of about 0.3–1.2 m generally decreased within a day. About two weeks following the end of rainfall, pressure heads slowly decreased to pre-storm levels (Fig. 14). Pressure response at the deeper (5.88 m) transducer in boring 24 during these storms was smaller than the shallow (3.35 m) transducer response (Fig. 14).

Analysis. Three questions are particularly important in understanding the role of rainfall in creating destabilizing pore pressure conditions: (1) does enough rain fall during a storm

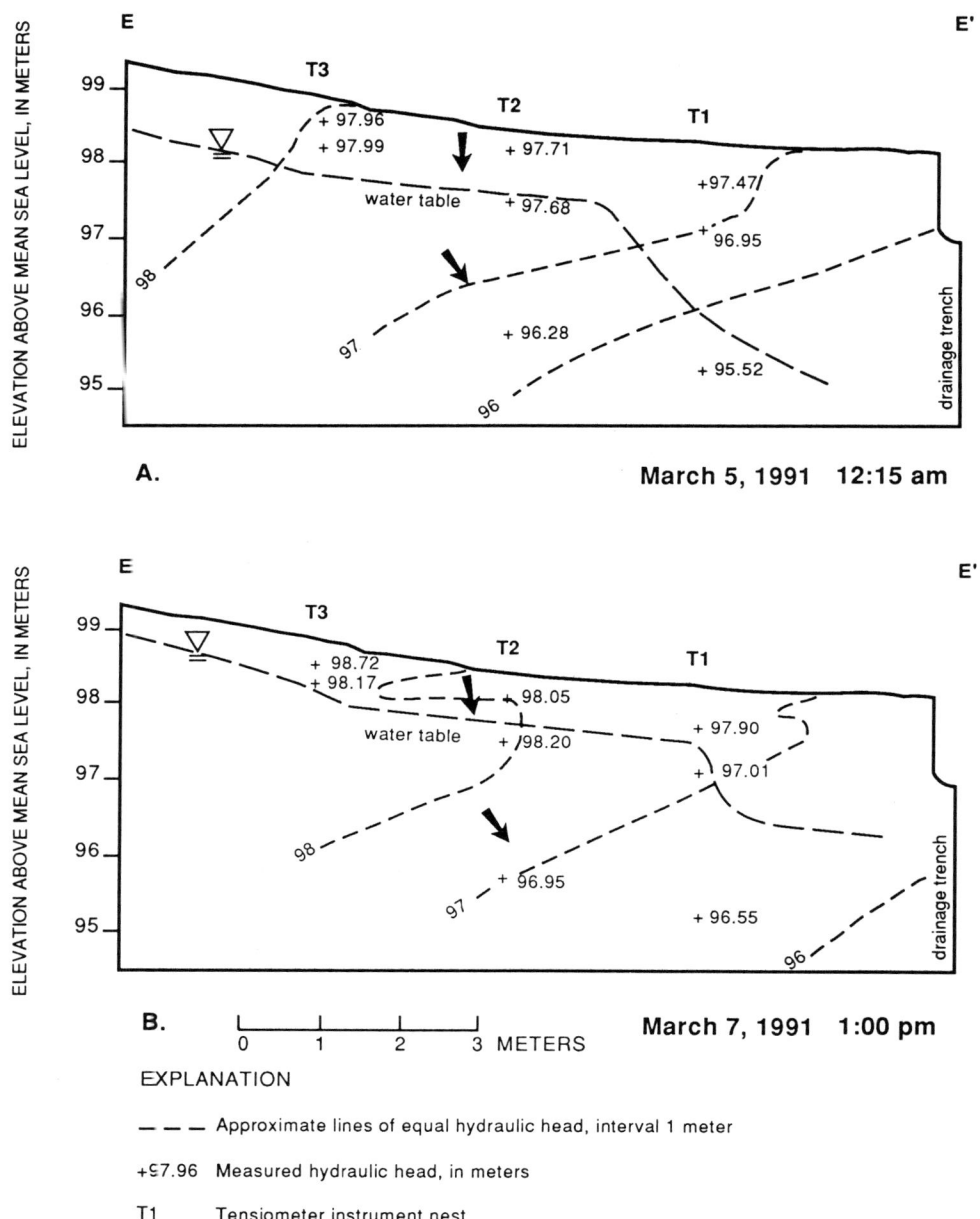

Figure 16. Cross section E–E' through one lot in the upslope part of landslide showing hydraulic head distribution (after Torikai, 1992). (A) Distribution on March 5, 1991, before rain. (B) Distribution on March 7, 1991, during rain. Rainfall shown in Figure 17. Location of cross section E–E' shown on Figure 2 (about 40 m southeast of the intersection of lines B–B' and C-C') Arrows indicate approximate direction of ground-water flow. Datum for hydraulic head is mean sea level.

to saturate or nearly saturate the landslide? (2) how long is the lag time between the onset of rainfall and pore pressure response at depth within the landslide? and (3) how large is the peak pore pressure response at depth to intense bursts of rain falling on the ground surface? Simple analyses quantify these important aspects of the hydrologic response during rainstorms.

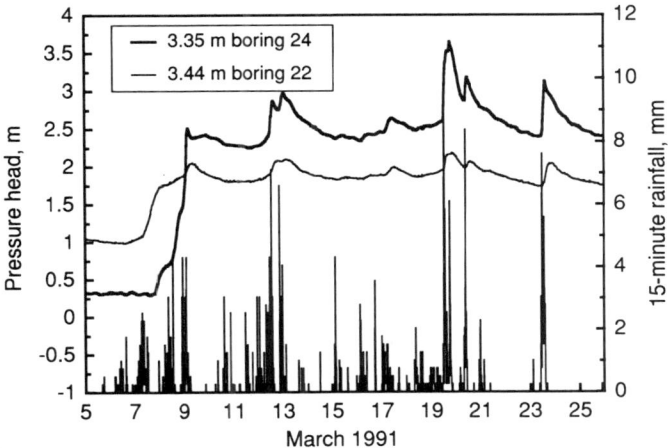

Figure 18. Response of shallow pressure-transducer piezometers during March 1991 rainstorms. Initial response occurs about 1–2 days after the onset of rainfall. During the storm, peak pressure responses occur within hours following high intensity rainfall bursts. Piezometer locations shown in Figures 2 and 13.

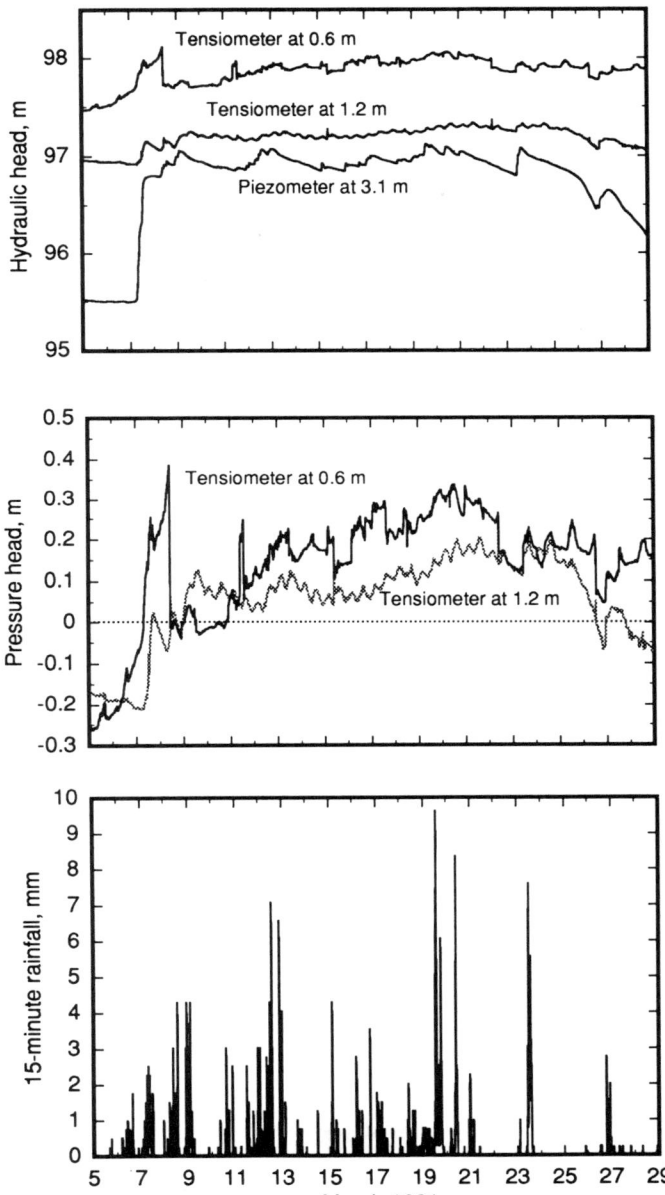

Figure 17. Rainfall and total hydraulic head measured at tensiometer nest T1 during the March 1991 rainstorms (Torikai, 1992). Transducer noise has been removed; sharp pressure drops in the tensiometer at 0.6 m depth are due to instrument servicing. Tensiometers initially measured soil tensions indicating unsaturated conditions; during most of the storm they measured positive pore pressures indicating saturated conditions.

Mass balance calculations indicate that enough rain falls during most storms to completely fill the available porosity in the materials above the water table, provided ground-water discharge and evapotranspiration are relatively minor during the storm. For an unsaturated thickness of 2 m (typical of the upslope one-fourth of the landslide prior to extended rainfall), the total amount of water required to fill the available porosity ranges from about 4 mm of rainfall for wet initial conditions to 140 mm for dry initial conditions (Baum and Reid, 1992). Many observed storms have rainfall in excess of this amount (see Table 3 for examples).

Infiltration of water from the ground surface can explain most of the observed lag between the onset of extended rainfall (Fig. 15A) and the pore pressure response at depth (Fig. 15C). Assuming incipient ponding at the ground surface, we estimated the elapsed time, t_D, for a wetting front to advance vertically downward to a depth, D, using a Green-Ampt analysis (Bouwer, 1978):

$$t_D = \frac{\theta_f}{K_s}\left[D - h_e \ln\left(\frac{h_e + D}{h_e}\right)\right] \quad (1)$$

where θ_f is the fillable porosity of the soil, K_s is the saturated hydraulic conductivity of the soil, and h_e is the effective pressure head beneath the wetting front. For wet initial conditions, the estimated time range is 2–50 hours for a saturated wetting front to infiltrate 2 m vertically through uniform materials to the water table, assuming K_s values between 4.6×10^{-7} m/s (near-surface soil) and 1.7×10^{-8} m/s (upper debris-apron deposits), $h_e = 0.3$ m, and $\theta_f = 0.002$ (for wet initial conditions with a soil tension of about -18 cm). The estimated time range is similar to the observed range of lag time, about 12–60 hours, for piezometers 3.3–3.4 m deep.

The effective pressure head, h_e, used in the preceding Green-Ampt analysis is difficult to measure directly, but Bouwer (1978) describes a method for estimating a related parameter called the critical pressure head, h_{cr}, that approximates h_e. The critical pressure head, h_{cr}, was estimated using the unsaturated characteristic curves shown in Figure 10 and the methods described in Bouwer (1978). Physically, $-h_{cr}$, approximates the height of a saturated capillary fringe (Bouwer, 1978). For the near-surface soils on the Alani-Paty landslide, $-h_{cr}$, and thus h_e, is about 0.3 m.

After the landslide materials are nearly saturated or tension saturated during extended rainy periods, the observed rapid pressure responses within the landslide (Fig. 15D) can be analyzed as the transmission and attenuation of rainfall-induced oscillating flux (infiltrating rainfall) at the ground surface. The attenuation of a sinusoidal oscillating pressure signal (in this situation approximating flux) through a homogeneous and isotropic, saturated, porous material can be estimated using a one-dimensional vertical diffusion model (Carslaw and Jaeger, 1959; Todd, 1980). Here, the oscillating signal represents oscillating bursts of intense rainfall infiltrating at the ground surface. Solutions to this model give the maximum amplitude of the pressure fluctuation, p, at a depth x from the signal source of

$$p = p_0 \exp\left(-x\sqrt{\frac{\pi S_s}{t_0 K_s}}\right), \quad (2)$$

where p_0 is the pressure of the source signal, S_s is the specific storage of the material, t_0 is the period of the oscillation, and K_s is the saturated hydraulic conductivity of the material. The time lag, t_L, between the oscillating source signal and the pressure response at depth is given by:

$$t_L = x\sqrt{\frac{t_0 S_s}{4\pi K_s}}. \quad (3)$$

This signal is superimposed on any existing pore pressure distribution, such as that occurring with a downslope hydraulic gradient. Similar solutions have been used to analyze the pressure response to rainfall in other landslides (Iverson and Major, 1987; Haneberg, 1991a).

Using this model with field-measured hydraulic properties, the estimated pressure response and time lag at depth are similar to that observed (Figs. 19 and 20) and indicate that pressure transmission from oscillating rainfall bursts is sufficient to cause the rapid pressure responses observed during extended rainy periods with intense rainfall bursts. Observed responses are from pressure sensors in the upslope quarter of the landslide; scatter in the observed responses may be the result of locally varying hydraulic properties. Both the pressure attenuation and time lag simulated using the higher K_s value better match the observations than those using the mean K_s value (Figs. 19 and 20). The better match using the higher K_s value indicates that the overall hydraulic conductivity of the upslope part of the landslide may be higher than the mean K_s, perhaps due to interconnected cracks. The maximum value of the source signal, p_0, was equal to the sensor depth multiplied by the unit weight of water and represents the maximum pressure in a completely saturated static condition. Thus, the atten-

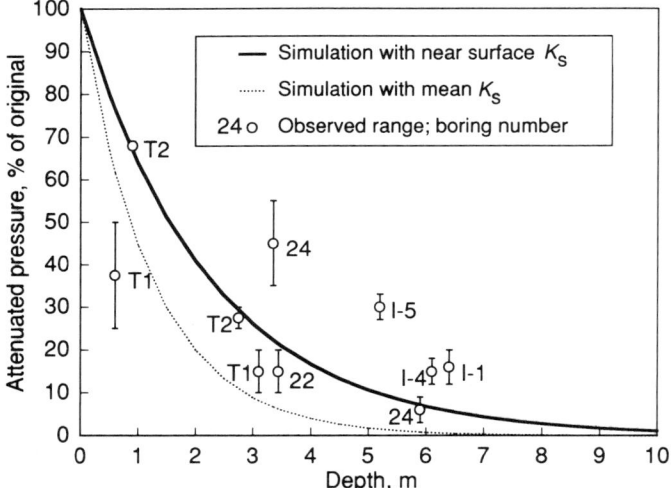

Figure 19. Observed and simulated pressure attenuation with depth. Observed values based on piezometer responses from selected December 1990 and March 1991 rainstorms and estimated as the change in pressure at depth divided by the change in pressure at the ground surface. Simulations assume an oscillating pressure signal with a period of one day at the ground surface and attenuation through a saturated, homogeneous material with $S_s = 0.0025$ m^{-1} (estimated as the slope of the moisture retention curve near saturation) and two K_s values: (1) the geometric mean of near-surface soil values (4.6 × 10^{-7} m/s) and (2) the geometric mean of all values determined between the ground surface and a depth of 6 m (1.4 × 10^{-7} m/s).

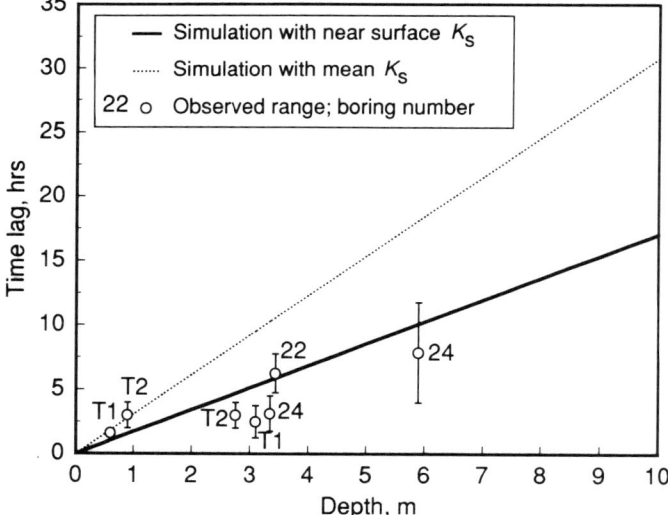

Figure 20. Observed and simulated time lag in pressure response at depth from a signal at the ground surface. Observed values based on piezometer responses from selected December 1990 and March 1991 rainstorms. Simulations use same parameters as in Figure 19.

uated pressure, p/p_0 is the actual pressure rise divided by this maximum possible pressure rise. The analysis assumes that the transient signal from rainfall is superimposed on a background of a steady pressure distribution. The intense rainfall oscillation period was estimated as about one day, typical of rainy periods such as that recorded during the March 1991 storms. Note that pressure increases at the depth of the basal slip surface are small for a one-day oscillation period (Fig. 19). Pressure increases that are sufficient to trigger movement result from longer periods (7–10 days) of rainfall (Fig. 11).

Thus, water enters the landslide primarily at the ground surface by infiltration of rainfall and surface runoff. Infiltration is sufficient to saturate the landslide during rainy periods and to create a pore pressure response, typically 1–2 days after the onset of rainfall. After the initial response and during prolonged rainfall, the landslide materials remain saturated or nearly saturated and pore pressures can increase rapidly in response to the downward movement of pressure waves from the ground surface during intense bursts of rainfall.

MECHANICS OF THE LANDSLIDE

Insufficient data about the initial movement of the landslide or about changes in topography and hydrology caused by urban development of the landslide area preclude detailed analysis of the causes of the landslide. Rather, the following analysis focuses on the conditions that existed at the time of our field investigation (January 1989–March 1991) as they relate to overall stability, periodic movement, and deformation of the landslide.

The low shear strength of silty clays from the basal slip surface and the high water pressures measured during wet periods when the landslide moved are consistent with sliding on the observed basal slip surface of the landslide. Baum and Fleming (1991) performed stability analyses (Janbu, 1973) using cross section A–A' of the landslide and their best estimate of pore pressure at the slip surface (during wet periods) based on measurements through April 1990. They computed a factor of safety, F, of 0.99 assuming the following parameters: saturated unit weight of the landslide material, $\gamma_t = 21.7$ kN/m^3; cohesion for effective stress, $c' = 6.1$ kPa; angle of friction for effective stress, $\phi' = 8°$. They repeated calculations over the range of uncertainty in the pore pressures (±1 m) and depth of the slip surface (±1.5 m at the ends, ±0.6 m elsewhere) and determined that the factor of safety varied by only ±4%. The assumed values are consistent with the measured shear strength parameters for samples of silty clay from the basal slip surface of the landslide, which range from $c' = 1.9$ to 6.7 kPa and from $\phi' = 6°$ to 10.9° (Table 2). For the average values of these parameters, $c' = 4$ kPa and $\phi' = 8.9°$, the factor of safety is 0.98, practically the same as computed by Baum and Fleming (1991). If the slip surface is assumed to have no cohesion, then ϕ' must be between 11° and 11.4° for the factor of safety to be unity for pore pressures observed during rainy periods. The factor of safety is fairly sensitive to the strength parameters, for example, if $c' = 6.65$ kPa and $\phi' = 6°$, then $F = 0.82$ (Baum and Fleming, 1991).

Average strength for cross section B–B' was estimated (back-calculated) using the method of Fellenius (Lambe and Whitman, 1969). Depth and water pressures for section B–B' have been measured only near the intersection of Alani Drive and Paty Drive. We used depths shown in Figure 3B. Using measurements in borings 9, 22, I2, and I4 and the observed downward hydraulic gradient, we estimated that pressure head at the slip surface might range from 4.6 to 5.5 m beneath Paty Drive, 5.2 to 7.3 m beneath the downslope edge of Alani Drive, and from 5.5 to 7.6 m near mid-length of section B–B'. Assuming these water pressures and depths, and constant $\phi' = 12°$, c' ranges from 3.8 to 7.7 kPa, or assuming no cohesion, ϕ' ranges from 14° to 16.5°. The average strength computed for section B–B' is slightly greater than that computed for section A–A'. The greater strength is not surprising because the average slope of section B–B', about 12°, is 3° steeper than the average slope of section A–A', about 9°. However, insufficient data exist to determine whether the strength difference is real or is an artifact of assumptions about the thickness and ground-water pressures in section B–B'.

The pattern of longitudinal stretching and shortening revealed by mapping was used to infer the state of stress and constrain the interslice forces used in stability analysis (Baum and Fleming, 1991). The local style of deformation was determined from about 140 individual observations of stretching, shearing, or shortening at the ground surface of the landslide (Baum et al., 1989, plate 2). Shearing was observed mainly at the flanks of the landslide where walls, streets, and sidewalks are displaced 1–3 m laterally without much differential vertical movement. Damage to human-made structures in the head consists of stretching accompanied by sagging (partial collapse of structures into voids left by stretching). Typically, concrete garage slabs are pulled apart and the driveways sag. Sagging is particularly common in the head of the Alani element. In the downslope part of each element, deformation indicates shortening or compression. Driveways have been pushed into the street, or the garage and masonry walls between neighboring lots are buckled. Horizontal stresses should approach the minimum, or active, state in the upslope part of the landslide where longitudinal stretching has occurred (Figs. 4 and 21). Similarly, horizontal stresses should approach the maximum, or passive, state in the downslope two thirds of the landslide where longitudinal shortening has occurred. Between the zones of shortening and stretching were areas where the styles of deformation were mixed. This zone of mixed styles is called the neutral zone because it does not seem to be shortening or stretching significantly (Baum and Fleming, 1991, fig. 2).

Baum and Fleming (1991) used the active and passive states to estimate the permissible range of interslice forces in stability analysis of the landslide in limiting equilibrium (factor of safety equal to one). The shaded area in Figure 21 shows the permissible range of interslice forces assuming active stress in the upslope part, at rest stress in the neutral zone, and passive stress in the downslope part. For reference, the pair of dashed lines on the left side of Figure 21 indicates the range of interslice forces consistent with passive failure (shortening) in the zone of stretching and the dashed lines on the right indicate the range of interslice forces consistent with active failure (stretching) in the zone of shortening. Use of constant strength parameters in limit-equilibrium stability analysis of section A–A' resulted in interslice forces that exceeded permissible values in the upslope third of the landslide and near midlength. The interslice forces computed using constant strength parameters are consistent with longitudinal shortening over the entire length of the landslide (Fig. 21). Interslice forces computed for $c' = 0$ and $\phi' = 11°$ were practically the same as those shown in Figure 21 for $c' = 6.1$ kPa and $\phi' = 8°$. The interslice forces computed for a nonuniform distribution of strength were consistent with the permissible range based on the observed pattern of strain and earth-pressure theory. The analysis indicates strength on the upslope 53 m of the slip surface and 125–156 m from the head may be roughly 40%–60% higher than average and strength on the remainder of the slip surface roughly 25%–35% less than average (Baum and Fleming, 1991).

Excess driving force in the upslope half of the Alani-Paty landslide (which includes the stretching, neutral, and part of the shortening zones) evidently drives movement of the entire landslide; the downslope half is inherently stable and resists movement (Baum and Fleming, 1991). The boundary between the driving and resisting elements occurs at or near the thickest part of the landslide where the interslice forces reach their maximum (Fig. 21). The interslice forces computed using the observed pattern of strain and earth-pressure theory helped identify the driving and resisting elements of the landslide. Driving and resisting forces act in all parts of the landslide but in driving elements the shear stress caused by the weight of the landslide (the driving force) exceeds the shear strength available at the slip surface (resisting force). The opposite occurs in resisting elements. The longitudinal (interslice) force transfers the excess driving force from the driving elements to the resisting elements. Driving and resisting elements can be identified from a graph of longitudinal force; driving elements occur where the force increases in the downslope direction, and resisting elements occur where the force decreases (Fig. 21).

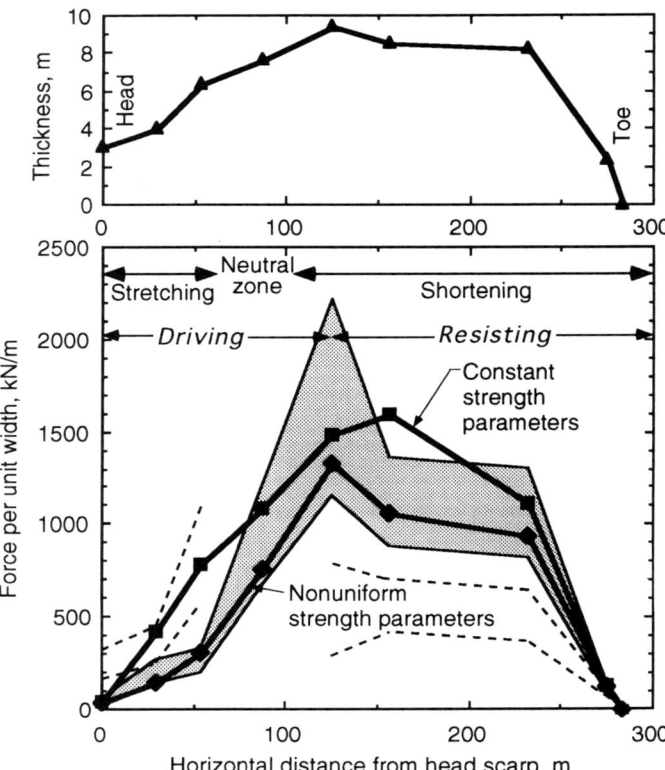

Figure 21. Longitudinal force per unit width computed for section A–A' of the landslide (adapted from Baum and Fleming, 1991, figs. 7 and 8). Thickness of the landslide is shown on the upper graph. Shaded area shows range of values computed using earth-pressure theory and assuming active failure in the zone of stretching, passive failure in the zone of shortening, and horizontal stresses consistent with a normally consolidated soil in the neutral zone. Heavy solid lines indicate values computed using limit-equilibrium stability analysis (Janbu, 1973) for a factor of safety of 1 and the average shear strength mobilized along the slip surface (constant strength parameters $c' = 6.1$ kPa, $\phi' = 8°$) and for a nonuniform distribution of shear strength also consistent with a factor of safety of 1. The nonuniform distribution assumes higher than average strength in the upslope quarter of the landslide and near mid length and lower than average strength elsewhere ($c' = 10.5$ kPa and $\phi' = 12°$ in the upslope 1/4 of the landslide, $c' = 10.5$ and $\phi' = 11°$ near mid length, and $c' = 6.7$ kPa and $\phi' = 6°$ elsewhere). The earth pressure computations used the same dimensional and water data as the stability analysis. A pair of dashed lines on the left side of the graph indicates the range of forces for passive failure in the zone being stretched; another pair on the right side shows the range of forces for active failure in the zone being shortened. The forces computed by stability analysis using constant strength parameters are too high in the zone of stretching and near mid length of the landslide. The forces computed by stability analysis using the nonuniform strength parameters are consistent with the range of forces estimated by earth pressure theory.

Absorbed storm water triggers movement of the Alani-Paty landslide. The landslide moved on a few occasions during the period of monitoring, but most of the time it was stationary (Table 3). The episodes of movement appear to have occurred after periods of intense rainfall during storms that lasted several days. The absorbed water affected the landslide by increasing subsurface water pressures in the upslope part of the landslide (Fig. 22) and by slightly increasing the driving force (the component of the weight acting parallel to the slip surface). However, the increased driving force is offset by the increased resistance due to normal component of the weight.

Elevated ground-water pressures help to trigger movement by decreasing the normal force across the basal slip surface of the landslide. Increases of water level/pressure head ranging from 0.3 to 3.3 m were observed in piezometers and pressure transducers in the upslope fourth of the landslide. We used measured water levels and pressures from various depths to estimate the pressure head that acted on the slip surface when the landslide was moving during wet periods and when it was stationary during dry periods (Figs. 22A, 22B). The shearing resistance of the slip surface decreases as the water pressure at the slip surface rises (Figs. 22B, 22C). The shearing resistance is related to the total normal stress according to Terzagh's principle of effective stress, and the Mohr-Coulomb failure criterion (Lambe and Whitman, 1969).

During dry periods movement stops except for a small amount of creep at the head of the Kahaloa element, so that the landslide appears to be near limiting equilibrium (analysis indicates the factor of safety is probably between 1.02 and 1.05). The landslide moves during rainy periods when water pressures increase in the head, causing local decreases in the shearing resistance that reduce the factor of safety slightly below 1. Available shearing resistance in the upslope part of the landslide decreases approximately 3%–15% locally during wet periods (Fig. 22C). The percent decrease is similar for both the distributions of shear strength assumed in Figure 22C; the decreases are as much as 25% locally if the slip surface has no cohesion. Shearing resistance in the downslope part of the landslide stays nearly constant because water pressures near the basal slip surface in the downslope part of the landslide appear to be steady throughout the year. The decreased shearing resistance in the upslope part of the landslide amounts to a 2% decrease in the total shearing resistance of the entire slip surface for the parameters assumed in Figure 22C. Repeating the calculations over the range of uncertainty in our water pressure data indicates that the decrease is between 2% and 5%. The total shearing resistance decreases by 7% if the slip surface has no cohesion, and water levels vary between the dry season lower bound and the wet season upper bound estimates (Fig. 22B). In any case, total resisting forces are only slightly greater than driving forces during dry periods so that small deformation, such as creeping movements observed at the head (Fig. 11), may be possible even when pore pressures are too low to cause significant movement of the landslide.

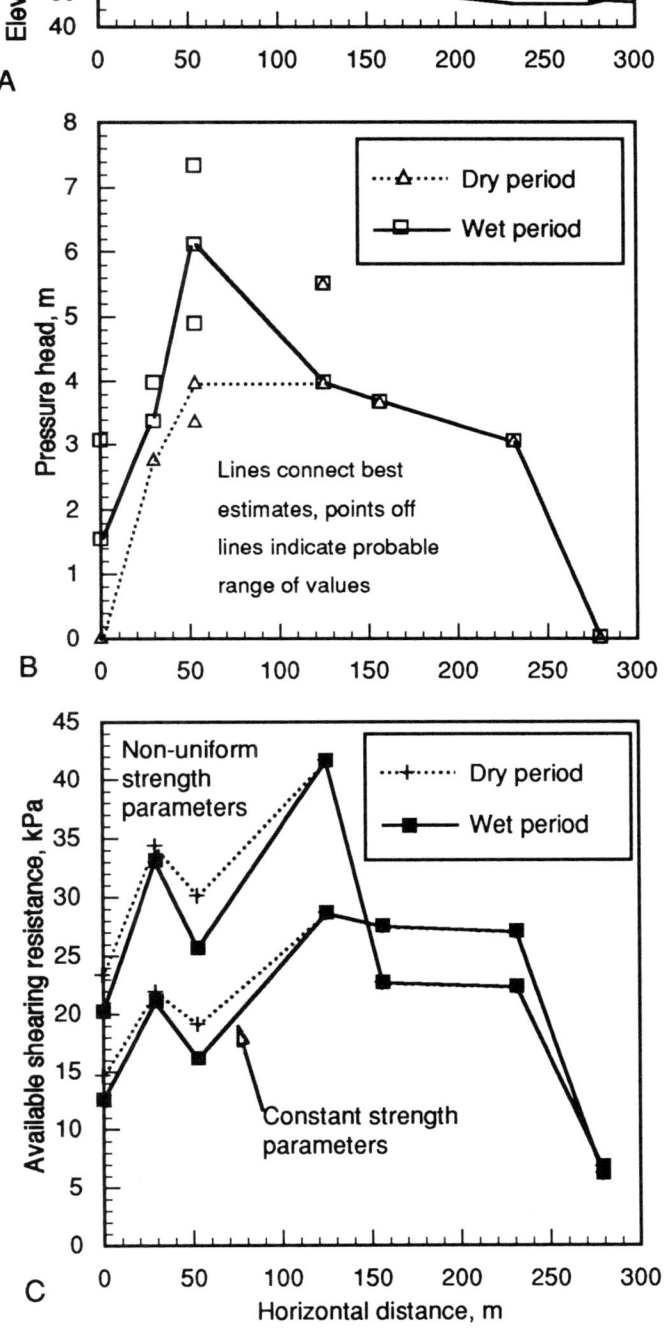

Figure 22. Estimated water pressures and shearing resistance at the basal slip surface for wet- and dry-period ground-water conditions in section A–A'. Lines connect best estimates; upper and lower limits indicate the range of uncertainty in the estimates. (A) Simplified cross section A–A' (B) Pressure head at the basal slip surface estimated using pressure heads from several depths in the boring or borings near the line of section. (C) Available shearing resistance of the basal slip surface at locations shown in B. The shearing resistance, τ, is determined by $\tau = c' + (\sigma_n - u) \tan \phi'$. The total normal stress, σ_n, is approximated by the component of the weight of overburden normal to the slip surface, $D\gamma_t \cos^2\theta$, where D is the vertical distance from the ground surface to the slip surface, γ_t is the unit weight of landslide material and θ is the slope of the slip surface. The following values were assumed for the parameters used in computing τ: $\gamma_t = 21.7$ kN/m^3; $u = \psi\gamma_w$, where ψ is the pressure head graphed in part B and γ_w is the unit weight of water, 9.8 kN/m^3; for average strength, $c' = 6.1$ kPa, $\phi' = 8°$; for nonuniform strength, $c' = 10.5$ kPa and $\phi' = 12°$ in the upslope quarter of the landslide, $c' = 10.5$ and $\phi' = 11°$ near midlength, and $c' = 6.7$ kPa and $\phi' = 6°$ in the downslope half of the landslide.

Water that fills tension cracks at the head of the landslide can increase driving force slightly. Water filling a 3-m-deep crack that terminates at the basal slip surface in the head of the landslide exerts a horizontal force per unit width of about 45 kN/m against the head of the landslide. The downslope driving force per unit width is about 7,300 kN/m for the Kahaloa element and 5,300 kN/m for the Alani element. Thus, water in the cracks can increase the driving force by about 0.6%–0.9%. The water can also reduce the amount of cohesion mobilized on parts of the slip surface that are shallower than the cracks (Lambe and Whitman, 1969); however, this reduction occurs on a minor part of the slip surface and is insignificant.

The weight of the water absorbed during storms increases the driving force insignificantly because the increased driving force is mostly offset by increased normal stress at the slip surface. The fillable porosity during the dry season is about 6.5%, and the pressure head in the upslope quarter of the landslide increases 1.5–3.0 m on average. If the entire landslide absorbs 0.15 m (0.065×2.3 m) of water during a wet period, the water would add 4.9×10^4 kN to the weight of the landslide. The weight of added rain would increase the driving forces that tend to destabilize the slope by about 7.6×10^3 kN or 0.9%, based on the average thickness of 7.6 m. However, the rain would also increase the total normal force by the same percentage and increase the resistance from 0% to 0.9% depending on the relative contributions of cohesion and friction to the shear strength.

Changes in ground water pressures in the upslope quarter of the landslide control movement of the entire landslide. Absorbed storm water triggers movement of the landslide mainly be reducing the shearing resistance of the upslope part of the slip surface. Increased pore pressures in the upslope part of the landslide reduce the shearing resistance of the upslope part by as much as 15% locally and the total shearing resistance over the entire slip surface by about 2%. Water filling cracks in the head of the landslide can augment the action of the increased pore pressures by increasing the driving force from 0.6% to 0.9%. The weight of the absorbed water has a negligible effect because most of the increased driving force is balanced by increased normal force acting at the slip surface.

DISCUSSION AND CONCLUSIONS

Debris-aprons on Oahu, Hawaii, like those in which the Alani-Paty landslide formed, have several features that make them prone to failure. The deposits, although gently sloping, are crudely stratified subparallel to the slope (Baum and Reid, 1992, fig. 7). In areas of Oahu where rainfall varies seasonally, smectite forms in the soils (Johnsson et al., 1991). Debris-aprons in these areas contain smectite-rich layers consisting of bouldery, highly plastic silty clay interfingered with weathered debris-flow deposits consisting of bouldery, plastic clayey silt. The smectite-rich layers appear to be pedogenically altered debris-flow deposits (buried soils). The smectite-rich layers are weak, (they appear to have properties similar to materials from the Alani-Paty landslide, Table 2) and have hydraulic conductivity 1–3 orders of magnitude lower than the silty layers. Thus, the debris-aprons contain weak layers parallel to the slope, and perched water tables can exist locally above the weak layers. Perching appears to be so widespread near the base of the Alani-Paty landslide that the slide stays mostly saturated perennially and ground-water flow is deflected downslope above the perching layer (Fig. 12). Perching of water on weak, slope-parallel layers makes the layers prone to failure.

Several other landslides on Oahu have features in common with the Alani-Paty landslide (Fig. 1). These have occurred in residential areas built on gently sloping debris aprons. Appearance and engineering properties (where known) of materials in the landslides are similar to those in the Alani-Paty landslide. Displacement is several meters or less, and some of the landslides are known to move slowly and episodically in response to rainfall (Peck and Wilson, 1968). Hydraulic conductivity of the Waiomao landslide is low but highly variable (Peck, 1968).

The Alani-Paty landslide moves as a coherent, deforming mass. Most deformation (and property damage) has occurred around the margins of the landslide, particularly the head and flanks of the landslide (Baum and others, 1989). Within the boundaries, deformation has consisted primarily of longitudinal stretching in the upslope third of the slide and longitudinal shortening in the downslope two-thirds (Baum and Fleming, 1991). Landslides in Ohio (Baum and Fleming, 1993; Fleming and Johnson, 1994), Utah (Fleming et al., 1993; Baum et al., 1993) and Colorado (Smith, 1993) have deformed similarly. The longitudinal deformation indicates that lateral earth pressures approach the minimum, or active, state in the zone of stretching and the maximum, or passive, state in the zone of shortening. Indeed, an internal toe, near mid-length of the slide, is consistent with passive failure (Fig. 4). Earth-pressure calculations based on these observations can be used to constrain the interslice forces in stability analysis (Fig. 21). The widespread occurrence of longitudinal deformation like that in the Alani-Paty landslide indicates that such calculation may be widely applicable and that field measurement of lateral stress distributions in active landslides merits considerable attention and effort. Improved understanding of the stress distributions could lead to improved design of remedial works.

The landslide moves when infiltration of storm water elevates pore pressures in the upslope quarter of the slide. Little water appears to come from bedrock upslope from the landslide. Tensiometric observations confirmed that infiltration alone, without contributions from bedrock, leaking pipes, or storm drains, was adequate to explain the increasing pressures. Ground-water pressures in the upslope quarter of the landslide increased 1–3 m during rainy periods, but water pressures in the downslope three-quarters of the slide remained relatively steady (Fig. 13). The main body of the landslide moved during extended rainy periods about 50–60 mm in January 1990 and 10–30 mm in March 1991 (Fig. 11, Table 3). No other changes that could have triggered movement were observed at these times, and so we attribute movement to increased ground-water pressures resulting from infiltration. Mechanical stability analysis indicates that the shearing resistance at the slip surface of the landslide decreases locally by 3%–15% and overall by only about 2% as a result of the increased ground-water pressure (Fig. 22). Thus a relatively small change from dry season ground-water conditions to wet-season conditions is sufficient to cause movement.

Ground-water response in the Alani-Paty landslide differs from response in other slow-moving landslides. For example, pore pressures near the basal slip surface of the Minor Creek landslide (about 6 m deep) increase about a month after the start of the rainy period, show little response to daily rainfall, and stay high throughout the wet season (Iverson and Major, 1987). At the Delhi-Pike landslide, the wetting front advances slowly (about 10^{-7} m/s) at the beginning of the wet season in November and early December (Haneberg, 1991b). After the entire 1-2 m thickness of material is nearly saturated, it stays uniformly moist for several months, and pore pressures respond in less than a day to rainfall. In contrast, shallow pore pressures in the upslope quarter of the Alani-Paty landslide increase in less than three days following the start of a wet period. Pressures stay elevated throughout the wet period and increase soon after intense rainfall. Shallow pressures increase within a few hours, and pressures near the basal slip surface increase slightly within 12 hours. These increases last about a day. Pressure increases that are great enough to trigger movement of the entire landslide result only from storms that last a week or more and have a peak value of the running ten-day-average rainfall greater than 25 mm/day (Fig. 11). In any case, pressures return to dry season levels in about two weeks following the end of a wet period. Pore pressures in the remainder of the Alani-Paty landslide change insignificantly throughout the year. The different pore-pressure response patterns in these three landslides appear to be related to differences in the depth of the basal slip surface, the magnitude and distribution of hydraulic conductivity, and the dry season moisture content of soils above the water table. The physically based pressure-diffusion model can explain these differing response patterns.

The landslide is near limiting equilibrium under wet-season ground-water conditions and residual strength. Limit-equilibrium stability analysis by the method of Janbu (1973), which uses ground-water pressures observed when the landslide was moving and measured thickness of the slide (Fig. 3A), indicates that the average shear strength parameters for the slip surface are about $\phi' = 8°$ and $c' = 6.1$ kPa or $\phi' = 11°$ with no cohesion. These parameters are consistent with measured residual strength parameters (Table 2).

In addition to conventional geotechnical methods, we used several other methods of data collection to enable us to make accurate analyses and to gain new insights about the mechanisms of landslides in clay-rich deposits. Our detailed mapping of damage to human-made structures (streets, curbs, houses, walls, etc.) was essential for locating bounding structures (scarps, toes, and flanks) of the landslide and determining the pattern of internal deformation (Baum et al., 1989). Displacement vectors obtained by applying photogrammetric methods to successive sets of historic aerial photographs confirm much of the pattern of deformation identified by the detailed mapping. Monitoring of ground-water conditions using 70 piezometers and pore-pressure transducers made it possible to recognize the patterns of ground-water flow and pressure response (Table 1, Figs. 12, 13). Monitoring only a few piezometers would not have allowed the accurate determination of the mechanisms for water entering the landslide and of the pore pressure conditions required to trigger movement. Monitoring of tensiometers equipped with differential pressure transducers revealed useful detail about infiltration and near surface build-up of positive pressure in landslide materials (Fig. 17). Finally, we used simple, physically based models to understand and quantify the hydrological and mechanical processes affecting the Alani-Paty landslide.

Although we lack detailed records or eye-witness accounts of the formation of the landslide, several observations suggest how such landslides form. Analysis of longitudinal strain and forces indicates that the upslope half of the landslide drives movement, and analysis of rainfall, landslide-movement, and water-level data indicates that water-level fluctuations in the upslope quarter of the landslide trigger movement of the entire landslide. Failure probably started somewhere in the upslope part of the landslide and then progressively spread downslope. Growth of the failure surface, like movement, was probably episodic over a period of many years. This kind of progressive failure is consistent with the observed enlargement of the landslide. The Alani element was well-developed before enlargement became evident in the ground downslope from it (area B, Fig. 4); thus, failure started in the steeper, upslope part of the debris apron and spread downslope. Detailed mapping in the vicinity of the landslide and throughout Manoa Valley has revealed incipient landslide scarps and toes around the margins of the Alani-Paty and Hulu-Woolsey landslides (Baum et al., 1989; Baum and Reid, 1992; R. W. Fleming, unpub. data, 1991). Detailed monitoring in areas of incipient landslide features should identify areas subject to accelerated movements. Remedial measures could then be performed in areas so identified before they enlarged to unmanageable sizes or caused significant damage.

ACKNOWLEDGMENTS

We thank C. Michael Street and his staff at the Department of Public Works, City and County of Honolulu, for coordinating access to sites for drilling and instrumentation. We also thank residents and owners of property in the landslide area for their cooperation in providing access to sites for drilling and instrumentation of the landslide. Geolabs-Hawaii performed drilling and sampling, and STV/Lyon Associates shared field measurements. Bob Fleming, Steven Ellen, Steve Spengler, Jill Torikai, Lori Lui, and Cynthia Wilburn helped with field work. James Messerich performed the photogrammetric measurements.

REFERENCES CITED

American Society for Testing and Materials, 1990, Annual Book of ASTM Standards: Philadelphia, American Society for Testing and Materials, v. 4.08, 1089 p.

Bates, R. L., and Jackson, J. A., editors, 1987, Glossary of geology (third edition): Falls Church, Virginia, American Geological Institute, p. 723.

Baum, R. L., and Fleming, R. W., 1991, Use of longitudinal strain in identifying driving and resisting elements of landslides: Geological Society of America Bulletin, v. 103, p. 1121–1132.

Baum, R. L., and Fleming, R. W., 1993, Patterns of movement in clay-rich landslides [abs.]: Eos, American Geophysical Union Transactions, v. 74, p. 300.

Baum, R. L., and Reid, M. E., 1992, Geology, hydrology and mechanics of the Alani-Paty landslide, Manoa Valley, Oahu, Hawaii: U.S. Geological Survey Open-File Report 92-501, 87 p., 6 oversize plates.

Baum, R. L., Fleming, R. W., and Ellen, S. D., 1989, Maps showing landslide features and related ground deformation in the Woodlawn area of the Manoa Valley, City and County of Honolulu, Hawaii: U.S. Geological Survey Open-File Report 89-290, 16 p., 2 oversize plates.

Baum, R. L., Spengler, S. R., Torikai, J. D., and Liu, L. A. S. M., 1990, Summary of geotechnical and hydrologic data collected through April 30, 1990, for the Alani-Paty landslide, Manoa Valley, Honolulu, Hawaii: U.S. Geological Survey Open-File Report 90-531, 67 p.

Baum, R. L., Reid, M. E., Wilburn, C. A., and Torikai, J. D., 1991, Summary of geotechnical and hydrologic data collected from May 1, 1990, through April 30, 1991, for the Alani-Paty landslide, Manoa Valley, Honolulu, Hawaii: U.S. Geological Survey Open-File Report 91-598, 102 p.

Baum, R. L., Fleming, R. W., and Johnson, A. M., 1993, Kinematics of the Aspen Grove landslide, Ephraim Canyon, central Utah, in Landslide processes in Utah—Observation and theory: U.S. Geological Survey Bulletin 1842, p. F1–F34.

Bouwer, H., 1978, Groundwater hydrology: New York, McGraw-Hill, 480 p.

Bouwer, H. and Rice, R. C., 1984, Hydraulic properties of stony vadose zones: Ground Water, v. 22, p. 696–705.

Carslaw, H. S. and Jaeger, J. C., 1959, Conduction of heat in solids: New York, Oxford University Press, 510 p.

City and County of Honolulu, 1990, Surveys of Manoa slide [unpublished]: Department of Public Works, City and County of Honolulu, Hawaii, June 1989–February 1990.

El-Kadi, A. I., 1987, SOIL—A computer program to estimate the parameters of soil hydraulic properties: Indiana, International Ground Water Modeling Center, Holcomb Research Institute, 47 p.

Fleming, R. W., and Johnson, A. M., 1989, Structures associated with strike-slip faults that bound landslide elements: Engineering Geology, v. 27, p. 39–114.

Fleming, R. W., and Johnson, A. M., 1994, Landslides in colluvium, in Landslides of the Cincinnati, Ohio, area: U.S. Geological Survey Bulletin 2059, p. B1–B24.

Fleming, R. W., Baum, R. L., and Johnson, A. M., 1993, Deformation of landslide surfaces as indicators of movement processes, in Proceedings, Seminar on landslide hazards, 2nd, Cosenza, Italy, March 5–6, 1990: Geografia Fisica e Dinamica Quaternaria, v. 16, p. 9–11.

Freeze, R. A. and Cherry, J. A., 1979, Groundwater: Englewood Cliffs, New Jersey, Prentice-Hall, 604 p.

Geolabs-Hawaii, 1984, Soil engineering investigation report for the study of earth movement in the vicinity of Kahaloa Drive, Kahaloa Place, Hulu Place, Woolsey Place, and Lanikaula Street, Manoa Valley, Honolulu, Hawaii [unpublished]: Report to the Department of Public Works, City and County of Honolulu, Hawaii, April, 23 p., appendices, 11 plates.

Geolabs-Hawaii, 1985, Geotechnical engineering investigation, remedial work for earth movement areas, vicinity of Hulu and Woolsey Place, Manoa Valley, Oahu, Hawaii [unpublished]: Report to the Department of Public Works, City and County of Honolulu, Hawaii, September, 19 p., appendices, 11 plates.

Geolabs-Hawaii, 1988, Field data report no. 1 earth movements in the Woodlawn area, Manoa Valley, Honolulu, Oahu, Hawaii [unpublished]: Report to the Department of Public Works, City and County of Honolulu, Hawaii, September, 2 p., appendices.

Geolabs-Hawaii, 1990, Field data report, earth movements in the Woodlawn area, Manoa Valley, Honolulu, Oahu, Hawaii [unpublished]: Report to the Department of Public Works, City and County of Honolulu, Hawaii.

Giambelluca, T. W., 1986, Land-use effects on the water balance of a tropical island: National Geographic Research, v. 2, p. 125–151.

Giambelluca, T. W., Lau, L. S., Fok, Y. S., and Schroeder, T. A., 1984, Rainfall frequency study for Oahu: Honolulu, Hawaii, State of Hawaii, Department of Land and Natural Resources, Report R73, 34 p.

Giambelluca, T. W., Nullet, M. A., and Schroeder, T. A., 1986, Rainfall atlas of Hawaii: Honolulu, Hawaii, State of Hawaii, Department of Land and Natural Resources, Report R76, 267 p.

Haneberg, W. C., 1991a, Pore pressure diffusion and the hydrologic response of nearly saturated, thin landslide deposits to rainfall: Journal of Geology, v. 99, p. 886–892.

Haneberg, W. C., 1991b, Observation and analysis of pore pressure fluctuations in a thin colluvium landslide complex near Cincinnati, Ohio: Engineering Geology, v. 31, p. 159–184.

Iverson, R. M., and Major, J. J., 1987, Rainfall, ground-water flow, and seasonal movement at Minor Creek landslide, northwestern California: Physical interpretation of empirical relations: Geological Society of America Bulletin, v. 99, p. 579–594.

Janbu, N., 1973, Slope stability calculations, in Hirschfield, R. C., and Poulos, S. J., eds., Embankment-dam engineering, Casagrande volume: New York, Wiley, p. 47–86.

Johnsson, M. J., Ellen, S. D., McKittrick, M. A., 1991, Clay mineralogy of soils in the southeastern Koolau Mountains, Oahu, Hawaii—Effects of varying chemical weathering intensity and impact on slope processes: Geological Society of America Abstracts with Programs, v. 23, no. 5, p. A71.

Lambe, T. W., and Whitman, R. V., 1969, Soil mechanics: New York, Wiley, 553 p.

Langenheim, V. A. M., and Clague, D. A., 1987, The Hawaiian-Emperor volcanic chain, Part II: Stratigraphic framework of volcanic rocks of the Hawaiian Islands, in Decker, R. W., Wright, T. L., and Stauffer, P. H., eds., Volcanism in Hawaii: U.S. Geological Survey Professional Paper 1350, p. 55–84.

MacDonald, G. A., Abbott, A. T., and Peterson, F. L., 1983, Volcanoes in the sea: Honolulu, University of Hawaii Press, 517 p.

Peck, R. B., 1959, Report on causes and remedial measures Waiomao Slide, Honolulu [unpublished]: Report to the City and County of Honolulu, Hawaii, 31 p.

Peck, R. B., 1967, Stability of natural slopes, in Proceedings, American Society of Civil Engineers: Journal of the Soil Mechanics and Foundation Division v. 93, p. 403–417.

Peck, R. B., 1968, Supplementary report on remedial measures Waiomao slide, Honolulu [unpublished]: Report to the City and County of Honolulu, Hawaii, 35 p.

Peck, R. B., and Wilson, S. D., 1968, The Hind Iuka landslide and similar movements Honolulu, Hawaii [unpublished]: Report to the City and County of Honolulu, Hawaii. 21 p.

Peterson, D. M., Ellen, S. D., and Knifong, D. L., 1993, Distribution of past debris flows and other rapid slope movements from natural hillslopes in the Honolulu District of Oahu, Hawaii: U.S. Geological Survey Open-File Report 93-514, 32 p., 2 plates, scale 1:24,000.

R. M. Towill Corporation, 1989, Aerial topographic map of portion of Manoa Valley: Honolulu, Hawaii, R. M. Towill Corporation, scale 1:1,200.

Smith, W. K., 1993, Photogrammetric determination of movement on the Slumgullion Slide, Hinsdale County, Colorado, 1985–1990: U.S. Geological Survey Open-File Report 93-597.

Stearns, H. T., 1939, Geologic map and guide to the island of Oahu, Hawaii: Territory of Hawaii, Division of Hydrography, Bulletin 2, 75 p., scale 1:62,500.

STV/Lyon Associates, 1990, Woodlawn area earth stabilization, Manoa Valley, Honolulu, Oahu, Hawaii, Landslides and Facilities Report [unpublished]: Report to the Department of Public Works, City and County of Honolulu, Hawaii, June, 2 volumes, appendices.

STV/Lyon Associates, 1991, Graphs of inclinometer and pore-pressure transducer data for the Paty-Alani landslide, Manoa valley, Honolulu, Oahu, Hawaii [unpublished]: Department of Public Works, City and County of Honolulu, Hawaii, various dates.

Todd, D. K., 1980, Groundwater hydrology (second edition): New York, John Wiley and Sons, 535 p.

Torikai, J. D., 1992, Rainfall infiltration and pore-pressure response in the active Alani-Paty landslide, Manoa Valley, Honolulu, Hawaii [M.S. thesis]: Honolulu, Hawaii, University of Hawaii, 122 p.

van Genuchten, M. T., 1980, A closed form equation for predicting the hydraulic conductivity of unsaturated soils: Soil Science Society of America Journal, v. 44, p. 892–898.

Walter Lum Associates, Inc., 1979, Waiomao slide area remedial work phase II, subsurface investigation Palolo Valley, Honolulu, Hawaii [unpublished]: Report to the city and county of Honolulu, Hawaii, 21 p.

Wentworth, C. K., 1951, Geology and ground-water resources of the Honolulu–Pearl Harbor area, Oahu, Hawaii: Honolulu, Hawaii, City and County of Honolulu, Board of Water Supply, 111 p.

William Cotton and Associates, 1992, Logs of trenches and large diameter borings [unpublished]: Department of Public Works, City and County of Honolulu, Hawaii, and William Cotton and Associates, Los Gatos, California.

Manuscript Accepted by the Society May 20, 1994

Landslides on clay and shale hillslopes in Tuscany, Italy

R. Bertocci, P. Canuti, and N. Casagli
Earth Science Department, University of Florence, Via G. La Pira, 4, 50121-Firenze, Italy
C. A. Garzonio
Urban and Regional Planning Department, University of Florence, Via Micheli, 7, 50121-Firenze, Italy
P. Vannocci
Earth Science Department, University of Florence, Via G. La Pira, 4, 50121-Firenze, Italy

ABSTRACT

This paper presents the work currently underway at the Earth Science Department of the University of Florence on the setting up of a landslide data base and information system for the Tuscan Region of Italy, with particular reference to the clay and shale slopes. Some examples of preliminary processing of the collected data are given. Landslides in Tuscany are mainly controlled by the lithological features of the soils and by the intensity and frequency of extreme rainfalls. Index properties can be used for the characterization of soils in different sedimentary basins and can be employed for the estimation of the intrinsic strength parameters, such as the residual and the critical state (fully-softened) friction angles. The combination of geomorphological and geotechnical parameters gives information on the causes and the evolution of slope movements. Two examples are given in this paper. The first deals with slope failures in marine sediments at a local scale; the comparison between the height and the inclination of the slopes that are affected by active or stabilized landslides can be used as back analysis providing that reasonable assumptions on pore water pressures are made. The second example consists of a regional analysis of mudslides. Two-dimensional and three-dimensional analyses are compared in order to explain the causes of the movements. The shorter mudslides on gentler slopes can be caused by undrained loading mechanisms. The interpretation of the longer mudslides is more difficult because they often occur at low slope angle, requiring high pore water pressure to explain the movement. In these cases, paleoclimatic factors probably need to be taken into account.

INTRODUCTION

Studies of landslides on the argillaceous hilly slopes of Tuscany have been going on for some years now at the Earth Science Department of the University of Florence: in the 1970s as part of the National Research Council Project "Soil Preservation" and from 1985 on with the National Group for the Prevention of Hydrogeological Disasters (GNDCI) in collaboration with the Civil Protection Department of the Italian Government. One of the four research lines of the GNDCI deals with the "forecast and prevention of high risk slope movements" and its main research project is the study of unstable built-up areas (SCAI Special Project). It involves more than 40 operative units all over Italy and includes among its tasks the inventory of the slope instability phenomena and the assessment of hazard conditions in towns and populated areas. In many cases, the towns date back to the medieval era and the Renaissance.

Tuscany is one of the Italian regions most subject to landslides. A recent report by the Ministry of Public Works (Catenacci, 1992) states that about 0.3% of the total regional surface area is involved in mass movements, for a total area of about 68.72 km², and these figures probably underestimate the real situation. In the course of this century landslides in Tuscany have caused about 50 deaths, more than 100 casualties and have left more than 1,000 people homeless.

According to the State Law 445/1908 and subsequent integrations, the land beneath 44 towns was classified as unstable, and funds were allocated to reinforce or transfer them. As part of the SCAI Project, the GNDCI Operative Units 2.14 (Florence) and 2.12 (Pisa) are working on the systematic collection of information on the geomorphological and geotechnical characteristics of landslides. This inventory, which is still being compiled, indicates that the total number of landslides affecting developed areas amounts to about 160, a figure that highlights a much more serious situation than the one estimated by the Ministry of Public Works. Moreover, so far more than 600 major slope movements affecting only marginally built-up areas but involving infrastructures, agricultural and industrial zones, have already been recorded at a regional scale. In this phase, the task of the Earth Science Department of Florence is to collect data which will then be fed into a geographical information system to be used for civil protection purposes.

This paper describes the work currently underway to develop criteria for data recording and on the specific slope instability problems of Tuscany. Therefore, it gives some examples of data processing, and it shows how the methods of engineering geology and geotechnics can be used to forecast regional patterns of slope instability.

GEOMORPHOLOGICAL AND GEOLOGICAL OUTLINE

Tuscany is placed on the western side of the Northern Apennine mountain belt. The belt consists of a sedimentary complex built up during Tertiary time by the thrusting of tectonic units of a western oceanic domain (Ligurian units) on top of the African continental margin (Tuscan units). This thrust system was subsequently affected by a phase of extensional tectonics from the late Tortonian on. This latter phase produced a horst and graben structure, with a northwest-southeast trend. The main grabens were filled with sediments of the Neogene and Quaternary marine and fluvio-lacustrine cycles (Fig. 1). The geomorphic features of the region closely reflect its geological structure: on a regional scale the landscape is characterized by ridges with a northwest-southeast trend, made up of Mesozoic and Tertiary units with folded structures, which are separated by basins developed on the main grabens.

Slope instability phenomena are mainly associated with the pelitic sediments, which can be divided into four main groups:

(1) Plio-Pleistocene fluvio-lacustrine clays: fine-grained sediments consisting of slightly over-consolidated clays and silts with sandy partings and peat (Ambrosetti et al., 1978; Bossio et al., 1994). They constitute the typical deposits of the eastern tectonic basins of Tuscany, not affected by the Neogene marine transgressions (Mugello, Casentino, Valdarno). Many of these basins were subject to a phase of tectonic uplift from the late Pleistocene on, which caused intense fluvial erosion, steepening of the slopes and, consequently, slope instability problems.

(2) Mio-Pliocene marine clays: made up of fissured over-consolidated clays, silts and marly clays of the Pliocene and Miocene marine cycles, sometimes with sandy lenses (Ambrosetti et al., 1978; Bossio et al., 1994). They outcrop in the western basins of Tuscany (Volterra, Elsa, Siena, Radicofani). Miocene sediments outcrop only over limited areas and most of these basins are filled with sediments of the Pliocene marine transgression. These deposits are made up mainly of blue clays that gradually shade into the yellow sands of the regressive sequence. The slopes on blue clays are particularly subject to surface mass movements and strong erosive processes (badlands and *biancane*) caused by fluvial erosion and often accelerated by poor agricultural practices.

(3) Argillaceous mélanges (Cretaceous-Eocene, belonging to the Ligurian units): made up of clay shales with chaotic structure caused by an intense tectonization (tectonites or tectonic mélanges) or by submarine gravitational processes (olistostromes, sedimentary mélanges, sedimentary breccias). They outcrop in mountain zones and are commonly formed by the assemblage of rock blocks (sandstones, ophiolites, calcarenites, limestones) embedded in a matrix of sheared clay shales (Abbate and Sagri, 1970; Abbate et al., 1970; Bettelli and Panini, 1987; Bortolotti, 1992). These materials are well known in the Apennines under the informal names of "chaotic complex" or *"Argille scagliose"* (Merla, 1952). The geomechanical behavior of these formations is complex and dependent on the relative proportions between strong rock and weak matrix. However, in the most typical outcrops in the mountain range of the Apennines, the clay shale component is predominant, hence they can be analyzed by a classical soil mechanics approach.

(4) Clay shales, siltstones and marls (mainly of the Cretaceous-Eocene systems, belonging both to Ligurian and Tuscan units): composed of pelitic oceanic sediments, involved in the Apennine orogenesis, disrupted to varying extents by tectonics. They are usually made up of alternating clay shales, siltstones and marlstones with intercalations of calcarenites, sandstones and limestones (Bortolotti et al., 1970; Abbate and Sagri, 1970, Bortolotti, 1992). They outcrop in the mountain range of the Apennines and are usually overlain with thick sandstones or limestones. Their geomechanical behavior lies in an intermediate field between partially lithified cohesive soils and jointed rocks.

LANDSLIDE iNFORMATION SYSTEM

The Landslide Information System currently being set up at the Earth Science Department at the University of Florence

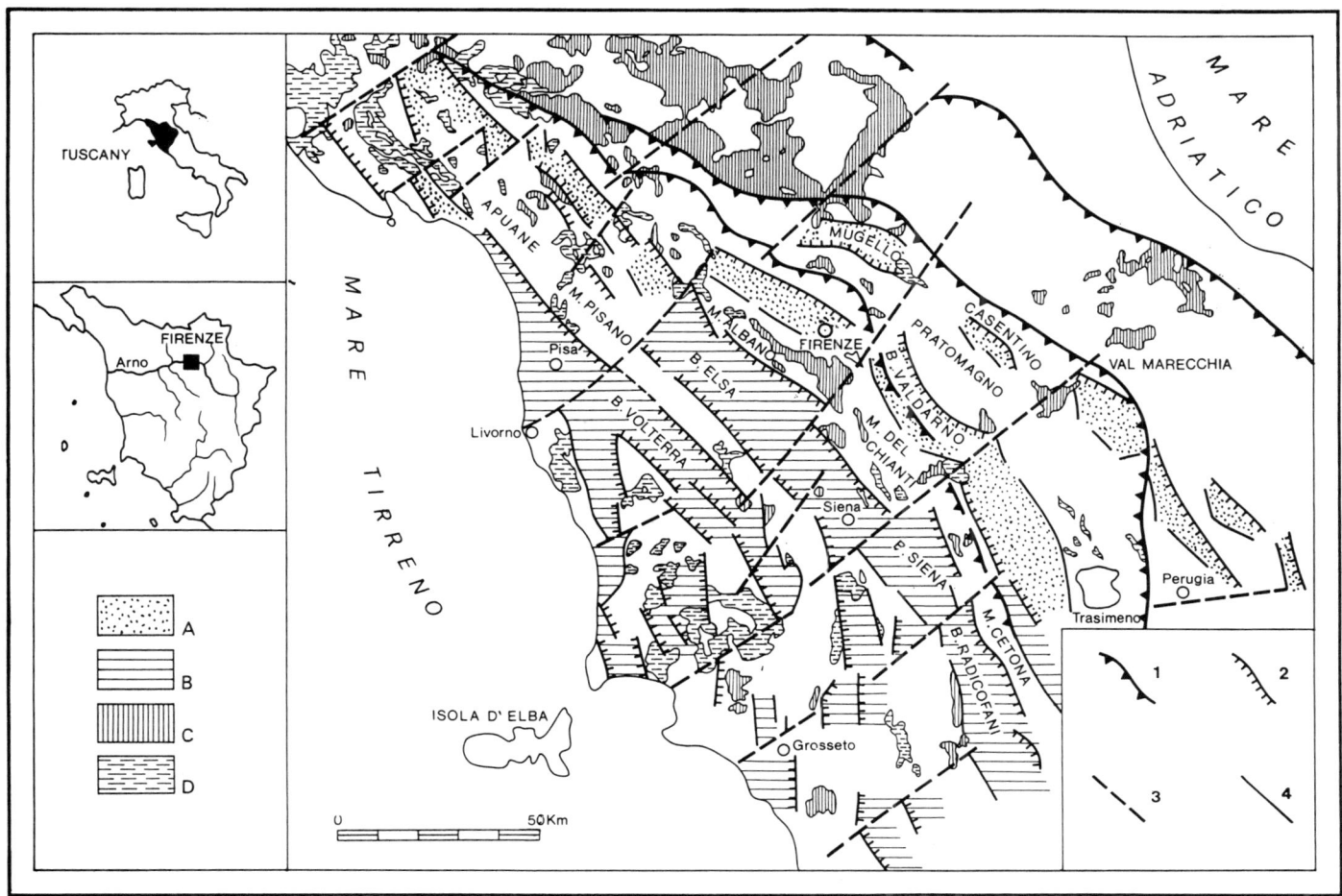

Figure 1. Geological-structural sketch of the Tuscan Apennines. (a) Plio-Pleistocene fluvio-lacustrine basins; (b) Mio-Pliocene marine basins; (c) main outcrops of shale mélanges; (d) Main outcrops of clay shales, siltstones and marls. (1) Main thrust; (2) main faults at the borders of grabens; (3) transverse tectonic lines; (4) minor faults. (Modified from Bossio et al., 1994).

is composed of two parts, which correspond to the causes and the effects of the landslides respectively:

(1) A geographic information system with a series of thematic maps, both in raster and vector format, on the main factors causing slope instability: topography, lithology, land use, climatic, and seismic factors.

(2) A data-bank of the geological, geomorphological, morphometrical, geotechnical, and historical characteristics of the landslides.

Within the context of the national SCAI Special Project, particular care was given to landslides affecting developed areas, as they pose a higher risk than those affecting undeveloped areas. Some of the slope instability scenarios in the Tuscan historical towns are depicted in Figure 2. Most of them are linked to landslides in argillaceous materials or in sandy or rocky layers overlying clays or clay shales.

LANDSLIDE DISTRIBUTION AND CAUSES

The distribution of the slope instability phenomena is dictated essentially by the lithological characteristics. Figure 3 shows contours of frequency of the major landslides for the River Arno drainage basin, the main watershed of Tuscany, which covers a total area of about 8,830 km^2. The greatest concentrations of slope failures are located over the terrains with a higher clay fraction. Therefore, lithological and geotechnical information provide the best tools for the spatial prediction of landslide danger.

Most of the slope instability phenomena are linked to the heavy rainfall that occurs in the months of November and December (Fig. 4). The maximum annual rainfalls for one to five consecutive day durations, published yearly by the National Hydrological Survey, are probably the best data avail-

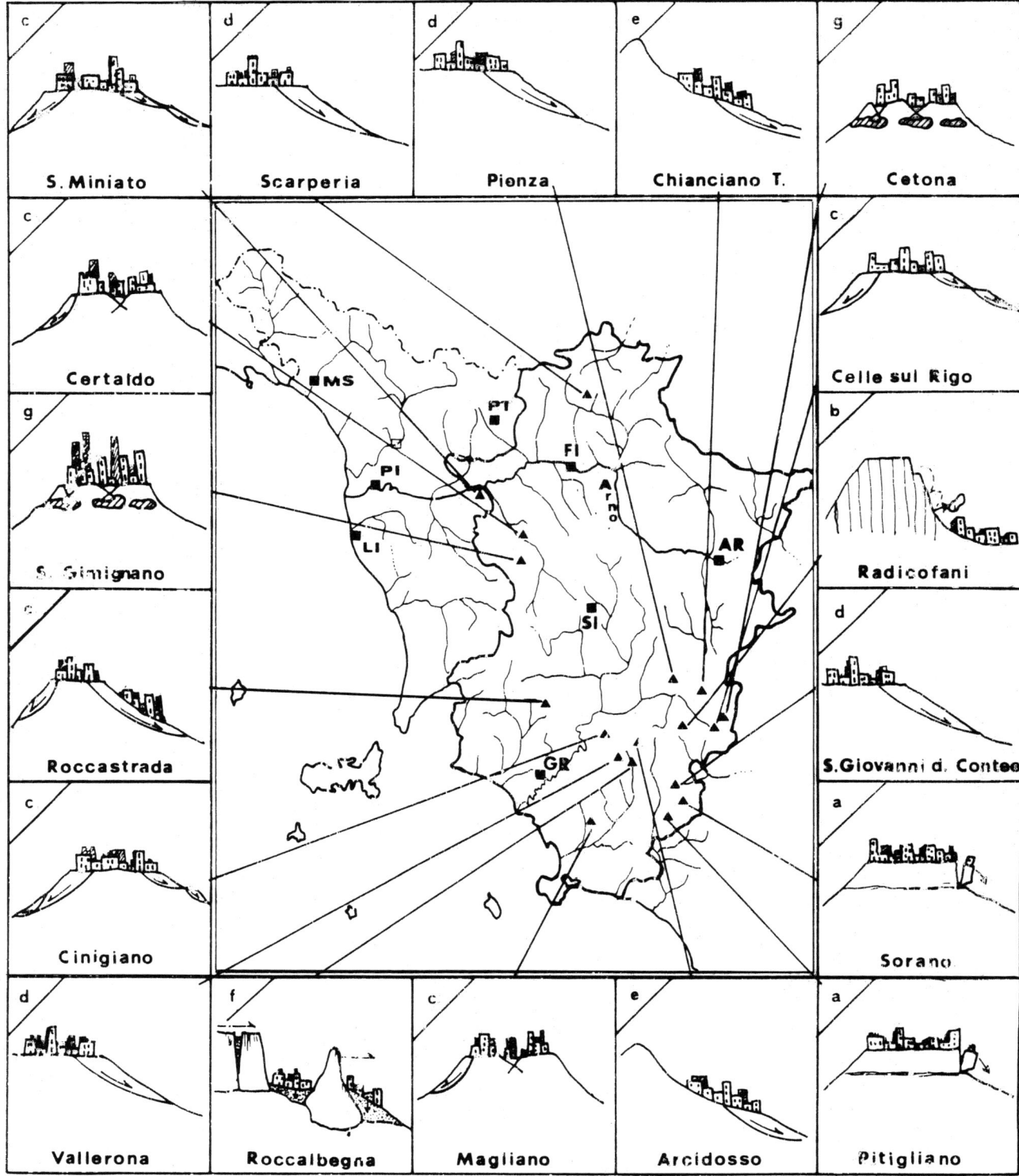

Figure 2. Main unstable built-up areas in eastern Tuscany, studied by the SCAI operative unit of Florence. (a) Hilltop town with slopes affected by rockfalls and toppling failures; (b) town at the base of a steep slope subject to rockfalls; (c) hilltop town with slopes affected by landslides or mudslides; (d) town at the border of a mesa subject to landslide; (e) hillslope town affected by landslides or mudslides; (f) town on unstable slopes for deep-seated complex mass movements; (g) hilltop town unstable for collapse of underground cavities.

able for hazard forecasting on a regional scale. For each rainfall gauge of the hydrologic network in the region, these data are elaborated statistically by means of Gumbel's law for extreme-value series (Fig. 5) according to the methodology presented by Benedetti et al. (1979). The heavy rainfall values for different durations with a given return period can be represented in cartographic form, as shown in Figure 6, and included in a data base linked to a geographic information system. Discrepancies between the map of extreme rainfall (Fig. 6) and the map of landslide distribution (Fig. 3) suggest that lithology has a stronger effect on the spatial distribution of the slope instability than does rainfall.

A large number of the surveyed landslides originated during wetter climatic conditions during late Pleistocene and Quaternary time. In the coldest periods, the mountain range of the Apennines was characterized by a very wet climate and, at the highest altitudes, even by periglacial conditions, as can be seen from the presence of stratified slope waste deposits (*grèzes litées*) and other periglacial features (Carton, 1990; Canuti, 1993).

Another important factor that caused a large number of ancient landslides is the intense tectonic uplift that the entire Apennine chain has been subject to since the early Pleistocene, with a maximum intensity in the period up to late Pleistocene time (Ambrosetti et al., 1978). This uplift was responsible for the closure of the main marine and fluvio-lacustrine basins and caused a deepening of the drainage pattern, thus triggering slope instability phenomena. Most of these ancient landslides can be reactivated, totally or only in part, during heavy rainfalls in present-day conditions. The detection of these reactivated slope movements is of crucial importance for their interpretation and analysis.

An example of mudslide reactivated by human activity and by heavy rainfalls is present at San Marcello Pistoiese in the Apennines northwest of Florence. Constructions were made in the 1970s on the upper part of the track of an ancient mudslide on clay shale mélanges. The constructions were allowed by the lack of adequate laws on ground investigations. With the heavy rainfalls of the winter 1976–1977 the mudslide was reactivated (Fig. 7). Tension cracks were observed after a rainfall of 200 mm in 10 days. The cracking stopped for about two weeks following a relatively dry period. Another rainfall of more than 200 mm in 10 days, during the first days of 1977, caused the reactivation of cracking and the start of a slow slid-

Figure 3. Distributions of major landslides in the River Arno drainage basin (upslope of Pontedera). Contours are drawn considering the number of major landslides (from the data bank of the Earth Science Department of Florence) in circles with areas equal to 1/100 the extension of the whole basin.

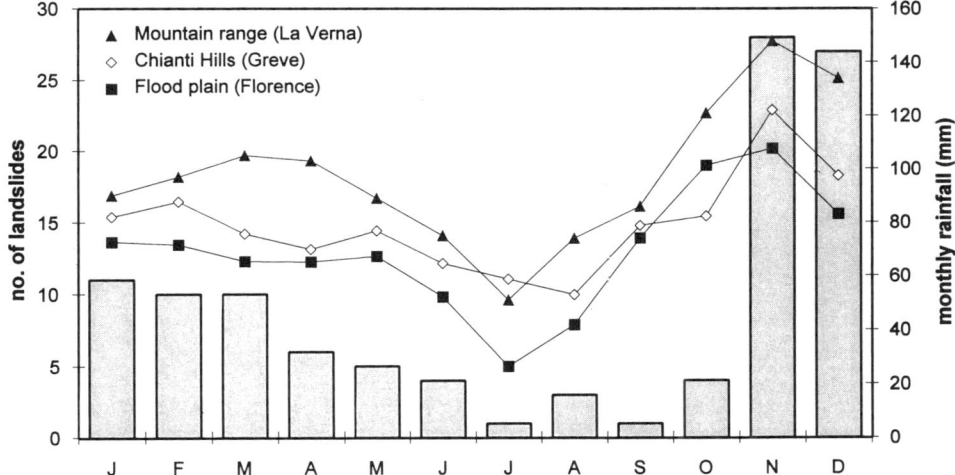

Figure 4. Number of landslides (bars) in Tuscany (1900–1992) compared with the monthly rainfalls (lines) (period 1957–1986) in different physiographic environments.

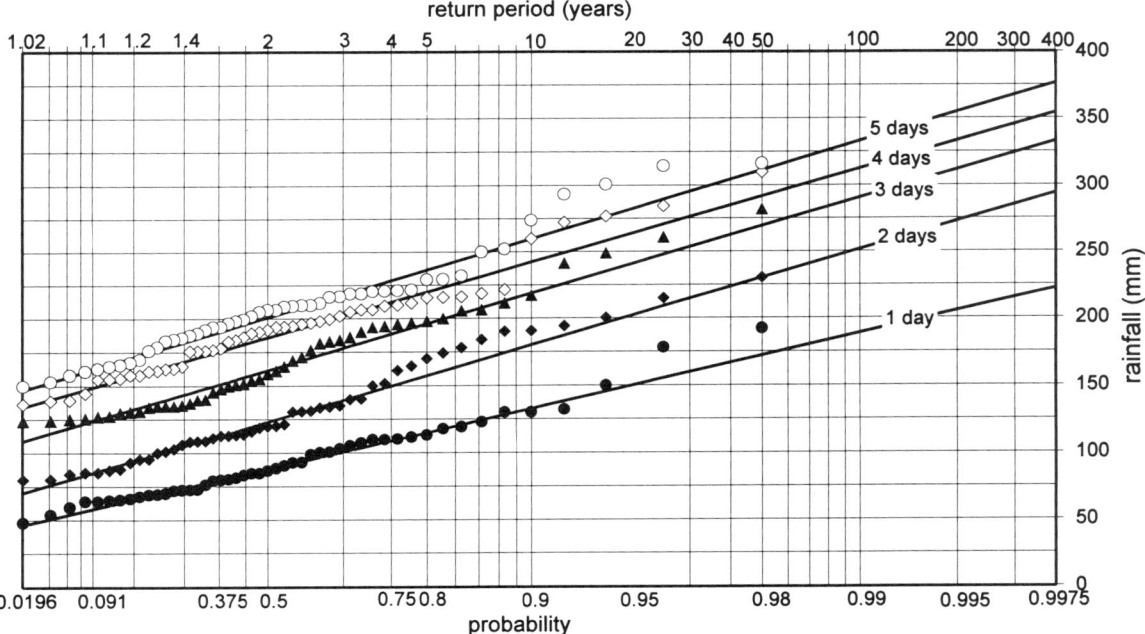

Figure 5. Maximum rainfall with a one- to five-day duration in the period 1957–1986 at the San Marcello Pistoiese rain gauge, represented in Gumbel's probability chart.

ing downslope. A rainfall of 150 mm in 10 days was sufficient to cause a sudden acceleration of the movement on February 17, 1977 causing the destruction of buildings.

With reference to Figure 5, the one- to five-day rainfalls in the period have a return period of about 1.4 years. However, the accumulated rainfall for the months of December and January is 763 mm; according to the Gumbel's law, it is associated with a return period of 10 years. Therefore, it can be argued that slides in low permeability soils are influenced by heavy rainfalls over long span of times (some months). Within these rainy periods, short (some days) concentrated rainfalls are sufficient to trigger the movements.

GEOTECHNICAL PROPERTIES OF CLAYS AND SHALES

The geotechnical characterization of the fine-grained soils, which causes most of the instability phenomena, is of prime importance for the understanding of the mechanics of the slope movements.

In setting up the data base, information on the geotechnical properties of the terrains, drawn mainly from research work carried out at the Earth Science Department at the University of Florence, was systematically recorded. Despite the relatively small amount of data with respect to the geological complexity and the extent of area under investigation, some simple conclusions can be still drawn.

Index properties

A large amount of information is available on the index properties (Atterberg limits, grain-size distribution) of clays in different localities of the region. Plasticity charts of silty-clayey soils belonging to different geological environments are shown in Figure 8.

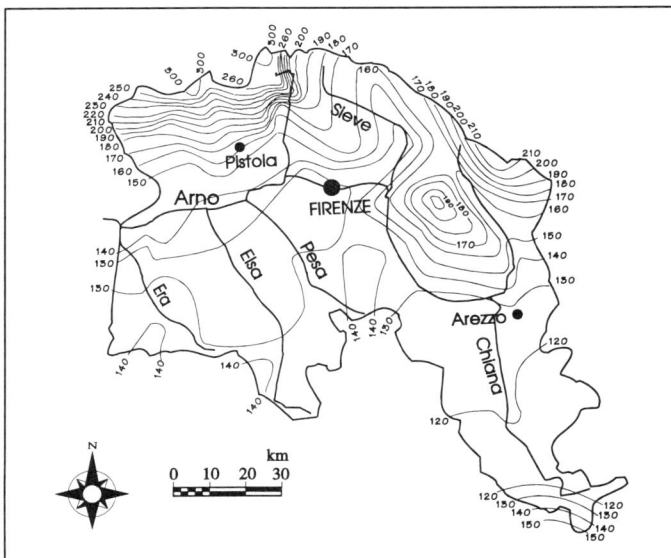

Figure 6. Cartographic representation of the forecast of the maximum heavy rainfalls, with a five-day duration and a 10-year return period in the River Arno drainage basin (after Canuti, 1985).

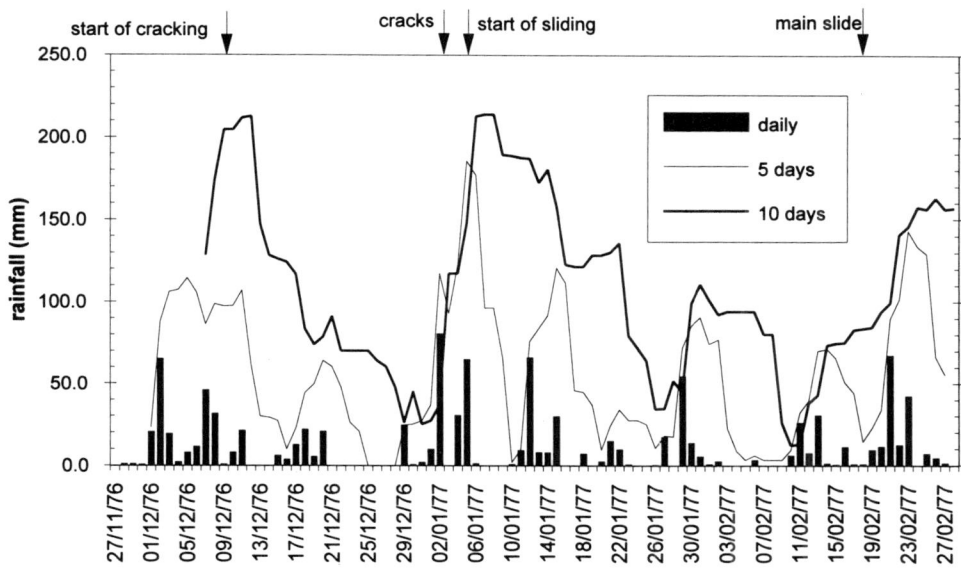

Figure 7. Rainfalls at San Marcello Pistoiese during the reactivation of the mudslide.

For the Cretaceous-Paleogene clay shales and mélanges (Fig. 8a), compositional characteristics vary considerably in different locations of the Apennine mountain range, from low plasticity clayey silts (CL) in the Chianti Mountains (Sillano Formation) to extremely high plasticity clays (CE) in the Val Marecchia (varicolored argillites). Moreover, these formations are all characterized by a structural and textural complexity, caused by the high degree of fissuring, the typical scaly structure, and the pervasive tectonic shears.

The fine-grained marine deposits of the Pliocene transgression (Fig. 8b), known as blue clays, are composed mainly of intermediate or high plasticity silty clays (CI and CH). The index properties are about the same in the different basins, thus showing that the sedimentary environment in the Pliocene sea was quite uniform over large areas. The textural and structural features are quite uniform too, and are characterized by a certain degree of overconsolidation and fissuring.

Conversely, the compositional characteristics of the soils formed in the Plio-Pleistocene fluvio-lacustrine basins (Fig. 8c) vary significantly from basin to basin: low plasticity clays (CL) in the early Pleistocene Valdarno basin, intermediate plasticity clays (CH) in the early Pleistocene Mugello basin, and very high plasticity silts in the late Pliocene Valdarno basin (MV). This fact suggests that each soil was deposited in an independent sedimentary domain. Moreover, some textural differences can be noted: the clayey silts of the Pliocene basin of the Valdarno are heavily over-consolidated, fissured, and jointed and are intercalated with peat levels; the silty clays of the early Pleistocene Valdarno basin show a high degree of fissuring and frequent sandy partings, whereas the silty clays of the early Pleistocene Mugello basin seem more homogeneous and present a lower degree of fissuring.

Drained shear strength

The amount of shear strength data available is not sufficient to allow us to make a spatial prediction of landslide danger on a regional scale. It is a well-known fact that the drained shear strength of soils depends on many compositional, textural, and structural factors. Therefore, in general, it is not possible to estimate the strength parameters of terrains over a large area on the basis of index testing. However, a large number of the landslides in the clay and shale hillslopes in Tuscany can be placed in one of the following two categories: (1) first-time slides in fissured over-consolidated materials where the long-term field strength can be very close to the laboratory strength of the reconstituted normally consolidated material (critical state strength) (Chandler, 1984); and (2) reactivated landslides where the mass strength is reduced to the residual value, caused either by the presence of tectonic pre-existing shears or by the presence of ancient slip surfaces which originated under colder and wetter paleoclimatic conditions.

Both the critical state and the residual strength are purely frictional and are linked only to the compositional characteristics of soils, regardless of their fabric and structure. Hence correlations between critical state and residual friction angles, in terms of effective stresses, can be attempted with some index properties.

The correlation between friction angles and clay fraction (Fig. 9a) coincides well with the experimental curves of Chandler (1984, modified from Lupini et al., 1981) and Blondeau (1973); as the clay fraction increases, we have a transition from rolling shear to sliding shear, caused by the re-orientation of the flat clay minerals (Lupini et al., 1981).

The correlation between residual strength and plasticity

index (Fig. 9b) coincides with the relationship proposed by Kanji (1974). With respect to the relationship proposed by Chandler (1984), there is a slight underestimation of the friction angle at low values of the plasticity index. The experimental boundaries for the critical state friction angle given by Jamiolkowski et al. (1979) are shown in the diagram, but the data collected are generally considerably lower. These discrepancies can be ascribed to the low precision and accuracy in the determinations of the Atterberg limits and to differences in the methodologies of soil disaggregation.

As far as the correlation with the liquid limit is concerned (Fig. 9c), the experimental curve of Jamiolkowski and Pas-

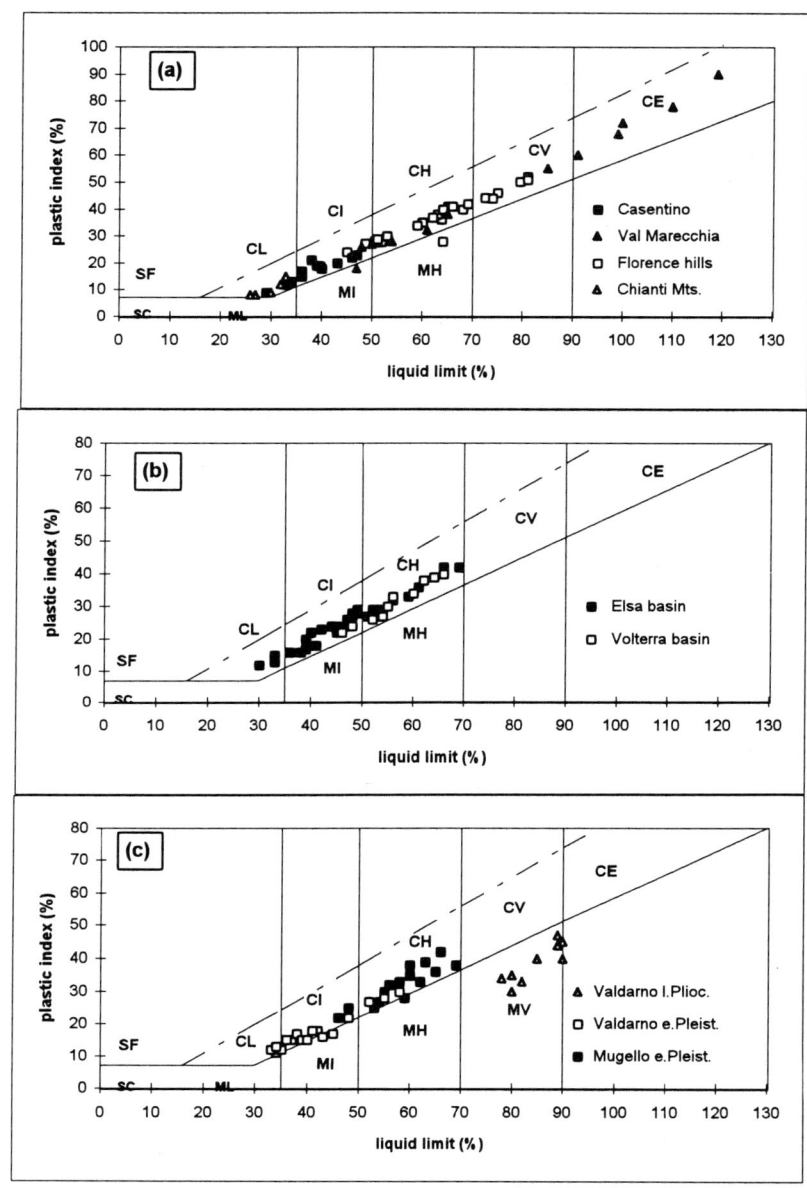

Figure 8. Plasticity charts (British Standard 5930, 1981) for some of the clay and shale deposits in Tuscany: (a) Cretaceous-Paleogene clay shales and mélanges; (b) Pliocene marine clays ("blue clays"); (c) Plio-Pleistocene fluvio-lacustrine clays. Oblique solid line: limit between clays and silts; oblique chain-dotted line: upper bound. SF = very silty and/or clayey sand; SC = very clayey sand; CL = low plasticity clays; CI = intermediate plasticity clays; CH = high plasticity clays; CV = very high plasticity clays; CE = extremely high plasticity clays; ML = low plasticity silts; MI = intermediate plasticity silts; MH = high plasticity silts; MV = very high plasticity silts.

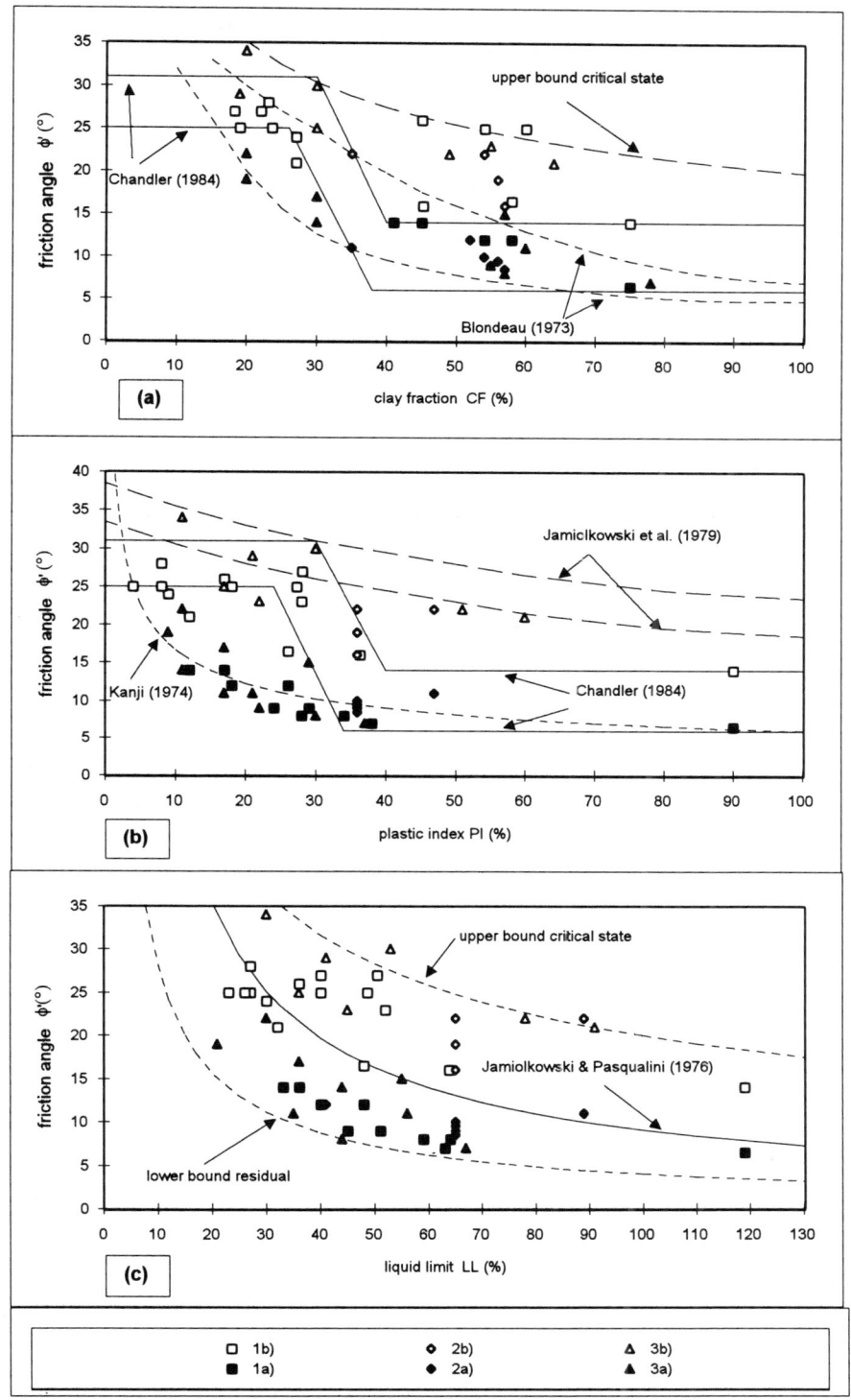

Figure 9. Correlation between critical state and residual friction angles with clay fraction and Atterberg limits, compared with experimental curves taken from literature. Black square, (1a) clay shales and mélanges—critical state strength; black diamond, (2a) Pliocene marine clays—critical state strength; black triangle, (3a) Plio-Pleistocene fluvio-lacustrine clays—critical state strength. Open square, (1b) clay shales and mélanges—residual strength; open diamond, (2b) Pliocene marine clays—residual strength; open triangle, (3b) Plio-Pleistocene fluvio-lacustrine clays—residual strength.

qualini (1976) for residual strength overestimates the collected data and seems to make an upper bound to them, separating the residual friction angle data from the critical state data.

These diagrams confirm the difficulties, already noted by Lupini et al. (1981), in the estimation of strength parameters on the basis of index properties. They can, however, be used as guidelines for future analyses. It seems that the best result for the residual friction angles can be obtained from the relations established by Blondeau (1973) with the clay fraction, and by Kanji (1974) with the plasticity index.

GEOMORPHOLOGICAL ANALYSES

The geomorphological and geotechnical data collected during a landslide inventory can be used to find simple trends and patterns of slope behavior, which in turn can be used for predictions both at a regional and local scale. This section presents some examples of simple data processing, drawn from literature and adapted to the specific situation of Tuscany, using some of the data collected so far at the Earth Science Department of Florence.

Maximum stable angle of natural slopes

Geomorphological analysis can be employed to find the maximum stable angle of natural slopes in homogeneous lithologies as shown by Skempton and Delory (1957) and Chandler (1984). Examples of this are the natural slopes in the Pliocene marine deposits forming unstable slopes near the unstable town of Certaldo in the Elsa basin. In this area two main lithologies are present:

Ps unit: clayey and sandy silts: LL = 32–37, PI = 12–17, $\phi' = 34°$ and $c' = 5$ kPa;
Pag unit: clays and clayey silts: LL = 45–50, PI = 23–27; $\phi' = 23°-25°$ and $c' = 15-20$ kPa;

where LL is the liquid limit, PI is the plasticity index, ϕ' is the friction angle, and c' is the cohesion, in terms of effective stress.

Figure 10 shows two diagrams where the slope height of several rotational and translational first-time landslides is plotted versus the inclination. Active and stabilized landslides are plotted with different symbols. Both diagrams show that the steeper slopes, formed by fluvial downcutting in a transient pore water pressure regime, are unstable and tend, in the long term, to be affected by landslides. Mass movements tend progressively to form gentler slopes and to become stabilized, i.e., to stop until new disturbances are caused by human activity or by extreme meteorological events. These simple data allow us to perform a back analysis of the slope movements, providing that reasonable assumptions are made regarding pore water pressures. Pore water pressures can be usefully expressed in terms of the dimensionless pore water pressure ratio, r_u, defined by Bishop and Morgenstern (1960). Field evidence and monitored landslides on similar lithologies in nearby zones (unstable slops near the towns of San Miniato and Monte-

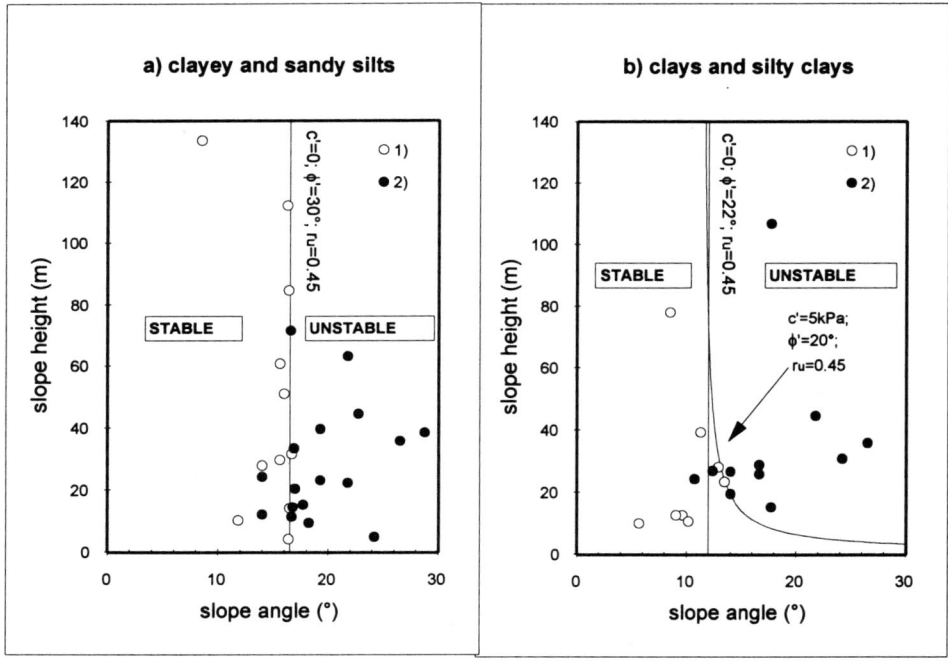

Figure 10. Slope height versus slope angle for active and dormant landslides in the Pliocene marine basin of the Elsa (near Certaldo), compared with theoretical curves. (a) Clayey and sandy silts; (b) clays and silty clays. Open circles, (1) stabilized landslides; black circles, (2) active landslides.

spertoli) show rather high values of r_u, ranging from 0.40 to 0.48. An average value of 0.45 has been taken for analysis purposes.

For frictional soils ($c' = 0$), slope instability phenomena consist mainly of sheet slides, which can be interpreted using infinite slope analysis (Skempton and Delory, 1957). Therefore, the maximum stable angle of the slopes can be computed from the friction angle, and it is not linked to the slope height.

The maximum stable angle of slopes for the unit Ps is equal to 16.5° (Fig. 10a), which, for $c' = 0$ and $r_u = 0.45$, corresponds to a field friction angle $\phi' = 30°$, slightly below the laboratory angle.

As far as the lithological unit Pag is concerned (Fig. 10b), a boundary between stable and unstable slopes can be drawn at a maximum stable angle of 12°, which can be obtained from $\phi' = 22°$ with $c' = 0$ and $r_u = 0.45$. Alternatively, a different boundary can be chosen by taking into account a cohesive component in the drained shear strength. In this case the maximum stable angle depends on the slope height. Curves of slope height versus slope angle can be obtained using the Bishop and Morgenstern (1960) stability charts or their extension made by Chandler and Peiris (1989). For $r_u = 0.45$, the best boundary between stable and unstable slopes is given by strength parameters $\phi' = 20°$ and $c' = 5$ kPa. These parameters are both lower than the laboratory values, which shows that the field behavior of slopes is strongly influenced by processes that cause a decrease of the shear strength of cohesive soils. In the case of both the Ps and the Pag units, the field strength parameters seem to be closer to the critical state values than to the peak values.

This kind of diagram, drawn up with simple morphometric data on landslides or slopes, can be used for planning purposes or for hazard mapping. However, this analysis can be applied only to slopes controlled by first-time slides. In all the cases where the reactivation of ancient slope movements is possible, failures can also occur at gentler slope angles, as reactivated landslides are governed by residual strength parameters.

Mudslide analysis

The most widespread case of ancient slides, which are periodically reactivated in the wet season, is represented by mudslides in clay and shale soils.

A comparison between morphometric data and theoretical curves obtained from three-dimensional analysis of mudslides is proposed by Hutchinson (1988). This comparison is shown in Figure 11 for examples from Tuscany. Hutchinson (1988) classifies as "lobate" the shorter mudslides in which the mechanism of movement is critically influenced by the passive thrust caused by undrained loading in the upper zone. Conversely, the mechanism of the longer mudslides, defined as "elongate," becomes independent of their length.

A large number of elongate mudslides are present in the mountain zones of the Tuscan Apennines, particularly on clay shale and mélange outcrops. They reach lengths of several kilometers and generally have slope angles ranging from 7° to

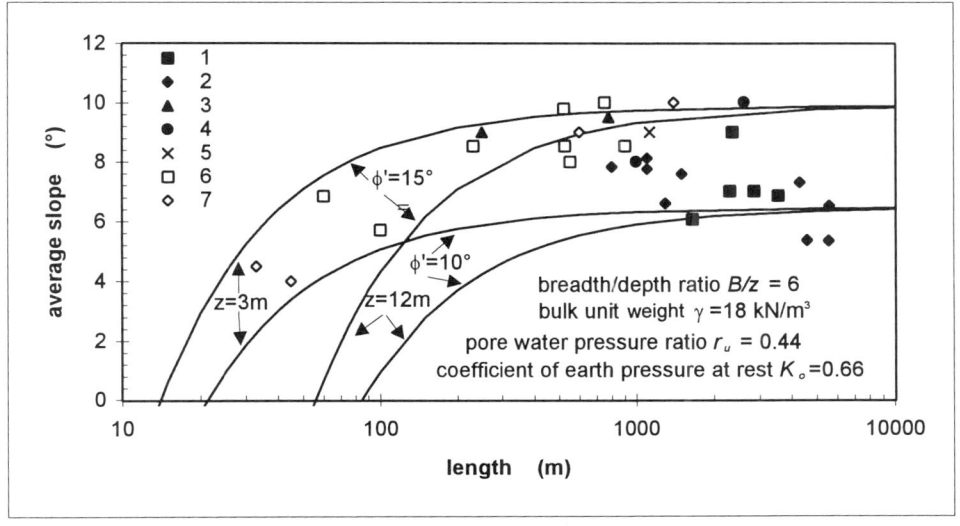

Figure 11. Average slope versus length for typical mudslides in Tuscany compared with theoretical curves from three-dimensional analysis of rectangular channels assuming a rear passive thrust from undrained loading (adapted from Hutchinson, 1988). *Clay shales and mélanges:* black square, (1) Casentino (La Verna); black diamond, (2) Val Marecchia (Sasso di Simone); black triangle, (3) Pistoia Mountains (San Marcello); black circle, (4) Florence hills (Pelago); X, (5) Southern Tuscany (Roccalbegna). *Pliocene marine clays:* open square, (6) Volterra and Elsa basins. *Plio-Pleistocene fluvio-lacustrine clays:* open diamond, (7) Valdarno.

10°; gentler slopes, even 5°, can only be found in particular conditions. Most of these mudslides are dormant or inactive in present-day conditions, and their origin is probably linked to the worst paleoclimatic conditions.

The mechanisms of the elongate mudslides in Tuscany, especially of those in the gentlest slopes, often require high values of pore-water pressure, as is shown by back analysis. A simple way of showing this is to plot the residual friction angle versus the mudslide slope angle. Theoretical curves of the pore water pressure ratio r_u required for the movement can be drawn using the equation of Skempton and Delory (1957):

$$F = \left(1 - \frac{r_u}{\cos^2\beta}\right)\frac{\tan\phi'}{\tan\beta},$$

where β is the slope angle, ϕ' the effective friction angle, and F is the factor of safety.

As can be seen in Figure 12 many of the elongate mudslides in the Apennines require r_u values ranging from 0.30 to 0.55, or even higher if we take into account the fact that the r_u curves are drawn without considering lateral stresses on the mudslide tracks. These data suggest the fact that high pore water pressures, in some cases even artesian, are necessary to explain the movement; perhaps this shows that the origin of these slope movements is to be linked to worse paleoclimatic conditions in the mountain area during the late Pleistocene or early Holocene.

CONCLUSIONS

The systematic collection of information on the slope movements currently carried out in Tuscany by the GNDCI operative unit of Florence allows us to make some observations on slope instability patterns and trends on a regional and a local scale. The collection of data on the geotechnical properties of soils can be used as a guideline for future laboratory analysis and for spatial predictions of specific types of slope movements. Morphometric data can be used for geomorphological analysis, with the support of the geotechnical information and the theoretical methods of stability analysis. Future work will be carried out on the regional analysis of the relationships between landslides and heavy rainfalls; in this case the availability of a time series of the slope movements is of crucial importance to make any correlation. An accurate landslide inventory is the best tool to obtain this information, despite the fragmentation and the problems in interpreting historical data. It is the hope of the writers to set up in Tuscany a permanent record service on landslides in order to produce landslide annals, analogous to the ones that are already edited for rainfall data.

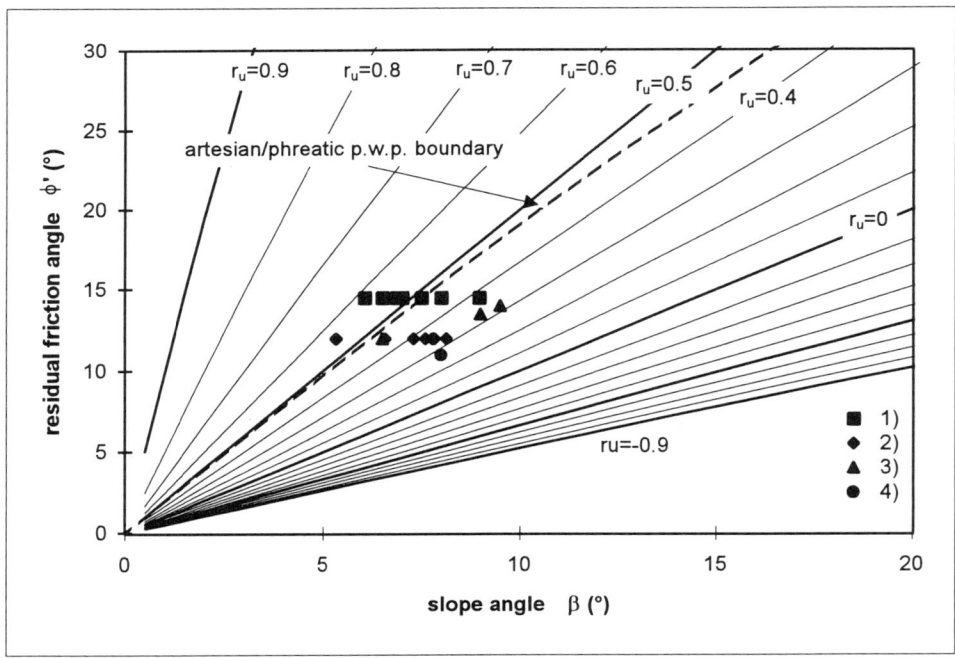

Figure 12. Residual friction angle versus slope angle for some elongate mudslides in clay shales of mountain zones of the Tuscan Apennines compared with pore water pressure ratio r_u curves required for failure according to the equation of Skempton and Delory (1957). Black square, (1) Casentino (La Verna); black diamond, (2) Val Marecchia (Sasso di Simone); black triangle, (3) Pistoia Mountains. (San Marcello); black circle, (4) Florence hills (Pelago).

ACKNOWLEDGMENTS

The work described in this paper was supported by the National Research Council–GNDCI line no. 2 "Forecast and Prevention of High Risk Slope Movements," coordinated by Paolo Canuti (Earth Science Department, Florence).

REFERENCES CITED

Abbate, E., and Sagri, M., 1970, The eugeosynclinal sequences, *in* Sestini G., ed., Development of Northern Apennines geosyncline: Sedimentary Geology, v. 4, p. 251–340.

Abbate, E., Bortolotti, V., and Passerini, P., 1970, Olistostromes and olistolithes, *in* Sestini G., ed., Development of Northern Apennines geosyncline: Sedimentary Geology, v. 4, p. 521–557.

Ambrosetti, P., and 11 others, 1978, Evoluzione paleogeografica e tettonica nei bacini tosco-umbro-laziali nel Pliocene e nel Pleistocene inferiore: Memorie Società Geologica Italiana, v. 19, p. 573–580.

Benedetti, R., Canuti, P., Moisello, U., and Tenti, G., 1979, Studi di protezione idrogeologica nella regione Toscana: Le precipitazioni massime prolungate (1-5 giorni) nel bacino dell'Arno: L'Ateneo Parmense—Acta Naturalia, v. 15, p. 99–115.

Bettelli, G., and Panini, F., 1987, I melanges dell'Appennino Settentrionale dal T. Tresinaro al T. Sillaro: Memorie Società Geologica Italiana, v. 39, p. 187–214.

Bishop, A. W., and Morgenstern, N. R., 1960, Stability coefficients for earth slopes: Géotechnique, v. 10, p. 129–150.

Blondeau, F., 1973, The residual shear strength of some French clays: measurement and application to a natural slope landslide: Proceedings, National Research Council–Istituto Ricerca Protezione Idroelogica, Cosenza, 1971: Geologia Applicata & Idrogeologia, v. 8, p. 125–141.

Bortolotti, V., ed., 1992, Appennino Tosco-Emiliano: Guide geologiche regionali: Società Geologica Italiana, 329 p.

Bortolotti, V., Passerini, P., Sagri, M., and Sestini, G., 1970, The miogeosynclinal sequences, *in* Sestini G., ed., Development of Northern Apennines geosyncline: Sedimentary Geology, v. 4, p. 341–444.

Bossio, A., and nine others, 1994, The Neogenic and Quaternary basins of Tuscany, *in* Proceedings, Geological Society of Italy Meeting, 76th: Memorie Società Geologica Italiana (in press).

British Standard 5930, 1981, Code of practice for site investigations: London, British Standards Institution.

Canuti, P., 1985, Indagini sui fenomeni franosi e stabilità dei versanti nella Regione Toscana: Geologia Applicata e Idrogeologia, v 20, p. 417–436.

Canuti, P., ed., 1993, IV Seminario Deformazioni Gravitative Profonde in Toscana: Florence National Research Council–National Group for the Prevention of Hydrogeological Disasters Special Publication, University of Florence, 73 p.

Carton, A., 1990, Evidence of Quaternary glacial and periglacial morphogenesis in the Tosco-Emilian Apennine, *in* The loess in northern and central Italy: A loess basin between the Alps and Mediterranean region": Milan, Quaderni Geodinamica Alpina Quaternaria, p. 74–77.

Catenacci, V., 1992, Il dissesto geologico e geoambientale in Italia dal dopoguerra al 1990: Memorie descrittive della Carta Geologica d'Italia, v. 47, 300 p.

Chandler, R. J., 1984, Recent european experience of landslides in over-consolidated clays and soft rocks *in* Proceedings, International Symposium on Landslides, 4th, Toronto, v. 1: Toronto, International Symposium on Landslides, p. 61–78.

Chandler, R. J., and Peris, T. A., 1989, Further extension to the Bishop and Morgestern stability charts: Ground Engineering, p. 33–38.

Hutchinson, J. N., 1988, General report: Morphological and geotechnical parameters of landslides in relation to geology and hydrogeology: Proceedings 5th International Symposium on Landslides, 5th, Lausanne, *in* Bonnard, C., Landslides, v. 1, p. 3–36.

Jamiolkowski, M., and Pasqualini, E., 1976, Sulla scelta dei parametri geotecnici che intervengono nelle verifiche di stabilità dei pendii naturali e artificiali: Atti Istituto di Scienza delle Costruzioni, Politecnico di Torino, v. 319, 53 p.

Jamiolkowski, M., Lancellotta, R., Marchetti, S., Nova, R., and Pasqualini, E., 1979, Design parameters for soft clays: European Conference on Soil Mechanics and Foundation Engineering, 7th, Brighton.

Kanji, M. A., 1974, The relationship between drained friction angles and Atterberg limits of natural soils: Géotechnique, v. 24, p. 671–674.

Lupini, J. F., Skinner, A. E., and Vaughan, P. R., 1981, The drained residual strength of cohesive soils: Géotechnique, v. 31, 181–213.

Merla, G., 1952, Geologia dell'Appennino Settentrionale: Bollettino Società Geologica Italiana, v. 70, p. 95–382.

Skempton, A. W., and Delory, F. A., 1957, Stability of natural slopes in London Clay: Proceedings, Conference of International Society of Soil Mechanics and Foundation Engineering, 4th, v. 2: p. 378–381.

Manuscript Accepted by the Society May 20, 1994

Characterizing durability of mudrocks for slope stability purposes

Jeffrey C. Dick
Department of Geology, Youngstown State University, Youngstown, Ohio 44555
Abdul Shakoor
Department of Geology and the Water Resources Research Institute, Kent State University, Kent, Ohio 44242

ABSTRACT

The lithologic characteristics and slake durability of mudrocks from 48 natural and human-made slopes were investigated with the objective of developing a classification of mudrock durability for slope stability purposes. Slopes from which samples were taken ranged from long-term stable to highly unstable. Samples included 10 claystones, 17 shales, and 21 mudstones. The durability of mudrocks was classified as high, medium, or low on the basis of relationships between lithologic characteristics, slake durability, and slope conditions, as observed in the field. The three classes of durability were then used to evaluate the likelihood of occurrence of the four commonly recognized types of instability that affect slopes in mudrocks: excessive erosion, slumps, debris flows, and undercutting-induced failures. High-durability mudrocks were found to be susceptible to undercutting only, medium-durability mudrocks to slumps, debris flows, and undercutting-induced failures, and low-durability mudrocks to all four types of slope instability.

INTRODUCTION

Mudrocks (shales, claystones, mudstones, and siltstones) are notorious in the construction and mining industries for their generally poor engineering properties. Many mudrocks weather rapidly so that the rock-like properties they exhibit during excavation often change to properties more characteristic of soil, resulting in countless slope, excavation, and mine-highwall failures. The susceptibility of mudrocks to weathering and the extent of changes in their engineering behavior depend upon their durability characteristics. Durability characteristics are particularly important in mudrock slope design because the rock often remains exposed to the weathering environment. Durable mudrocks tend to resist weathering resulting in stable slopes for long periods of time. Nondurable mudrocks, however, tend to weather rapidly and are prone to slope instability problems. Common slope instability problems attributable to nondurable mudrocks include excessive erosion, slumping, debris flow, and undercutting leading to rock falls.

Mudrock durability research was initiated in the 1950s by the British coal mining industry in response to highwall and mine roof failure problems (Taylor, 1988). The research that followed has greatly enhanced our understanding of mudrock weathering processes, produced standardized durability tests, and established relationships between durability and lithologic characteristics of mudrocks. Although several durability tests have been developed, the test that most accurately characterizes mudrock durability is the slake durability test developed by Franklin and Chandra (1972). It is the most widely used test of mudrock durability, is recommended by the International Society for Rock Mechanics (1979), and has been standardized by the American Society for Testing and Materials (ASTM, 1990). Relationships between durability and the lithology of mudrocks are well established. The combined works of Chandler (1969), Fookes et al. (1971), Spears and Taylor (1972), Van Eeckhout (1976), Russell and Parker (1979), Russell (1981), Smart et al. (1982), Grainger (1983), Steward and Cripps (1983), Shakoor and Brock (1987), Taylor (1988), and Dick and Shakoor (1992b) demonstrate that mudrock durability is controlled by clay content, proportion of expandable clay minerals, rock fabric, sedimentary structures (laminae, slickensides, and microfractures), and cementation.

Dick, J. C., and Shakoor, A., 1995, Characterizing durability of mudrocks for slope stability purposes, *in* Haneberg, W. C., and Anderson, S. A., eds., Clay and Shale Slope Instability: Boulder, Colorado, Geological Society of America Reviews in Engineering Geology, v. X.

A mudrock durability classification system based on lithologic characteristics has been proposed by Dick et al. (1994).

The role of durability in evaluation and design of slopes in mudrocks is an active area of research. Franklin (1983) has developed a rating system for Ontario shales wherein slake durability, strength, and plasticity characteristics are used to evaluate the field performance of shale slopes. Shakoor and Weber (1988) used the second cycle slake durability index (Id_2) as a laboratory measure of differential weathering in the study of shale undercutting in upper Pennsylvanian and lower Permian age rocks along portions of Interstate 77 in southern Ohio and northern West Virginia. In related research, Shakoor and Rodgers (1992) used Id_2 as a predictor of the rate of shale undercutting.

The research presented here is part of an ongoing investigation regarding the roles and interrelationships of durability and mudrock lithology in characterizing and predicting long-term slope behavior. It is based on a study of lithology, durability, and slope characteristics of 48 different mudrock slopes in various parts of the United States and Canada.

FIELD METHODS

The 48 mudrock slopes sampled for this study were selected on the basis of published information and exploratory field work. Information regarding age, location, and lithology of the geologic formations sampled is provided in Table 1. A general assessment of slope stability and rock mass properties was made at each site. Mudrock samples were usually collected from hand-excavated pits. The state of weathering was assessed and thickness of the regolith was measured. The mudrocks were distinguished using a slightly modified form of the Blatt et al. (1980) classification. The classification (Table 2) recognizes six different classes of mudrocks based on the percentage of clay-size particles (<0.004 mm) and the presence or absence of laminations. Modifications to the Blatt et al. (1980) classification include use of the term "laminated" in place of fissile, and placement of the textural division between mudstones and claystones at 50% clay-sized particles. According to this classification, the samples consisted of 10 claystones, 21 mudstones, and 17 shales (10 mudshales, 6 siltshales, and 1 clayshale).

LABORATORY INVESTIGATIONS

Laboratory investigations included determinations of slake durability index, lithologic characteristics, and selected engineering properties. The durability of mudrocks was characterized using the two-cycle slake durability index test (ASTM, 1990, D4644). Lithologic characteristics analyzed included clay content, particle size distribution, expandable clay mineral content, fabric, and degree of microfracturing. The engineering properties determined included plasticity index, adsorption, absorption, dry density, and void ratio. Clay content and particle size distribution were determined using sieve and hydrometer analyses (ASTM, 1990, D422). The relative proportion of expandable clay minerals was estimated using the X-ray diffraction method described by Schultz (1964). The plasticity index (ASTM, 1990, D4318) and adsorption (96 hours at 95% humidity) were determined because these properties are sensitive to the total clay content and the relative proportion of expandable clay minerals. Dry density (ASTM, 1990, C97), absorption (ASTM, 1990, C97), and void ratio were used to characterize fabric. These properties are useful indicators of fabric because they are dependent on the relative proportion of void space and minerals in the mudrocks. Scanning electron microscopy (SEM) was used to confirm fabric and textural characteristics indicated by dry density, absorption, void ratio, and particle size analysis. Frequency of microfractures and slickensided surfaces, I_{mf}, for the mudstones and claystones was measured using the method described by Dick and Shakoor (1992b). In this method, mudrocks are dry cut into orthogonal shapes having a minimum dimension of 5 cm. The number of open microfractures (including slickensided surfaces) along linear traverses oriented parallel to bedding on each surface is counted. The I_{mf}, expressed as the number of microfractures per unit length, is then calculated by dividing the number of microfractures along each traverse by the length (cm) of the traverse. An average I_{mf} is obtained using at least two cut chunks of each mudrock.

LITHOLOGIC CHARACTERISTICS AND ENGINEERING PROPERTIES AS INDICATORS OF MUDROCK DURABILITY

The extent to which individual lithologic characteristics control the durability of mudrocks depends upon which lithologic characteristic is dominant in the rock. An earlier study of durability and lithologic characteristics of 61 mudrock samples (Dick, 1992) indicated that for mudrocks in general, no single lithologic characteristic or combinations of lithologic characteristics can reliably predict durability. Discriminant analysis of lithologic variables, however, successfully distinguished between mudrocks of high and low durability 78% of the time. Although this is unacceptable for most practical applications, the discriminant analysis demonstrated that mudrocks of high durability are associated with an increase in the degree of consolidation (as measured by dry density and void ratio) and mudrocks of low durability are associated with increasing percentage of clay minerals (as measured by adsorption and bulk specific gravity).

The absence of a meaningful relationship between lithology and durability of mudrocks in general can be attributed to the lithologic diversity of mudrocks. The number of lithologic characteristics contributing to mudrock durability can be reduced for the purpose of evaluating the relationships between lithology and durability by subdividing the mudrocks into claystones, mudstones, and shales. Following the modified

TABLE 1. MUDROCK FORMATIONS SAMPLED

Formation	Age	Location	Lithology
Chagrin Shale	Devonian	Lake Co., Oh.	Mudshale
Chagrin Shale	Devonian	Lake Co., Oh.	Mudshale
Chagrin Shale	Devonian	Lake Co., Oh.	Siltshale
Allegheny Group	Pennsylvanian	Lawrence Co., Pa.	Siltshale
Allegheny Group	Pennsylvanian	Lawrence Co., Pa.	Mudshale
Fairview Formation	Ordovician	Kenton Co., Ky.	Mudstone
Fairview Formation	Ordovician	Kenton Co., Ky.	Mudstone
Kope Formation	Ordovician	Hamilton Co., Oh.	Mudstone
Kope Formation	Ordovician	Hamilton Co., Oh.	Claystone
Conemaugh Group	Pennsylvanian	Noble Co., Oh.	Mudstone
Monongahela Group	Pennsylvanian	Galia Co., Oh.	Mudstone
Conemaugh Group	Pennsylvanian	Athens Co., Oh.	Mudstone
Conemaugh Group	Pennsylvanian	Allegheny Co., Pa.	Claystone
Conemaugh Group	Pennsylvanian	Allegheny Co., Pa.	Mudstone
Dunkard Group	Permian	Wood Co., W. Va.	Mudstone
Dunkard Group	Permian	Wood Co., W. Va.	Mudstone
Conemaugh Group	Pennsylvanian	Noble Co., Oh.	Siltshale
	Jurassic	Riverside Co., Ca.	Mudstone
Monterey Formation	Miocene	Riverside Co., Ca.	Mudstone
Georgian Bay Formation	Ordovician	Ontario, Canada	Claystone
Georgian Bay Formation	Ordovician	Ontario, Canada	Mudshale
Queenston Shale	Ordovician	Ontario, Canada	Mudstone
Queenston Shale	Ordovician	Ontario, Canada	Mudstone
Chagrin Shale	Devonian	Lake Co., Oh.	Mudshale
Chagrin Shale	Devonian	Lake Co., Oh.	Mudshale
Pierre Shale	Cretaceous	Huerfano Co., Co.	Claystone
	Pennsylvanian	Garfield Co., Co.	Mudstone
Benton Formation	Cretaceous	Eagle Co., Co.	Mudshale
Mancos Shale	Cretaceous	Garfield Co., Co.	Mudstone
Kirtland Formation	Cretaceous	Garfield Co., Co.	Claystone
Mancos Shale	Cretaceous	Garfield Co., Co.	Mudstone
Morrison Formation	Jurassic	Grand Co., Utah	Claystone
Lewis Formation	Cretaceous	Grand Co., Utah	Mudstone
Mowry Shale	Cretaceous	Uintah Co., Utah	Claystone
Pierre Shale	Cretaceous	Jefferson Co., Co.	Claystone
Dakota Group	Cretaceous	Ellsworth Co., Ks.	Mudstone
Dakota Group	Cretaceous	Ellsworth Co., Ks.	Claystone
Morrison Formation	Cretaceous	Cimmarron Co., Ok.	Mudstone
Chanute Formation	Pennsylvanian	Johnson Co., Ks.	Claystone
Allegheny Group	Pennsylvanian	Lawrence Co., Pa.	Siltshale
Allegheny Group	Pennsylvanian	Lawrence Co., Pa.	Siltshale
Olentangy Shale	Devonian	Franklin Co., Oh.	Mudshale
Olentangy Shale	Devonian	Franklin Co., Oh.	Mudshale
Bedford Shale	Mississippian	Cuyahoga Co., Oh.	Mudstone
Dunkard Group	Permian	Wood Co., W. Va.	Siltshale
Conemaugh Group	Pennsylvanian	Athens Co., Oh.	Mudshale
Conemaugh Group	Pennsylvanian	Athens Co., Oh.	Clayshale
Conemaugh Group	Pennsylvanian	Noble Co., Oh.	Mudstone

Blatt et al. (1980) classification (Table 2), the 48 mudrock samples were subdivided into 10 claystones, 17 shales (1 clayshale, 10 mudshales, and 6 siltshales), and 21 mudstones.

The results of the geological analyses and engineering tests for individual mudrock classes are statistically summarized in Table 3 and graphically illustrated by histograms in Figures 1, 2, and 3. When subdivided into individual classes, the standard deviation values of the measured properties are restricted and the mean values are indicative of dominant lithologic characteristics. For example, the high clay content of claystones is reflected in the relatively high plasticity index (mean = 29.4) and adsorption (mean = 8.59%) values. Similarly, the higher degree of consolidation of the shales is reflected in the relatively low absorption (mean = 5.98%) and void ratio (mean = 0.08) values. Most of the lithologic properties of the mudstones are intermediate between those of the shales and claystones. The values also suggest a general relationship between lithology and durability. The shales are com-

TABLE 2. CLASSIFICATION OF MUDROCKS*

	Percent Clay-Sized Particles		
	0–32%	33–49%	50–100%
Nonlaminated	Siltstone	Mudstone	Claystone
Laminated	Siltshale	Mudshale	Clayshale

*Modified from Blatt et al., 1980.

TABLE 3. RESULTS OF GEOLOGICAL AND ENGINEERING ANALYSES

		Claystone n = 10	Shale n = 17	Mudstone n = 21
Slake Durability Index (%)	Mean	16.9	77.4	54.3
	Std. Dev.	16.7	20.1	30.6
Percent <0.004 mm	Mean	66.5	35.5	40.5
	Std. Dev.	14.8	8.4	6.3
Expandable Clay (%)	Mean	28.4	4.8	7.1
	Std. Dev.	25.5	5.1	5.3
Plasticity Index (PI)	Mean	29.4	11.0	10.8
	Std. Dev.	18.6	3.2	2.9
Adsorption (%)	Mean	8.59	3.16	3.72
	Std. Dev.	4.89	1.29	0.72
Absorption (%)	Mean	35.4	5.98	10.0
	Std. Dev.	41.2	2.25	3.9
Dry Density (g/cc)	Mean	2.25	2.52	2.42
	Std. Dev.	0.15	0.12	0.14
Void Ratio	Mean	0.18	0.08	0.12
	Std. Dev.	0.05	0.03	0.04
Microfracture Frequency (#/cm)	Mean	1.22	n.a.	0.86
	Std. Dev.	0.48	n.a.	0.56

paratively durable (mean $Id_2 = 77.4\%$); the claystones are nondurable (mean $Id_2 = 16.9\%$); and the durability of mudstones is intermediate (mean $Id_2 = 54.3\%$).

LITHOLOGIC CONTROLS OF CLAYSTONE DURABILITY

The lithologic characteristics of claystones that distinguish them from other mudrocks are their high proportion of expandable clay minerals, grain size, and poor state of induration. The high proportion of expandable clay minerals is reflected in the comparatively high plasticity index and adsorption values (Table 3; Fig. 1). The poor state of induration of the claystones is reflected by comparatively high values of void ratio and adsorption and low values of dry density. Void ratio, absorption, and dry density results are supported by SEM analyses (Dick, 1992), which show a flocculated texture for all the claystones.

The low durability of claystones can be attributed to their poor state of induration and the presence of expandable mixed-layer and montmorillonite clay minerals. These lithologic characteristics make claystones particularly sensitive to changes in moisture content and release of confining stress that accompany excavation. A strong relationship between Id_2 and the percentage of expandable clay minerals (Dick and Shakoor, 1992b) supports this finding. Similar correlations between Id_2 and adsorption and Id_2 and plasticity index provide alternative indicators of durability (Dick and Shakoor, 1992a), and these indicators have the added benefit of being quick and inexpensive compared to X-ray analyses.

LITHOLOGIC CONTROLS OF SHALE DURABILITY

Shales are distinguished from other mudrocks by their laminated sedimentary structure. The laminations are largely the result of parallel to subparallel alignment of constituent clay minerals. This ordered texture is reflected in the comparatively low absorption, low void ratio, and high dry density values of the shales (Table 3 and Fig. 2). The shales used in this study included 10 mudshales, 6 siltshales, and 1 clayshale. Because of the relatively small number of siltshales and clayshales, all the shales were grouped together for the purpose of durability evaluation and classification. The comparable percent clay-size particles, percent expandable clay minerals, plasticity index, and adsorption values between the mudstones and the shales is attributed to the high proportion of mudshale samples (Table 3; Figs. 2 and 3).

Slake durability test results show that the shales are the most durable of all the mudrocks studied. The comparatively high mean Id_2 value (77.4%) is in part attributed to the six siltshale samples. SEM analysis shows the siltshales are characterized by a comparatively high degree of cementation, which translates to higher durability. The higher degree of cementation is also reflected by relatively low absorption, low void ratio, and high dry density values of the siltshales. Cementation of the siltshales, combined with the laminated structure of the shales, controls the durability of shales. Regression analysis of Id_2 and absorption shows a very strong relationship (Dick and Shakoor, 1992b). A similar relationship exists between Id_2 and void ratio and Id_2 and dry density (Dick and Shakoor, 1992b).

LITHOLOGIC CONTROLS OF MUDSTONE DURABILITY

Mudstones are intermediate between shales and claystones in all the properties measured (Table 3; Figs. 1, 2, and 3). The absorption, void ratio, and dry density values suggest a higher state of induration than the claystones, but a less-ordered structure than the shales. SEM analysis of mudstones shows a tighter packing of mineral grains than claystones, whereas relatively coarse-grained mudstones exhibit a degree of cementation comparable to that of siltshales (Dick, 1992).

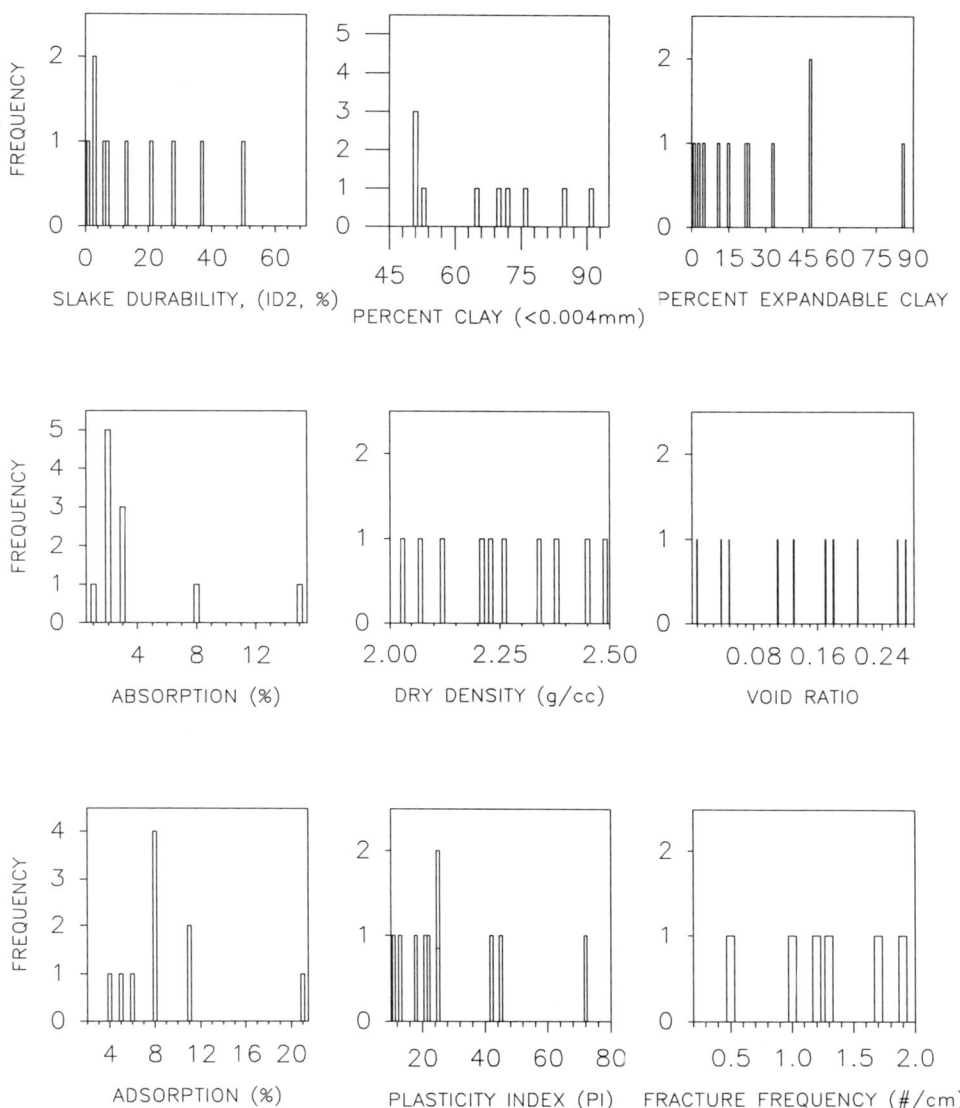

Figure 1. Frequency histograms of lithologic characteristics and engineering properties for claystones.

A distinctive characteristic of mudstones is the common occurrence of microfractures. The microfractures occur as slickensided surfaces and irregular fractures. The durability of mudstones is controlled by these microfractures as suggested by a strong correlation between Id_2 and microfracture frequency index (I_{mf}) (Dick and Shakoor, 1992b). The six slickensided mudstones tested had exceptionally low Id_2 values (less than 20%), suggesting that the presence of slickensides alone is a very useful lithologic indicator of mudstone durability.

MUDROCK DURABILITY CLASSIFICATION

The relationships between durability and lithologic characteristics of the claystones, shales, and mudstones were used to develop a lithologic classification of mudrock durability (Table 4). Three classes of durability are recognized. *High-Durability* mudrocks have Id_2 values >85%. *Medium-Durability* mudrocks have Id_2 values between 85% and 50%. *Low-Durability* mudrocks have Id_2 values <50%.

The durability of claystones is based on the weight percentage of expandable clay minerals. The only applicable class of durability for claystones is *Low Durability* because all the claystones in this study had Id_2 values <50%. The durability of mudstones ranged from *Low Durability* to *High Durability* and is based on the frequency of microfractures. Slickensided mudstones are classified as *Low Durability*. The durability of shales ranges from *High Durability* to *Low Durability*. Shale durability is based on absorption which is an indicator of min-

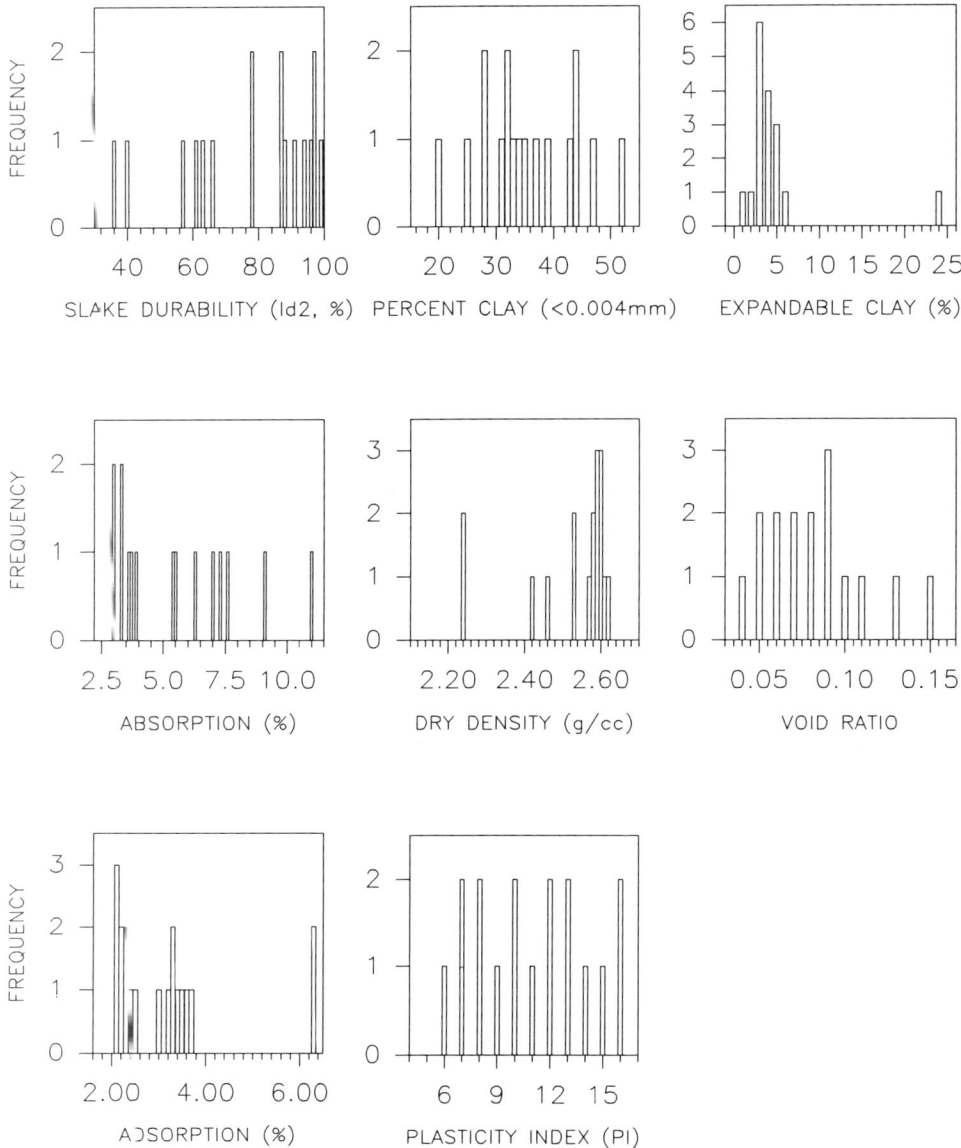

Figure 2. Frequency histograms of lithologic characteristics and engineering properties for shales.

eral grain alignment and degree of cementation. *High-Durability* shales are predominantly siltshales and tend to be well cemented. The durability ranking of the shales tends to decrease with increasing clay content so that mudshales tend to have *Medium Durability* and clay shales have *Low Durability*.

DURABILITY AS AN INDICATOR OF MUDROCK SLOPE INSTABILITY

Durability is the most important intact rock property to consider in the design of slopes in mudrocks, and it is often the most important general rock mass property. Our experience shows that most unstable mudrock slopes are associated with nondurable mudrocks and most stable mudrock slopes are either associated with durable mudrocks or the slope was specifically designed to minimize anticipated durability-related slope instability problems.

Slope instability problems associated with nondurable mudrocks tend to be surficial. Surficial instability generally involves weathered mudrock and is confined to the upper 2 m of the slope fan (Franklin, 1983). Surficial instability includes excessive erosion, slumps, debris flows, and undercutting-induced failures (rock falls, plane failures, wedge failures). These types of failures do not generally pose a substantial hazard, but may require costly routine maintenance. The problem

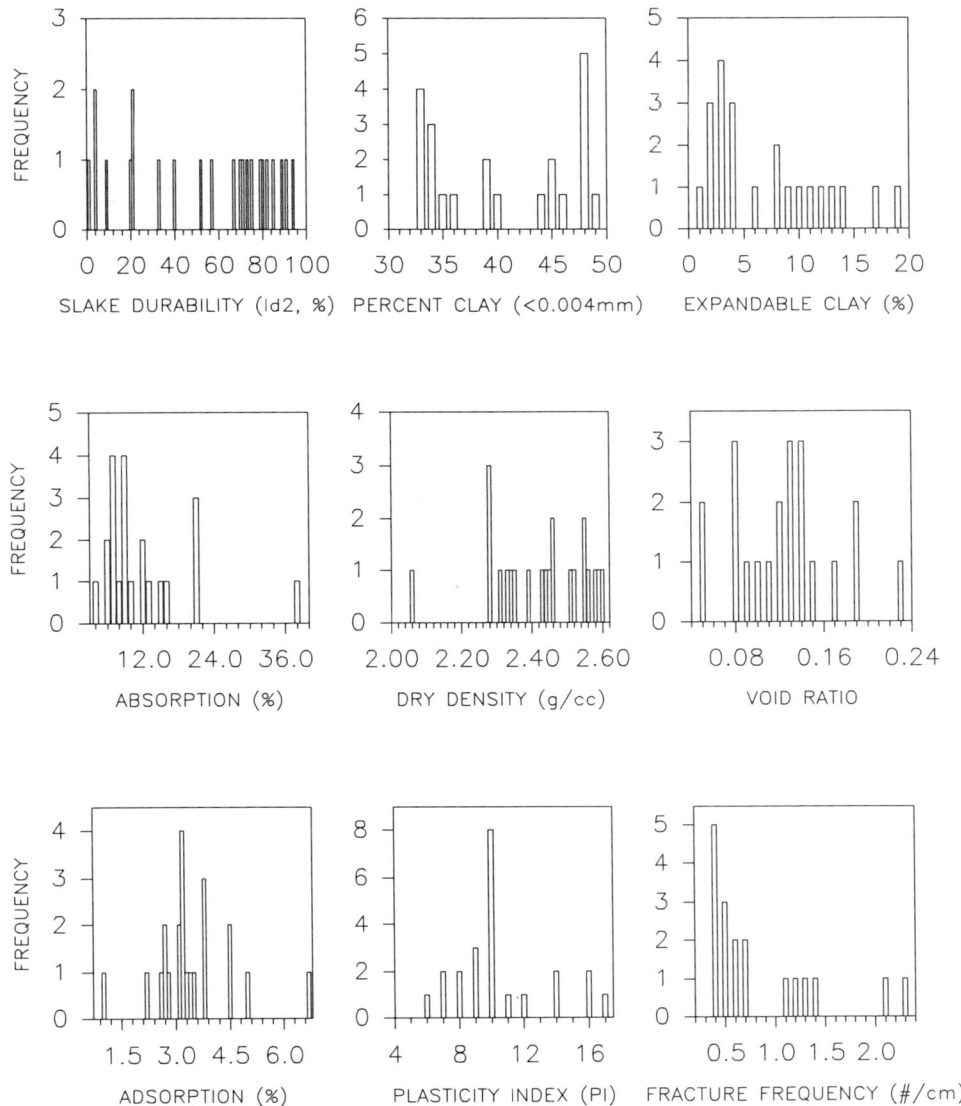

Figure 3. Frequency histograms of lithologic characteristics and engineering properties for mudstones.

Durability	Lithologic Characteristics			
	Claystone Exp. Clay (%)	Mudstone Slickensided	Mudstone I_{mf} (#/cm)	Shale Absorption (%)
High Id_2 >85%	n.a.	n.a.	<0.4	<5.5
Medium 50%< Id_2 <85%	n.a.	n.a.	0.8 - 0.4	10.0 - 5.5
Low Id_2 <50%	All	All	>0.8	>10.0

TABLE 4. LITHOLOGIC CLASSIFICATION OF MUDROCK DURABILITY

can largely be avoided with careful consideration of durability characteristics while designing slopes in mudrocks.

The principle objective of this research was to develop a geological classification of mudrock durability for slope stability purposes. In order to accomplish this objective, the laboratory based durability classification system (Table 4) was combined with observed durability behavior of mudrocks for a variety of unstable slope situations. The intent is not to provide mudrock slope design criteria, but to alert engineers and geologists to potential durability-related instability problems.

A tentative classification of mudrock slope instability based on durability is presented in Table 5. The classification is based on field observations involving multiple examples of mudrock slope instability; mudrocks involved were tested for

TABLE 5. DURABILITY CLASSIFICATION OF MUDROCK SLOPE INSTABILITY

Durability	Slope Instability			
	Excessive Erosion	Slump	Debris Flow	Undercutting (cm/yr)
High	Unlikely	Unlikely	Unlikely	2–3
Medium	Unlikely	Potential	Potential	3–5
Low	Probable	Probable	Probable	5–10

durability and subjected to lithologic analyses. The three classes of durability (high, medium, and low) are compared to the likelihood (unlikely, potential, and probable) of each of four common types of mudrock slope instability: excessive erosion, slump failures, debris flows, and undercutting-induced failures. All mudrocks are susceptible to undercutting. The projected rates of undercutting, as listed in Table 5, are based on the findings of Shakoor and Rodgers (1992). High-durability mudrocks are unlikely to result in excessive erosion, slumps, or debris flows. It is unlikely that medium-durability mudrocks will produce excessive erosion problems; however, there is potential for slumping and debris flow instability, particularly when wet conditions and relatively steep slopes (>40°) are involved. Low-durability mudrocks are prone to all four modes of slope instability.

Excessive erosion

Instability resulting from excessive erosion is generally limited to claystones and slickensided or heavily fractured low-durability mudstones. The combination of expandable clay minerals and poor induration makes these rocks particularly sensitive to water and highly susceptible to erosion.

Figure 4 shows an example of excessive erosion of a highway cut made in the Pierre Shale (Upper Cretaceous) along Interstate 25 near Walsenburg, Colorado. The Pierre Shale at this locality is a massive claystone that contains an appreciable amount of expandable clay minerals (~30%), has a low dry density (2.20 g/cc), and has low durability (Id_2 = 6%). The slope faces the west with an angle of ~35° and a height of ~5 m. The erosional gullies on the slope face coincide with joints spaced ~1 m. The claystone weathers rapidly, producing a thick (1.4 m) mantle of regolith. Excessive erosion under these conditions could be minimized by regrading the slope to 20° and establishing a vegetative cover or installing protective riprap or geotextile.

Slump failures

Slump failures are a potential problem in mudrocks of medium and low durability. All the slump failures observed

Figure 4. Excessive erosion of the Pierre Shale near Walsenburg, Colorado.

Figure 5. Slump failure in a Dunkard Group mudstone, Wood County, West Virginia.

during the course of this investigation involved mudstones and claystones. However, it is likely that slumping occurs in medium- to low-durability shales as well. Slumps tend to occur in oversteepened cut slopes during prolonged wet periods. The lithologic characteristics that contribute to this type of failure are the presence of water-sensitive expandable clays and a relatively high frequency (>0.4/cm) of microfractures.

An example of a slump in a microfractured mudstone is shown in Figure 5. The pictured slope is a west-facing highway cut made in mudstone of the Dunkark Group (Permian) along Interstate 77 in Wood County, West Virginia. The mudstone is massive with ~15 m of the rock exposed above highway grade level. The mudstone contains ~10% expandable clay minerals, is poorly indurated (dry density = 2.31 g/cc), has a moderate I_{mf} value (0.56/cm), and medium durability (Id_2 = 80%). Reliable rock attitude measurements could not be made at this site, however, nearby exposures gently dip southeast. The slope angle is ~35°, and the slope height is ~15 m. The estimated depth of the failure surface is 2 to 3 m.

The slump failure (Fig. 5) occurred in April 1991, following a prolonged period of heavy rain. Slump failures within mudstones of the Dunkard Group are common, but these failures could be minimized by benching the cut, regrading to a lower slope angle, providing internal drainage, or adding retaining-wall support.

Debris flow

Debris flows are a potential mode of failure in mudrocks of medium and low durability and tend to occur when developed regolith fails during extended wet periods or heavy rain. Debris flows were observed in claystones, mudstones, and shales having slope angles <40°. The regolith on steeper slopes generally does not develop in sufficient amounts to form debris flows.

An example of a debris flow is shown in Figure 6. The site consists of a benched cut in mudstones and shales of the Conemaugh Group (Pennsylvanian) near Athens, Ohio. The exposed section is a deltaic sequence consisting of medium- and low-durability mudstones and shales interbedded with resistant limestone and sandstone.

The debris flow (Fig. 6) originated in the massive calcareous mudstone layer near the top of the bench. The mudstone is ~10-m thick, has low durability (Id_2 = 39%), contains 45% clay-size particles, of which about 10% are expandable clay minerals, and has a relatively high I_{mf} value (1.25/cm). In order to maximize parking space, the slope was cut at ~35°. This is much steeper than natural slopes in the same rock, which form at ~15°. The low durability of the mudrocks, combined with an oversteepened slope, ground-water seepage, and heavy runoff, results in frequent debris flows and undercutting-induced failures at this location. Frequent failure of the slope limits regolith accumulation to ~30 cm at this site.

Undercutting

Undercutting-induced failures are common in slopes consisting of resistant limestones and sandstones underlain by relatively weak mudrocks. Rock falls and wedge failures are the most common forms of slope failure resulting from undercutting (Fookes and Sweeney, 1976; Rib and Liang, 1978; and Shakoor and Weber, 1988). Mudrocks of high, medium, and low durability are all susceptible to undercutting (Table 5). The rate of undercutting, however, is dependent on durability.

Shakoor and Rodgers (1992) have proposed the use of slake durability index (Id_2) as a predictor of mudrock undercutting rates. Based on their research, undercutting rates of 2 to 3 cm per year are expected for high-durability mudrocks, 3 to 5 cm per year for medium-durability mudrocks, and 5 to 10 cm per year for low-durability mudrocks. Other factors that influence rate of undercutting are ground-water seepage, surface

Figure 6. Debris flow (left of center) in mudstone of the Conemaugh Group near Athens, Ohio.

Figure 7. Rock fall resulting from undercutting of Dunkard Group mudstone, Wood County, West Virginia.

runoff, jointing, vegetation, talus accumulation, and slope angle (Shakoor and Rodgers, 1992).

An example of a rock fall resulting from mudrock undercutting is shown in Figure 7. The site is an east-facing road cut along Interstate 77 in Wood County, West Virginia. The slope is cut into rocks of the Dunkard Group (Permian) and consists of a 3-m thick low-durability mudstone (Id_2 = 20%) overlain by a massive and resistant calcareous sandstone. The mudstone is poorly indurated and slickensided. It contains ~15% expandable clay minerals, which in combination with ground-water seepage along the base of the overlying sandstone accelerates the rate of undercutting.

CONCLUSION

Durability is the most important intact rock property to consider in evaluating potential slope stability problems in mudrocks. The most frequently used method of evaluating mudrock durability is the slake durability test. However, considering the strong relationships between slake durability and lithologic characteristics of mudrocks, an alternative geological classification of mudrock durability may be useful for slope stability considerations.

The geological classification of mudrock durability provides the framework for the proposed durability classification of mudrock slope instability. In this classification, three classes of durability are compared to the likelihood of slope instability problems in claystones, shales, and mudstones. Four types of failures are recognized: excessive erosion, slumps, debris flows, and undercutting-induced failures. The purpose of the classification is to alert geologists and engineers to potential mudrock slope instability problems. The classification is applicable to surficial slope instability problems only and does not consider rock mass properties which are more relevant to deep-seated failures.

REFERENCES CITED

American Society for Testing and Materials, 1990, Soil and rock; dimension stone; geosynthetics: Annual book of the American Society for Testing and Material Standards 4.08: Philadelphia, American Society for Testing and Materials, 1092 p.

Blatt, H., Middleton, G. V., and Murray, R. C., 1980, Origin of sedimentary rocks: Englewood Cliffs, New Jersey, Prentice-Hall, 782 p.

Chandler, R. J., 1969, The effect of weathering on the shear strength properties of the Keuper Marl: Géotechnique, v. 19, p. 321–334.

Dick, J. C., 1992, Relationships between durability and lithologic characteristics of mudrocks [Ph.D. dissert.]: Kent, Ohio, Kent State University, 243 p.

Dick, J. C., and Shakoor, A., 1992a, Engineering properties as indicators of mudrock lithology [abs.]: Boulder, Colorado, Geological Society of America Abstracts with Programs, v. 24, no. 7, p. 294.

Dick, J. C., and Shakoor, A., 1992b, Lithologic controls of mudrock durability: Quarterly Journal of Engineering Geology, v. 25, p. 31–46.

Dick, J. C., Shakoor, A., and Wells, N. A., 1994, A geological approach toward developing a mudrock durability classification system: Canadian Geotechnical Journal, v. 31, p. 17–27.

Fookes, F. G., and Sweeney, M., 1976, Stabilization and control of rock falls and degrading slopes: Quarterly Journal of Engineering Geology, London, v. 9, p. 37–55.

Fookes, F. G., Dearman, W. G., and Franklin, J. A., 1971, Some engineering aspects of rock weathering with field examples from Dartmoor and elsewhere: Quarterly Journal of Engineering Geology, London, v. 4, p. 139–185.

Franklin, J. A., 1983, Evaluation of shales for construction projects—An Ontario shale rating system: Ottawa, Ontario Ministry of Transportation and Communications, 98 p.

Franklin, J. A., and Chandra, P., 1972, The slake durability test: International Journal of Rock Mechanics and Mining Sciences, v. 9, p. 325–341.

Grainger, P., 1983, Aspects of the engineering geology of mudrocks, with reference to the Crackington Formation of southwest England [unpublished Ph.D. thesis]: Exeter, England, University of Exeter, 407 p.

International Society for Rock Mechanics, 1979, Suggested method for the determination of the slake durability index: International Journal of Rock Mechanics and Mining Science and Geomechanical Abstracts, v. 16, p. 154–156.

Rib, H. T., and Liang, T., 1978, Recognition and identification, in Schuster, R. L., and Krizek, R. J., eds., Landslides: Analysis and control: Washington, D.C., National Academy of Science, Transportation Research Board, Special Publication 176, p. 34–80.

Russell, D. J., 1981, Controls on durability: The response of two Ordovician shales in the slake durability test: Canadian Geotechnical Journal, v. 9, p. 107–116.

Russell, D. J., and Parker, A., 1979, Geotechnical, mineralogical, and chemical interrelationships in weathering profiles of an overconsolidated clay: Quarterly Journal of Engineering Geology, London, v. 12, p. 107–116.

Schultz, L. G., 1964, Quantitative interpretation of mineralogical composition from X-ray and chemical data for the Pierre Shale: U.S. Geological Survey Professional Paper 391-C, p. 1–28.

Shakoor, A., and Brock, D., 1987, Relationship between fissility, composition, and engineering properties of selected shales from northeast Ohio: Bulletin of the Association of Engineering Geologists, v. 24, p. 363–379.

Shakoor, A., and Rodgers, J. P., 1992, Predicting rate of shale undercutting along highway cuts: Bulletin of the Association of Engineering Geologists, v. 29, p. 61–75.

Shakoor, A., and Weber, M. W., 1988, Role of shale undercutting in promoting rock falls and wedge failures along Interstate 77: Bulletin of the Association of Engineering Geologists, v. 25, p. 219–234.

Smart, G. D., Rowlands, N., and Isaac, A. K., 1982, Progress towards establishing relationships between the mineralogy and physical properties of coal measures rocks: International Journal of Rock Mechanics, Mining Sciences, and Geomechanical Abstracts, v. 19, p. 81–89.

Spears, D. A., and Taylor, R. K., 1972, The influence of weathering on the composition and engineering properties of in situ coal measures rocks: International Journal of Rock Mechanics and Mining Sciences, v. 9, p. 729–756.

Steward, H. E., and Cripps, J. C., 1983, Some engineering implications of chemical weathering of pyritic shale: Quarterly Journal of Engineering Geology, London, v. 16, p. 201–209.

Taylor, R. K., 1988, Coal measures mudrocks: Composition, classification, and weathering processes: Quarterly Journal of Engineering Geology, London, v. 21, p. 85–89.

Van Eeckhout, E. M., 1976, The mechanisms of strength reduction due to moisture in coal mine shales: International Journal of Rock Mechanics, Mining Sciences, and Geomechanical Abstracts, v. 13, p. 61–67.

MANUSCRIPT ACCEPTED BY THE SOCIETY MAY 20, 1994

Slope stability considerations in differentially weathered mudrocks

Abdul Shakoor
Department of Geology and the Water Resources Research Institute, Kent State University, Kent, Ohio 44242

ABSTRACT

Mudrock sequences usually occur as harder, more competent strata (siltstones, sandstones, limestones) alternating with softer, less competent strata (claystones, mudstones, shales). This type of stratigraphy is highly susceptible to differential weathering that results in undercutting of the competent layers by the incompetent layers. Undercutting promotes a variety of slope movements such as rock falls, plane failures, and wedge failures that may not occur otherwise. Examples of such failures from selected sites in Ohio, Pennsylvania, and West Virginia are presented. Because of their high speed, suddenness of occurrence, and occasionally large volume of rock involved, undercutting-induced failures can be quite hazardous.

A variety of remedial measures are employed to reduce the potential for undercutting-induced failures. For a timely implementation of these remedial measures, it is essential to estimate the anticipated rate of undercutting for a given site. In order to develop a method that could be used to predict the rate of undercutting, the amount of undercutting was measured for 14 road-cut sites in Ohio, Pennsylvania, and West Virginia. Information regarding the excavation dates of these cuts was obtained from the highway department records. The maximum amount of undercutting at each site was divided by the time since excavation to obtain the rate of undercutting. The rate of undercutting was then correlated with the slake durability index values of the undercutting units to develop prediction equations for estimating the maximum expected rate of undercutting for different types of mudrock. The paper also reviews various methods of treating slopes subject to differential weathering.

INTRODUCTION

The term "mudrocks" is collectively used to include fine-grained argillaceous rocks such as claystones, mudstones, siltstones, shales, and argillites. These rocks occupy approximately one-third of the land area (Franklin, 1983) and, therefore, are the most frequently encountered rocks in engineering construction. Mudrocks usually occur as alternating series of harder and softer strata. The harder units consist of siltstones, sandstones, or limestones whereas claystones, mudstones, or shales form the softer layers. This type of stratigraphy is particularly prone to differential weathering. When slopes cut through such sequences are left exposed to weathering for extended periods of time, the softer and weaker units erode more rapidly than the overlying stronger, more resistant units, resulting in undercutting of the resistant units. As undercutting proceeds, vertical support of the overlying resistant units diminishes and the slope becomes unstable. Undercutting can lead to a variety of slope movements including rock falls, wedge failures, and plane failures (Fookes and Sweeney, 1976; Rib and Liang, 1978; Young and Shakoor, 1987; Shakoor and Weber, 1988). Undercutting-induced slope failures can be quite hazardous because of their instantaneous occurrence, high speed, and sometimes large volume of rock involved (Fig. 1).

The extent of differential weathering, i.e., the amount of undercutting, depends upon two types of factors:

(1) Engineering properties and lithologic characteristics of the weaker, undercutting unit. These include slake durability, freeze-thaw resistance, unconfined compressive strength, type and amount of clay minerals, and rock fabric (Deo, 1972;

Shakoor, A., 1995, Slope stability considerations in differentially weathered mudrocks, *in* Haneberg, W. C., and Anderson, S. A., eds., Clay and Shale Slope Instability: Boulder, Colorado, Geological Society of America Reviews in Engineering Geology, v. X.

Figure 1. Large-size failure of a sandstone ledge along Sawmill Run Road in Pittsburgh, Pennsylvania. The failure occurred in 1984, killing two persons, and completely destroying a car and a bulldozer. Undercutting of sandstone by claystone appears to have contributed to this failure which can be categorized as both a rock fall and a plane failure.

pre-existing discontinuity whose strike is within 20° of the strike direction of the slope face and whose dip is less than that of the slope face but more than the angle of friction along the discontinuity surface (Hoek and Bray, 1981).

Rock falls

Rock falls predominate where rock discontinuities form orthogonal blocks in competent strata which are underlain by easily erodible incompetent strata. The undercutting by the weaker layer allows the loose blocks from the upper layer to fall under the influence of gravity (Figs. 2 and 3). The frequency and size of rock falls depend upon the joint spacing within the competent unit and the extent by which it has been undercut. However, undercutting is not always required for the rock falls to occur. Closely jointed rocks can lead to rock falls even if there is no undercutting involved.

The Oregon Department of Transportation (ODOT) has developed a Rock Fall Hazard Rating System (RHRS) that can be used to evaluate the hazard potential of rock falls (Pierson, 1991). The system, which is quite subjective, is based on an

Spears and Taylor, 1972; Stollar, 1976; Russell and Parker, 1979; McClure, 1981; Grainger, 1983; Oakland and Lovell, 1985; Shakoor and Brock, 1987; Taylor, 1988; Dick and Shakoor, 1992). The term "fabric," as used here, refers to void space, slickensides, laminations, and microfractures.

(2) Slope characteristics including slope aspect, slope angle, slope vegetation, surface runoff, ground-water seepage, talus accumulation, and fracture frequency within the undercut unit (Sowers and Royster, 1978; Rib and Liang, 1978; Shakoor and Rodgers, 1992).

This paper documents the types of slope failures promoted by undercutting, presents a method for predicting the rate of undercutting, and reviews some of the remedial measures that can be used to reduce the potential for undercutting-induced failures.

UNDERCUTTING-INDUCED FAILURES

The common forms of slope failure associated with undercutting include rock falls, wedge failures, and plane failures, as defined by Varnes (1978). In falls, a mass of rock is detached from a steep surface, along which little or no movement occurs, and descends most of its distance through the air. A wedge failure is a rapid, downward and outward movement of a wedge-shaped block of rock along the line of intersection of the two discontinuities forming the block, or along the steeper of the two discontinuity surfaces. A wedge failure occurs when the inclination of the line of intersection (plunge) is less than that of the slope face (i.e., the line daylights on the slope face) but greater than the friction angle (Hoek and Bray, 1981). A plane failure involves a sliding movement along a

Figure 2. Rock falls resulting from differential weathering of the slope at Rockport, Interstate 77, Ohio. The sandstone block in the foreground is more than 4.5 m (15 ft) in height.

Figure 3. Undercutting-induced rock falls along Interstate 77 near Dexter City, Ohio.

evaluation of 12 different factors that contribute to the overall hazard. Based on the final ratings, the frequency of rock falls is categorized as "new falls" (3 points), "occasional falls" (9 points), "many falls" (27 points), and "constant falls" (81 points). Fookes and Sweeney (1976) have discussed the general methods of stabilizing slopes subject to local rock falls and degradation. Barrett and White (1991) have reviewed rock fall prediction and control methods with reference to Colorado highways. A methodology for rock fall evaluation by computer simulation using a probablistic approach has been developed by Wu (1982).

Wedge failures

As stated previously, the condition for a wedge failure to occur is that the plunge of the line of intersection (ψ_i) should be greater than the friction angle (ϕ) along the discontinuity surfaces but less than the dip of the slope face (ψ_f); i.e., the line of intersection must "daylight" on the slope face (Hoek and Bray, 1981). This condition is fulfilled if the point of intersection of the two discontinuities, when plotted on a stereonet, falls within the stippled area referred to as the critical zone (Fig. 4). The critical zone is the area enclosed by the intersection of the great circle representing the slope face and the circle representing the friction angle.

Differential weathering can cause wedge failures to occur in the more competent layers even when the lines of intersection do not initially daylight on the slope face; i.e., the points of intersection do not fall within the critical zone. When the line of intersection is steeper than the slope face, the wedge has a greater tendency to move downward than outward. When the line of intersection is nearly vertical, as will be the case for a wedge formed by near vertical discontinuities, the wedge could fail only by downward movement. Such a movement can occur only if the underlying rock is removed through the process of differential weathering (undercutting). Figures 5

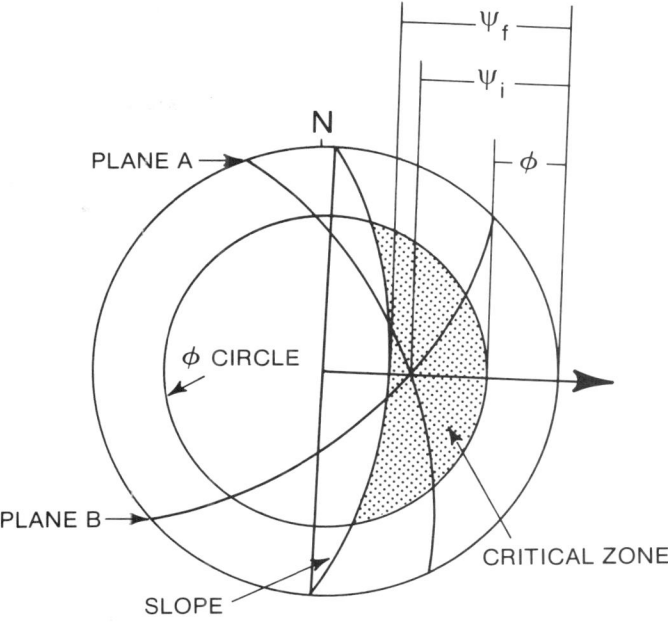

(After Hoek and Bray, 1981)

Figure 4. Stereographic projection showing conditions for wedge failure.

Figure 5. Example of a small wedge failure in a limestone ledge near Senecaville, Interstate 77, Ohio. Notice the steeply plunging line of intersection which did not daylight on the original slope face.

and 6 show examples of small-scale wedge failures resulting from the intersection of steeply dipping discontinuities in a relatively thin layer of limestone that has been undercut by a highly erodible shale layer. Because of their steeply dipping lines of intersection, these wedge failures could not have occurred without backwasting of the shale layer. Figure 7 shows a series of wedge failures in a resistant silty shale layer underlain by a less resistant claystone layer. Again, it is the undercutting action of the claystone layer that has led to the failure of the steeply plunging wedges in the upper layer. It should be noted that depending upon the locations of the competent layers on the slope face, the movement that initiates as a wedge slide may culminate as a rock fall. The term "wedge falls" has been proposed to describe such combinations of wedge failures and rock falls (Shakoor and Weber, 1988).

Plane failures

Like wedge failures, undercutting can promote plane failures even if the discontinuities running parallel to the slope face do not daylight on the slope face. Figures 1 and 8 illustrate how undercutting by claystone layers has led to plane failures in the overlying sandstone layers. It is evident from Figure 1 that such failures can be both large and hazardous. Figures 9 and 10 are good examples of plane failures that occurred in a resistant shale layer underlain by a weak claystone layer. It is highly unlikely that these failures would have occurred without removal of the support beneath the failed blocks through the process of undercutting.

PREDICTING THE RATE OF UNDERCUTTING

The time interval between the excavation of a road cut in a mudrock sequence and the initiation of undercutting-induced failures will depend upon the rate at which various mudrock formations are likely to undercut. For a timely implementation of remedial measures, it is essential to have a good estimate of the rate of undercutting for a given site. In order to devise a method that could be used to predict the rate of undercutting, 14 roadcut sites that exhibited significant amounts of undercutting were studied (Shakoor and Rodgers, 1992). Essential to the selection of these sites was the presence of pre-split blast hole markings on the undercut resistant units (Fig. 11). The presence of pre-split markings ensured that the exposed face of the resistant unit was the original surface exposed to weathering since the time of excavation. The amount of undercutting at each site was carefully measured at equally spaced intervals along the entire length of the cut. The measurements were taken at the contact between the undercut and the under-

Figure 6. Wedge failures in a limestone ledge near Senecaville, Interstate 77, Ohio, with near vertical lines of intersection. These wedges could not have fallen without undercutting of the shale layer below.

Figure 7. Steeply dipping wedge failures in a resistant shale layer near East Steubenville, State Route 2, West Virginia. Undercutting of the shale layer by a less resistant claystone layer near the base of the slope seems to have aided in the occurrence of these wedge failures.

Figure 8. Plane failure of a sandstone slab subsequent to undercutting, Mingo Junction, State Route 2, Ohio. Scar of a previous plane failure can be seen to the right of the sandstone slab.

Figure 9. This failure of a resistant shale material near East Steubenville, State Route 7, West Virginia, can be categorized as a plane failure and has been promoted by differential weathering of the slope.

Figure 10. Another example of a plane failure in a resistant silty shale underlain by a less resistant claystone near East Steubenville, State Route 7, West Virginia. Again undercutting of the resistant shale layer seems to have triggered this failure.

Figure 11. Pre-split blast hole markings on a sandstone ledge undercut by mudstone, Interstate 79, Roane County, West Virginia.

cutting units, using a tape measure. The extent of the overhang of the undercut unit relative to the undercutting unit at a given point of measurement defined the amount of undercutting. The number of measurements taken, and the interval between individual measurements, depended upon the length of the road cut. Information regarding the dates of excavation of the selected road cuts was obtained from highway department records. The measured amount of undercutting at each road cut and the age of the cut were used to estimate the rate of undercutting at each site.

Samples of the undercutting mudrock units were obtained from each site and tested for Atterberg limits, clay mineralogy, and slake durability. In order to obtain representative samples, each cut was divided into three equal segments and representative samples were obtained from the midpoints of the three segments. The samples were stored in air-tight containers to prevent slaking and preserve their natural water content. The American Society for Testing and Materials (ASTM) procedure D4318 (ASTM, 1990) was used to determine the liquid limit and plastic limit on samples disintegrated by alternate cycles of wetting and drying. The plasticity index was computed as the numerical difference between the liquid limit and the plastic limit. Clay mineralogy was determined by X-ray diffraction analysis and clay content by hydrometer analysis (ASTM, 1990, D422). Based upon the amount of clay content and the presence or absence of laminations, the samples were categorized as mudstones, clayshales, siltshales, and argillaceous limestones in accordance with the classification proposed by Potter et al. (1980).

The slake durability index test (ASTM, 1990, D4644) was used to determine the slake durability index (SDI) value for each sample. Three samples from each site were tested, each sample being subjected to two cycles of treatment. The test simulates the natural degradation of clay-bearing rocks to alternate cycles of wetting and drying (Franklin and Chandra,

1972). Slaking is considered to be the principal mechanism of weathering of soft rocks (Oakland and Lovell, 1985).

The average values of SDI for different sites were correlated with the corresponding values of the maximum rate of undercutting using bivariate regression analysis. The correlation coefficient (r) was used to evaluate the strength of the relationship. The results of regression analysis, shown in Figure 12, indicate that the maximum rate of undercutting has a fairly strong (r = 0.83) relationship with SDI for SDI values >30%, and a very strong (r = 0.99) relationship when SDI values are <30%. Figure 12 also provides equations for the two segments of the regression line that can be used to predict the rate of undercutting once the second-cycle slake durability index for the undercutting unit has been determined. It should be pointed out, however, that these equations are based on a limited number of sites investigated and a limited number of samples tested. Additional research is needed to validate and further refine these equations, especially with data from other climatic regions. Furthermore, the equations assume that the rate of undercutting has been constant since the date of excavation which may not necessarily be the case.

In addition to correlating slake durability index with the rate of undercutting, an attempt was also made to correlate Atterberg limits with the rate of undercutting, but no significant relationship was found to exist between these properties.

REMEDIAL MEASURES FOR SLOPES SUBJECT TO DIFFERENTIAL WEATHERING

The remedial measures commonly employed to reduce the frequency and hazard potential of undercutting-induced failures can be divided into two categories:

(1) Those that retard the rate of undercutting or prolong the initiation of undercutting.

(2) Those that minimize the potential for occurrence of undercutting-induced failures and the hazard associated with such failures.

Shotcreting of weak incompetent layers, along with the provision of drainage holes, is the most frequently used method to retard the rate of undercutting (Fig. 13). Specifications and details of shotcrete applications are provided in a special publication by the American Concrete Institute (1974). Generally a 7.5–10-cm (3–4-in) thick layer, containing fine-grained aggregate (<1.25 cm or ½ in), is considered adequate.

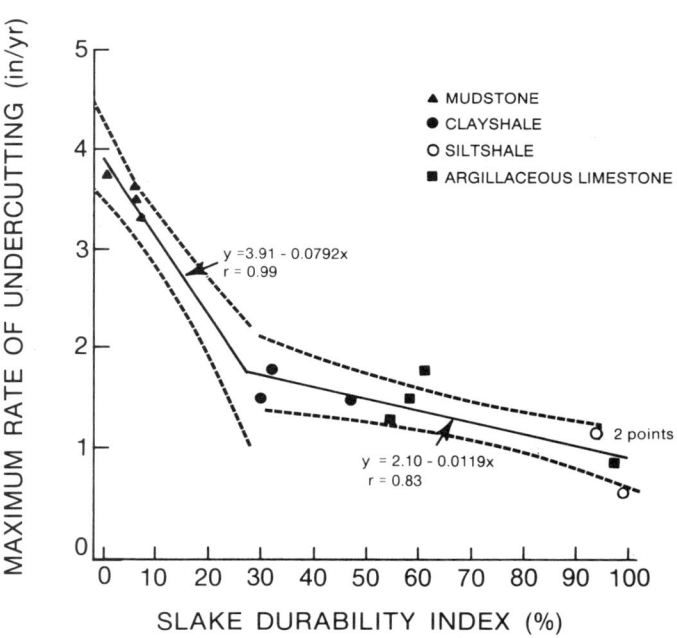

Figure 12. Correlation between the maximum rate of undercutting and the second-cycle slake durability index. In the figure, y = maximum rate of undercutting (in/yr); x = slake durability index (%); and r = correlation coefficient.

Figure 13. Shotcreting of a weak claystone layer that lies between a sandstone layer above and a relatively durable shale layer below, Duquesne Bluff, Parkway East, Pittsburgh, Pennsylvania.

For erodible layers near the toe of the slope, placement of gabbions is a viable option. A recent and ingenious method of delaying the initiation of undercutting is to provide a horizontal bench, ~5–7 m (~15–20 ft) wide, along the contact between the incompetent and competent layers (Fig. 14). The provision of such a bench does not eliminate differential weathering but does prolong the time by which the incompetent layer starts undercutting the competent layer. Once the vegetation takes place on such a bench, it can serve as an excellent delaying technique.

Periodic removal of overhangs (scaling), stabilization of unstable wedges and rock blocks using rock bolts, placement of anchored wire-mesh nets over portions of the slope, provision of properly designed and lined catchment ditches, provision of benches, and erection of sturdy catch fences are some of the techniques available to reduce the potential for undercutting-induced failures and to minimize the hazard associated with such failures. Recently, rock-fall attenuators and rock-fall impact barriers have been incorporated as energy absorbing devices (Barrett and White, 1991). Details of catchment-ditch design have been discussed by Ritchie (1963) and Peckover (1975). Lowell (1987) has developed design tables and charts based on Ritchie's criteria for catchment ditches. Mearns (1972) has evaluated the use of sand in catchment ditches as an energy absorber to prevent rock falls from rolling and bouncing. Smith and Duffy (1990) have described field tests and evaluation procedures for rock-fall restraining nets. Evans (1989) has investigated the use of rock-fall simulation programs for bench design. Generally, a combination of these techniques may be necessary for a given site.

SUMMARY

Differential weathering of road cuts located in mudrock sequences can quickly lead to undercutting of the more competent layers. The undercutting, in turn, promotes rock falls, wedge failures, and plane failures, which could not have occurred if the support below the competent layers had not been removed through the process of undercutting. Undercutting-induced failures can be quite hazardous because of their high speed and their sudden occurrence. For timely implementation of remedial measures necessary to reduce the potential for undercutting-induced failures, it is essential to be able to predict the rate at which the incompetent layers at a given site will undercut the competent layers. This can be done using the equations developed in this study which relate the second-cycle slake durability index of the undercutting rock unit with the maximum expected rate of undercutting. If the rate of undercutting happens to be very high, as is the case with claystones and mudstones, remedial measures can be adopted during the construction stage. In case of slow weathering rocks (e.g., silty shales), remedial action can be taken at an appropriate time after the cut has been excavated.

Figure 14. Remediation of the Sawmill Run Road failure of 1984 (Fig. 1). The remediation consists of cutting back the sandstone face, using presplitting, and providing a horizontal bench in the claystone layer along its contact with the overlying sandstone layer. The bench is designed to delay the undercutting of the sandstone layer.

REFERENCES CITED

American Concrete Institute, 1974, Use of shotcrete for underground structural support: Detroit, American Concrete Institute, Publication No. SP-45, 467 p.

American Society for Testing and Materials (ASTM), 1990, Soil and rock; dimension stone; geosynthetics: Annual Book of ASTM Standards, Volume 4.08: Philadelphia, ASTM, 1092 p.

Barrett, R. K., and White, V. L., 1991, Rockfall prediction and control, in Hambley, D. F., ed., Proceedings, National Symposium on Highway and Railroad Maintenance Annual Meeting, 34th: Chicago, Association of Engineering Geologists Symposium Series No. 6, p. 23–40.

Deo, P., 1972, Shales as embankment materials [Ph.D. dissert.]: West Lafayette, Indiana, Purdue University, Department of Civil Engineering, 202 p.

Dick, J. C., and Shakoor, A., 1992, Lithologic controls of mudrock durability: Quarterly Journal of Engineering Geology, v. 25, p. 31–46.

Evans, C., 1989, The design of catch bench geometry in surface mines to control rock fall [M.S. thesis]: Tucson, University of Arizona, Department of Mining and Geological Engineering.

Fookes, P. G., and Sweeney, M., 1976, Stabilization and control of local rock falls and degrading rock slopes: Quarterly Journal of Engineering Geology, v. 9, p. 37–55.

Franklin, J. A., 1983, Evaluation of shales for construction project—An Ontario shale rating system: Toronto, Ontario, Ministry of Transportation and Research, Research Development Branch, RR229, 99 p.

Franklin, J. A., and Chandra, A., 1972, The slake durability test: International Journal of Rock Mechanics and Mining Sciences, v. 9, p. 325–341.

Grainger, P., 1983, Aspects of the engineering geology of mudrocks, with reference to the Crackington Formation of Southwest England [Ph.D. dissert.]: Exeter, England, University of Exeter, 407 p.

Hoek, E., and Bray, J. W., 1981, Rock slope engineering: London, The Institute of Mining and Metallurgy, 358 p.

Lowell, S., 1987, Development and application of Ritchie's rock fall catch ditch design, in Proceedings, Federal Highway Administration Rock Fall Mitigation Seminar: Portland, Oregon, Federal Highway Administration.

McClure, J. G., 1981, Physiochemical investigations of shale slaking [Ph.D.

dissert.]: Berkeley, University of California, Department of Civil Engineering, 314 p.

Mearns, R., 1972, Solving a rock fall problem in Nevada County, California: Highway Research News, no. 49, p. 14–17.

Oakland, M. W., and Lovell, C. W. 1985, Building embankments with shale, *in* Ashworth, E., ed., Research and engineering applications in rock masses: Rapid City, South Dakota, 26th Symposium on Rock Mechanics, p. 305–312.

Peckover, F. L., 1975, Treatment of rock falls on railway lines: American Railway Engineering Association Bulletin 653, p. 471–503.

Pierson, L. A., 1991, Rock fall prediction and control, *in* Hambley, D. F., ed., Highway and railroad slope maintenance: Annual Meeting, 34th: Chicago, Association of Engineering Geologists Symposium Series No. 6, p. 1–22.

Potter, P. E., Maynard, J. B., and Pryor, W. A., 1980, Sedimentology of shale: New York Springer-Verlag, 306 p.

Rib, H. T., and Liang, T., 1978, Recognition and identification, *in* Schuster, R. L., and Krizek, R. J., ed., Landslides: Analysis and control: Washington, D.C., National Academy of Sciences, Transportation Research Board Special Publication No. 176, p. 34–80.

Ritchie, A. M., 1963, Evaluation of rock fall and its control: Washington, D.C., Highway Research Board, Highway Research Record 17, p. 13–28.

Russell, D. J., and Parker, A., 1979, Geotechnical, mineralogical, and chemical interrelationships in weathering profiles of an overconsolidated clay: Quarterly Journal of Engineering Geology, v. 12, p. 107–116.

Shakoor, A., and Brock, D., 1987, Relationship between fissility, composition, and engineering properties of selected shales from Northeast Ohio: Bulletin of the Association of Engineering Geologists, v. 24, p. 363–379.

Shakoor, A., and Rodgers, 1992, Predicting the rate of shale undercutting along highway cuts: Bulletin of the Association of Engineering Geologists, v. 29, p. 61–75.

Shakoor, A., and Weber, M. W., 1988, Role of shale undercutting in promoting rock falls and wedge failures along Interstate 77: Bulletin of the Association of Engineering Geologists, v. 25, p. 219–234.

Smith, D., and Duffy, J., 1990, Field tests and evaluation of rock fall restraining nets: California Department of Transportation, Report No. CA/TL-90/05, 138 p.

Sowers, G. F., and Royster, D. L., 1978, Field investigations, *in* Schuster, R. L., and Krizek, R. J., Landslides: Analysis and control: Washington, D.C., National Academy of Sciences, Transportation Research Board Special Report 176, p. 81–111.

Spears, D. A., and Taylor, R. K., 1972, The influence of weathering on the composition and engineering properties of in situ coal measures rocks: International Journal of Rock Mechanics and Mining Sciences, v. 9, p. 729–756.

Stollar, R. L., 1976, Geology and some engineering properties of near-surface Pennsylvanian shales in northeast Ohio [M.S. thesis]: Kent, Ohio, Kent State University, Department of Geology, 36 p.

Taylor, R. K., 1988, Coal measures mudrocks—Composition, classification, and weathering processes: Quarterly Journal of Engineering Geology, v. 21, p. 85–89.

Young, B. T., and Shakoor, A., 1987, Stability of selected road cuts along the Ohio River as influenced by valley stress relief joints, *in* Proceedings, 38th Annual Highway Geology Symposium: Pittsburgh, Highway Geology Symposium, p. 15–23.

Varnes, D. J., 1978, Slope movement: Types and processes, *in* Schuster, R. L., and Krizek, R. J., eds., Landslides: Analysis and control: Washington, D.C., National Academy of Sciences, Transportation Research Board Special Report 176, p. 11–33.

Wu, Shie-Shin, 1982, Rock fall evaluation by computer simulation: Washington, D.C., Transportation Research Record 131, p. 1–5.

MANUSCRIPT ACCEPTED BY THE SOCIETY MAY 20, 1994

Effect of argillic alteration on rock mass stability

Robert J. Watters
Department of Geological Sciences, Mackay School of Mines, University of Nevada, Reno, Nevada 89557
Warren D. Delahaut
FMC Gold Company, Paradise Peak Mine, Gabbs, Nevada 89409

ABSTRACT

The role of hydrothermal alteration in producing clay-rich rocks is discussed. Hydrothermal fluids derived from magmatic sources change rock lithologies to argillic rocks as a result of temperature, pressure and chemical effects. The grade of the argillization can vary from one in which only trace amounts of clay minerals are present to one in which there has been complete alteration into clay. Detailed geologic surface mapping and subsurface drilling are required to accurately delineate the extent and grade of alteration. Suggestions are presented for assessing the grade and amount of alteration. Two examples illustrate the effects of argillic alteration on the engineering design of excavated slopes. One example demonstrates a successful design where alteration effects were incorporated into the initial design stage. The second example illustrates where an inadequate geologic model underestimated the grade and extent of alteration, and a landslide of over four million tons of material resulted. Reappraisal of the geologic model enabled a successful mitigation and incorporation of the clay-rich lithologies to be designed, and permitted continued safe excavation at the site.

INTRODUCTION

The deleterious effect of rock alteration through weathering processes (Lumb, 1962; McFeat-Smith et al., 1989), with the consequent reduction in rock mass strength and increased potential for slope instability is well known to engineering geologists. Less familiar is the reduction in rock mass strength produced from rock alteration in response to hydrothermal fluids. Hydrothermal alteration (Guilbert and Park, 1986) differs from chemical and physical weathering in that it is independent of surface weathering processes and can be both vertically and laterally widespread. Alteration develops in response to hydrothermal fluids affecting the parent rock through a combination of chemical, temperature, and pressure effects. Argillic alteration is only one of several types of hydrothermal alteration (Fig. 1); other types of hydrothermal alteration (e.g., greisenization and fenitization) may develop and can increase or decrease the strength characteristics of the rock mass.

Slopes composed of unaltered rock material(s) may be perfectly stable, whereas the argillic altered rock material can exhibit major slope distress, depending on slope geometries and groundwater conditions. Argillic alteration is a type of rock alteration that changes the parent or country rock from a strong rock to a clay-rich weak rock assemblage. When the argillic alteration is intense, the original rock essentially can be compared to an overconsolidated clay.

Argillic alteration differs markedly from weathering induced alteration. Weathering, in most instances, is a ground surface phenomena with the greatest weathering generally occurring at the ground surface and decreasing with depth. Hydrothermal alteration is independent of surface effects and consequently can vary from hundreds to thousands of feet in depth (Fig. 2). It develops from hot, pressurized, chemically charged hydrothermal fluids. These fluids migrate from a magmatic system, along fault surfaces, shear zones, or a permeable rock unit, to a suitable host rock which will chemically react with the fluid. Hydrothermal fluids are the main source of mineralization in many economic ore deposits throughout

---

Watters, R. J., and Delahaut, W. D., 1995, Effect of argillic alteration on rock mass stability, *in* Haneberg, W. C., and Anderson, S. A., eds., Clay and Shale Slope Instability: Boulder, Colorado, Geological Society of America Reviews in Engineering Geology, v. X.

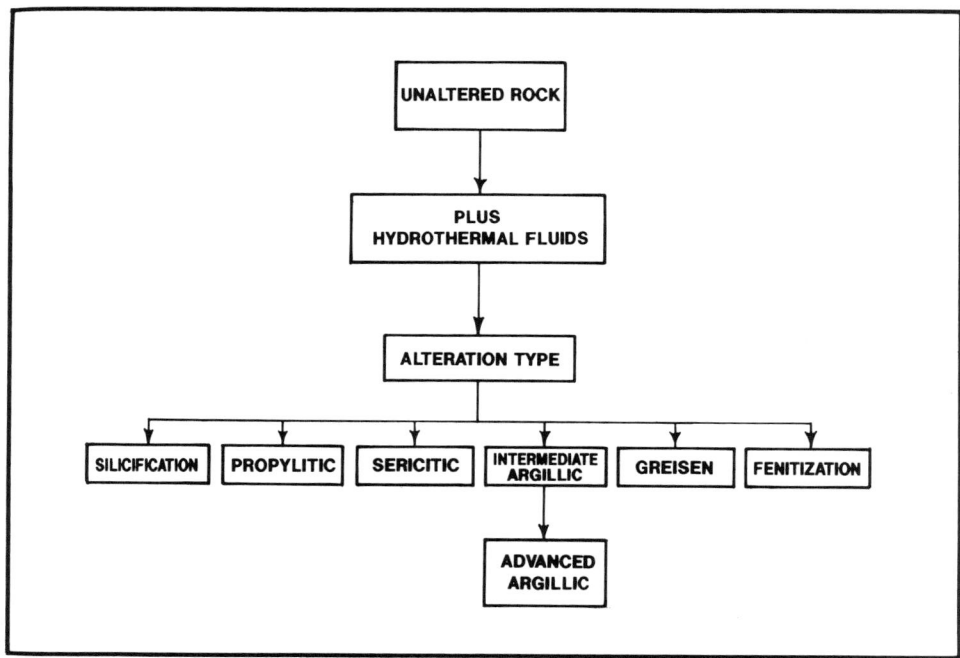

Figure 1. Types of hydrothermal alteration.

the world. In addition to changing the rock mass from a rock of no economic value to that of an ore, the strength characteristics also are affected by the fluids. Therefore, the mining of economic deposits, through large open pits, requires an understanding of hydrothermal alteration, its extent, engineering significance, and the selection of suitable slope design criteria based on the engineering behavior of the argillic rock mass.

The purpose of this paper is to examine the formation of argillic rock masses resulting from hydrothermal alteration, the classification systems describing the alteration, the strength changes the rock mass undergoes, and the major slope instability that can result.

EXTENT AND TYPES OF ALTERATION

As hydrothermal fluids migrate away from the magmatic source along fracture systems, wall rock alteration develops along the conduits (fractures), with shear strength reduction occurring on the fractures. This change is well recognized as joint alteration and has been addressed by many authors with regard to weathering (Barton, 1973; ISRM, 1981). The most dramatic and wide-ranging effects of hydrothermal alteration are on the intact rock and rock mass, in which the rock is chemically changed to a new rock type with different engineering behavior characteristics. These changes develop when the country rock is chemically unstable in the presence of hydrothermal fluids. Physical and chemical changes continue to occur to the rock material until physical and chemical equilibrium is re-established. Temperature, pressure and chemical gradients are the principal mechanicms by which alteration occurs. However, chemical gradients are most important in inducing change in the parent rock. The chemical reactions affecting change include hydrolysis, hydration-dehydration, decarbonization, silicification, and oxidation-reduction effects. Overall, the most significant is hydrolysis. Hydrolysis is the

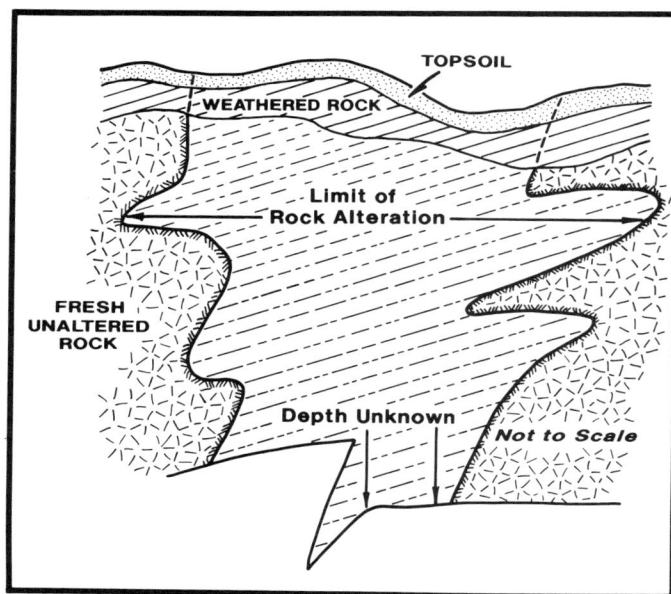

Figure 2. Cross section comparing vertical extent of surface weathering and hydrothermal alteration.

involvement of the hydrogen ion in changing or converting anhydrous silicates (e.g., feldspar) to hydrologized silicates (e.g., mica) and the clay minerals. When a carbonate and silica are present, the hydrogen ion can produce selective silicification of the rock mass, consequently producing a stronger rock mass. Hence, a localized strong silicified rock mass may be produced beside a very weak argillized rock mass. These radically different rock masses may co-exist for a few feet, tens of feet, or hundreds of feet.

The parent rock types most susceptible to alteration are those comprised of significant proportions of alumino-silicate minerals, e.g., granite, andesite, basalt, etc. (Meyer and Hemley, 1967). The main types of alteration that develop can be classified as porpylytic, intermediate argillic, advanced argillic, sericitic, potassic, silicic, greisen, and fenite are presented in Table 1. The different clay alteration types that can be developed in response to hydrothermal fluids are illustrated in Figure 3. The type and percentage of clay minerals formed and the amount of quartz present classify the type of alteration (Sales and Meyer, 1948).

CLASSIFICATION SYSTEMS FOR ALTERATION

Unlike rock weathering, where the classification of the grade of weathering has been somewhat standardized (Ruxton and Berry, 1957; Lumb, 1962; Krank and Watters, 1983; Lee and de Freitas, 1989), alteration classifications are more variable. This variability reflects the demands of the economic geologist, who needs to know minor amounts of alteration

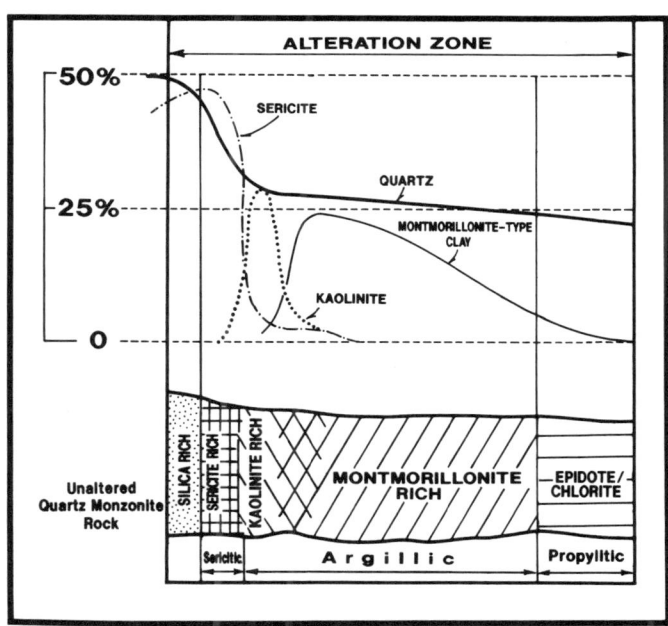

Figure 3. Change in the relative abundance of major mineral constituents produces sericitic, argillic, or propylitic alteration, in response to hydrothermal solutions (modified from Sales and Meyer, 1948).

because certain mineral assemblages may be indicators of economic mineralization. Different points of view exist with regard to mapping, classifying, and describing alteration from the standpoint of the economic geologist. However, all alteration systems follow one or two classification formats. One approach incorporates the name reflecting the most abundant or clearly identifiable mineral in the altered rock. This classification system results in a combination of rock texture, mineralogy, and the original rock, as the basis of describing the alteration (Creasey, 1959). The second approach is to classify based upon the major chemical change(s) during the alteration process (Hemley and Jones, 1964).

For any engineering classification system, the absence or presence of clay dramatically affects the engineering behavior of the material. Hence, once the type of alteration has been determined, e.g., argillic, the amount and grade of argillic alteration will control the engineering behavior of the rock mass. Table 2 details a suggested classification system for argillic alteration similar to that used with rock weathering. The determination of the type of alteration, grade, and vertical and lateral extent of the alteration is dependent on detailed surface geologic mapping and subsurface drilling. Both the type and grade of alteration may change rapidly over short distances. Consequently, a dense drilling pattern with quality geotechnical logs (or chip boards) is required if the effects of the alteration are to be incorporated into a slope design.

TABLE 1. MAIN TYPES OF ALTERATION

Alteration Type	Mineralogy
Propylitic	Epidote and/or chlorite minerals
Intermediate argillic	Significant amounts of kaolinite, montmorillonite, and amorphous clay
Advanced argillic	All feldspar minerals converted to dickite, kaolinite, alunite, and other aluminum rich argillic phases
Sericitic	Potassium feldspar and plagioclase converted to sericite
Potassic	Potassium and/or biotite formed as an alteration of mafic minerals
Silicic	Proportion of quartz or opaline silica increased
Greisen	Similar to advanced argillic with more sericite or muscovite.
Fenite	Occurs around carbonatite intrusions and represents sodium metasomatism with hydrous silicates and hematite

TABLE 2. ARGILLIC ALTERATION CLASSIFICATION

Alteration	Amount	Grade	Description
None	Zero	0	No alteration
Argillic	Trace	1	Scattered occurrence, clay <1%
Argillic	Slight	2	Mainly along fractures, scattered matrix occurrence
Argillic	Moderate	3	Along fractures and replacement of phenocrysts
Argillic	Complete	4	Complete replacement of phenocrysts and matrix

ALTERATION EFFECTS ON ROCK MASS STRENGTH

To assist in the development of slope design recommendations for three gold mines in the United States, alteration effects had to be assessed. Detailed field investigations were conducted in conjunction with rock strength testing, rock mass classification, and classification of alteration type and grade. Only the three rock lithologies which reflected the major rock units at the mine sites were studied. Each lithology was in excess of 300 ft (90 m) thick and consisted of (1) phonolite, (2) andesite, and (3) a sedimentary sequence of siltstones, mudstone, and sandstone. One of the three lithologies was dominant at each mine site and had experienced moderate to complete alteration. Geotechnical testing at each site concentrated on parameters that were utilized in characterizing the rock mass for slope design purposes. These tests included unconfined strength testing, rock shear testing, density and unit weight determination, rock discontinuity data collection, petrographic, scanning electron microscope (SEM), and X-ray diffraction (XRD) analysis. The SEM and XRD techniques are the most useful in classifying the type of clay alteration based on clay type and approximate percentage of clay present in the sample. Montmorillonite clay is easily identified on the XRD plot of Figure 4, where the spikes on the plot are at the two-theta values for montmorillonite. Further analyses can be performed by use of the SEM where the structure of the clay can be observed and photographed (Fig. 5) and the sample chemistry can be determined.

The testing results were incorporated into calculations to obtain the Rock Mass Rating (Bieniawski, 1973) and the rock mass strength from semi-empirical techniques (Hoek and Brown, 1980), for each of the lithologies at different grades of alteration. Table 3 contains a summary of the average values obtained from the testing and computation.

Illustrated in the table is the dramatic drop in unconfined compressive strength, joint cohesion, and friction values. The low values for joint friction reflect the formation of montmorillonitic clay in varying proportions in the tested samples. Unit weight shows declines of about 15%–20%, a function of lower specific gravity of the new clay minerals and increased rock porosity. The increase in porosity indicates that the rock

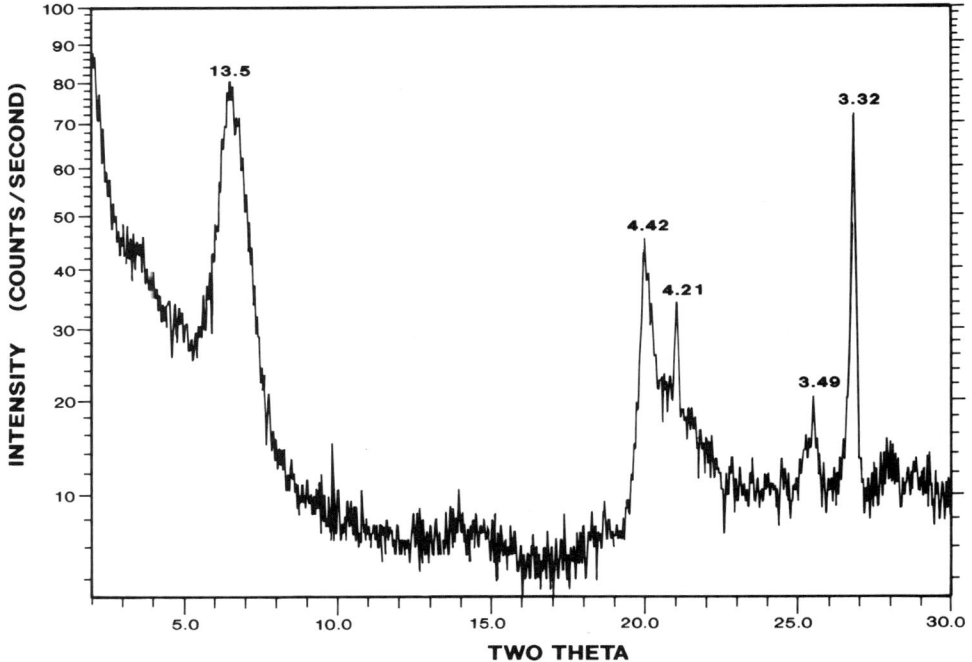

Figure 4. XRD plot for montmorillonite clay (field example 1).

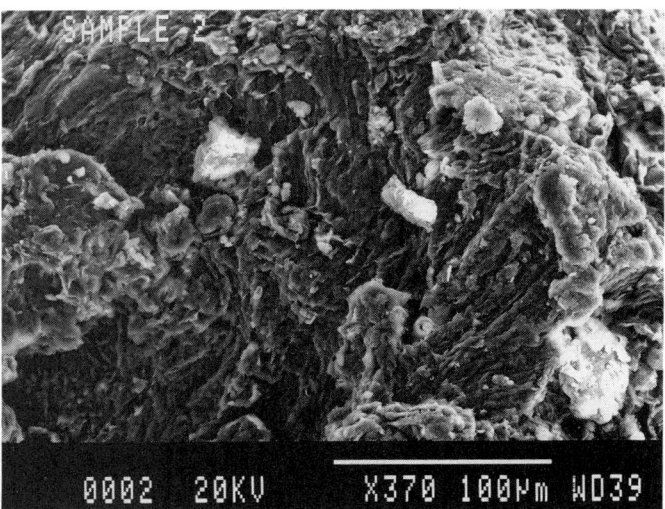

Figure 5. SEM photograph of montmorillonite clay (field example 1).

has greater capacity for storage of water within its matrix, which leads to lower strength behavior.

FIELD EXAMPLES

Two field examples illustrating the effects of clay alteration on slope stability are taken from two disseminated gold mining operations. The first example details the effects of both argillization and silicification on slope design. The second example illustrates only argillization. However, the effects were so widespread in the slope that a slope failure in excess of four million tons was triggered, requiring a major slope redesign. Computerized stability analyses for both examples were performed using Modified Bishop, Janbu, and block-surface-generated failure surface techniques using software developed by Sharma (1992).

EXAMPLE 1

The first example illustrates a successful recognition of alteration effects on slope design. It was recognized at the initial stage of the investigation that the two dominant rock units portrayed vastly different strength characteristics. One rock unit could be excavated easily by hand, the other rock unit resisted efforts to obtain small samples for study.

Site geology

The geologic environment at the mine site consists of upper Miocene rocks, which consisted, prior to alteration, of siltstone, sandstone, and minor subaqueous basalt flows. The siltstone and sandstone is interfingered and locally dips at between 15° to 17° to the east. The area is traversed by several large north-striking normal faults. A minor conjugate fault system striking N30°W and N30°E cuts the mine site. Two joint sets are prevalent throughout both sedimentary units. A strongly striking north-south set, with steep dips (70°–90°) both east and west, and a less developed east-west striking joint set. Pervasive hydrothermal alteration of the rocks occurred over the mine site area from the Pliocene to the present and is responsible for the economic mineralization. The high-angle faults acted as conduits for geothermal fluids resulting in intense silicification along faults and the many receptive stratiform zones.

Strong argillization (grade 4) occurred contemporaneous with the silicification, producing montmorillonite (Figs. 4 and 5), illite and kaolinite clays adjacent to the silicified zones. The siliceous rock mass is significantly stronger jointed than the argillic lithology, with the siliceous rock containing more persistent and continuous joints. Figure 6 is an east-west cross section through the deepest part of the pit, prior to excavation, and demonstrates how varied and contorted the alteration profiles are. Borehole data positions and orientations, together

TABLE 3. GEOTECHNICAL EFFECTS OF ALTERATION

Material	Alteration Grade	Type	Unconfined compressive strength (lb/in²)	Joint Cohesion (lb/ft²)	Joint Friction (°)	Unit Weight (lb/ft³)	Rock Mass Rating
Phonolite	0	Argillic	8,500	6,200	40	150	Good
	3		3,000	3,500	30	141	Fair
	4		500	2,000	22	133	Poor
Andesite	0	Argillic	18,000	5,500	38	155	Good
	2		8,000	4,000	33	151	Fair
	3		2,500	2,000	21	130	Poor/fair
	4		1,000	1,450	13	125	Poor
Sediments	0	Silicified	23,000	6,900	31	154	Good
	3	Argillic	2,000	2,600	26	135	Poor
	3	Mixed	5,500	4,000	26	145	Fair

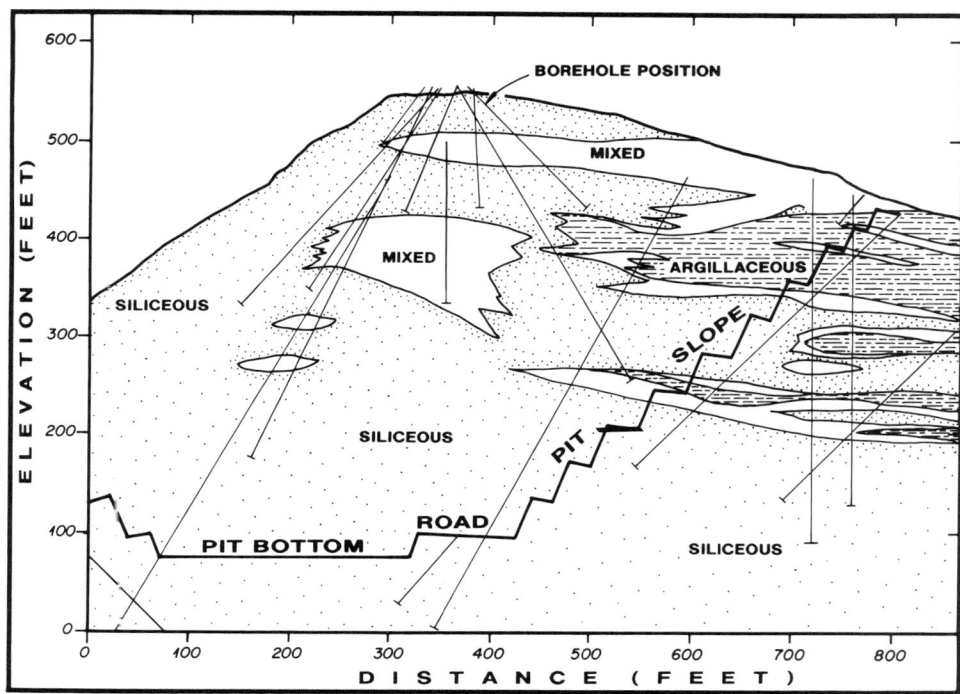

Figure 6. Cross section for example 1 with alteration types, borehole positions, and slope geometry.

with geologic units, are contained in the figure. The pit slope shown on the cross section is for a 45° slope, which was later increased to 55° after the analysis. No ground water table was detected in drilling, and it was assumed, for stability purposes, that the slopes would remain in the drained state for the lifetime of the mine.

Engineering analysis

The geologic model of Figure 6 was analyzed for a pit depth of 400 ft (120 m). The cross sections used for the slope stability appraisal were analyzed while assuming different proportions of argillic and siliceous rocks existed in the slope. Stability calculations were performed assuming the slope was (a) completely argillic (i.e., clay), (b) completely siliceous, (c) comprised of a mixed argillic/siliceous unit, and (d) consisted of a variable thickness of argillic rocks (100–200 ft; 30–60 m) overlying siliceous material. Given the rock structure and lithologic character of the rock units, stability calculations obtained the optimum design angles using the relevant lithology values for the sedimentary units of Table 3. Factors of safety varied as a function of the position of the argillic unit and the ratio of the argillic to silicified unit in the slope (Table 4). The geologic cross sections through the pit, similar to that of Figure 6, showed that the bottom 200–250 ft (60–75 m) of the pit slopes would be in siliceous rocks. The overlying 150–200 ft (45–60 m) would be in either argillic or mixed rocks.

Inspection of Table 4 demonstrates that a slope angle of 70° would cause failure in both argillic and mixed materials. A slope angle of 60° for thicknesses of argillic rocks up to 200 ft (60 m) overlying siliceous material would have a factor of safety of about 1.2 and greater for lesser thicknesses of argillic rock. Based on the analyses, the mine slopes were excavated at between 55° and 60°, depending on the lithologic composition of the slope in a given part of the mine, and the location of the haul road cutting across the slope. No significant failures developed during the operation of the pit, apart from localized bench failures. Such failures are expected where slopes are designed to lower factors of safety as compared to slopes

TABLE 4. FACTORS OF SAFETY FOR DIFFERENT SLOPE GEOMETRIES AND LITHOLOGIES

Slope Composition	Factor of Safety	Angle (°)
A. Entire slope		
Siliceous	1.36	60
	1.16	70
Argillic	0.86	60
	0.72	70
Mixed	1.14	60
	0.96	70
B. Argillic over siliceous		
100 ft argillic	1.32	60
150 ft argillic	1.27	60
200 ft argillic	1.19	60
200 ft mixed	1.3	60

designed for civil engineering purposes that have higher assigned factors of safety.

EXAMPLE 2

This example demonstrates what may develop in an argillic slope, if alteration effects are more pervasive in the slope than were anticipated in the design. The initial design did not incorporate the completely argillic altered rock mass adequately in the geologic model. As a result strength parameters higher than were appropriate were utilized. Consequently, a major landslide developed, mobilizing over four million tons of material. A major slope redesign and real-time continuous slope monitoring were required in order to permit the continued safe operation of the open pit.

Site geology

The rocks at the mine site are composed of upper Oligocene and Miocene volcanic rocks. The volcanic rocks can be divided into three volcanic sequences: an older sequence of andesitic lavas, a middle sequence of pyroclastic silicic ash-flow tuffs, and a younger sequence of andesitic lavas. The older andesite unit consists of a thick series of andesite to rhyodacite flows and flow breccia that are locally intercalated with volcaniclastic sedimentary rocks. The middle sequence is a series of silicic ash-flow tuffs, interstratified with intermediate lavas and epiclastic sedimentary rocks. The middle sequence overlies the older andesites along a low-angle normal fault (thrust?) contact. The upper andesitic sequence unconformably overlies the middle tuff sequence. The upper andesites are intermediate to felsic lavas, dominantly coarsely porphyritic. They include minor epiclastic and volcanic sedimentary rocks. Several sets of moderate to steep faults cut and locally offset the different volcanic sequences with north-south strikes. Two conjugate joint sets were determined to exist through discontinuity mapping. One set strikes essentially north-south with steep dips of between 70° and 90° to the east and west. The second joint set was less strongly developed with approximately a west northwest strike with a variable dip of between 50° and 80° to the northeast and southwest. Regionally the dip of the strata is to the east northeast at about 35°, but local variations to both the dip and dip direction are apparent. Hydrothermal alteration of the volcanic sequences was intense, with mixed layer illite-montmorillonite and kaolinite clays formed. The major alteration types of argillic, propylitic, and silicic were distinguished using XRD and SEM analyses.

Engineering analysis

The geologic model for the slope was developed from cored boreholes and rotary percussion drilling, which permitted numerous geologic cross sections to be prepared. A simpli-

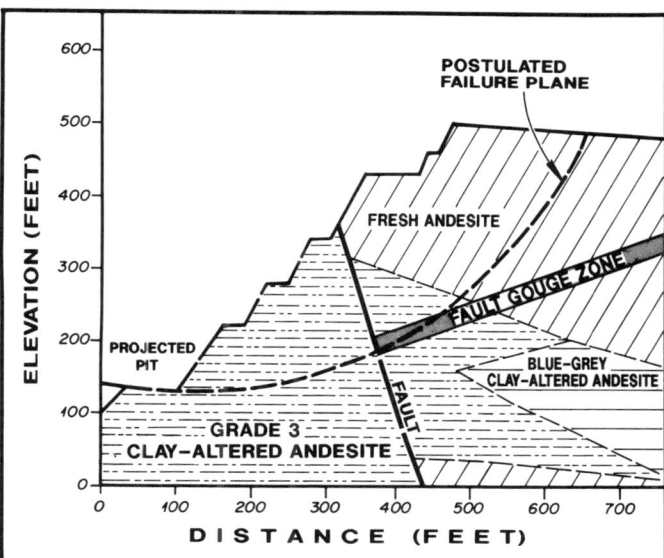

Figure 7. Initial geologic cross section and critical failure surface position (FS >1.0) for example 2.

fied cross section illustrating the two major rock units and the geologic structure is presented in Figure 7. The rock units comprising the slope consist of an upper fresh andesite overlying argillically altered andesite. One low angle fault gouge zone dipping toward the slope face was identified from the drilling records. Fresh rhyolite occurs in the middle and near the bottom of the pit but was not intersected in the initial slope excavations and was not involved in the slope instability. Four joint sets exist within the fresh andesite and rhyolite together with a large north-south striking fault. The natural ground water table, ~100 ft (30 m) below the ground surface, had been temporarily depressed below the proposed pit bottom through the use of submersible borehole pumps and horizontally bored drains.

Vertically oriented boreholes were drilled around the periphery of the pit and were installed with submersible pumps. The highly fractured rhyolite and fresh andesite were easily dewatered because of the high fracture flow permeability of both rock units. However, the clay-altered andesite, as a result of low permeability, produced locally a perched water table within the fresh andesite, where the dip of the clay andesite impeded flow through the fresh andesite toward the pit or the dewatering boreholes. The perched water table(s) were intersected by horizontally drilled boreholes, up to 200 ft (60 m) in length; the boreholes were drilled from the slope face. Flow rates up to 150 gal/min were measured from some boreholes, but within one to two days the flow rates dropped to 1–2 gal/min.

The initial slope design utilized geotechnical data for the rock units as detailed in Table 3. Analyses indicated that a stable slope would be accommodated with an angle of 45°. The stability analysis assumed that the most appropriate failure surface, or lowest calculated factor of safety, would be circular in nature. The circular failure surface was deemed appropriate because of the generally closely spaced jointing and low-angle bedding in the unaltered andesite. The failure surface was assumed to follow the jointing in the fresh andesite into part of the gouge zone and through the altered grade 3 andesite before daylighting in the pit bottom.

Slope excavations proceeded uneventfully in the area until a tension crack was noted some 200–250 ft (60–75 m) from the slope crest. Figure 8 is a plan view of the pit with cross-section positions, monitoring locations, and tension crack detailed. At the time of the failure, the pit excavations were ~200 ft in depth, with ~125 ft (40 m) of vertical excavation remaining. The tension crack increased in size, ultimately exhibiting >60 ft (20 m) of horizontal and 40 ft (13 m) of vertical movement (Fig. 9). Monitoring prisms were installed on the slope in order to safeguard mine personnel and establish the type of slope movement. Figure 10 contains data recorded during an approximately nine-month period. This data illustrate the total displacement of different parts of the landslide as slope remediation was performed and the slide controlled. At the

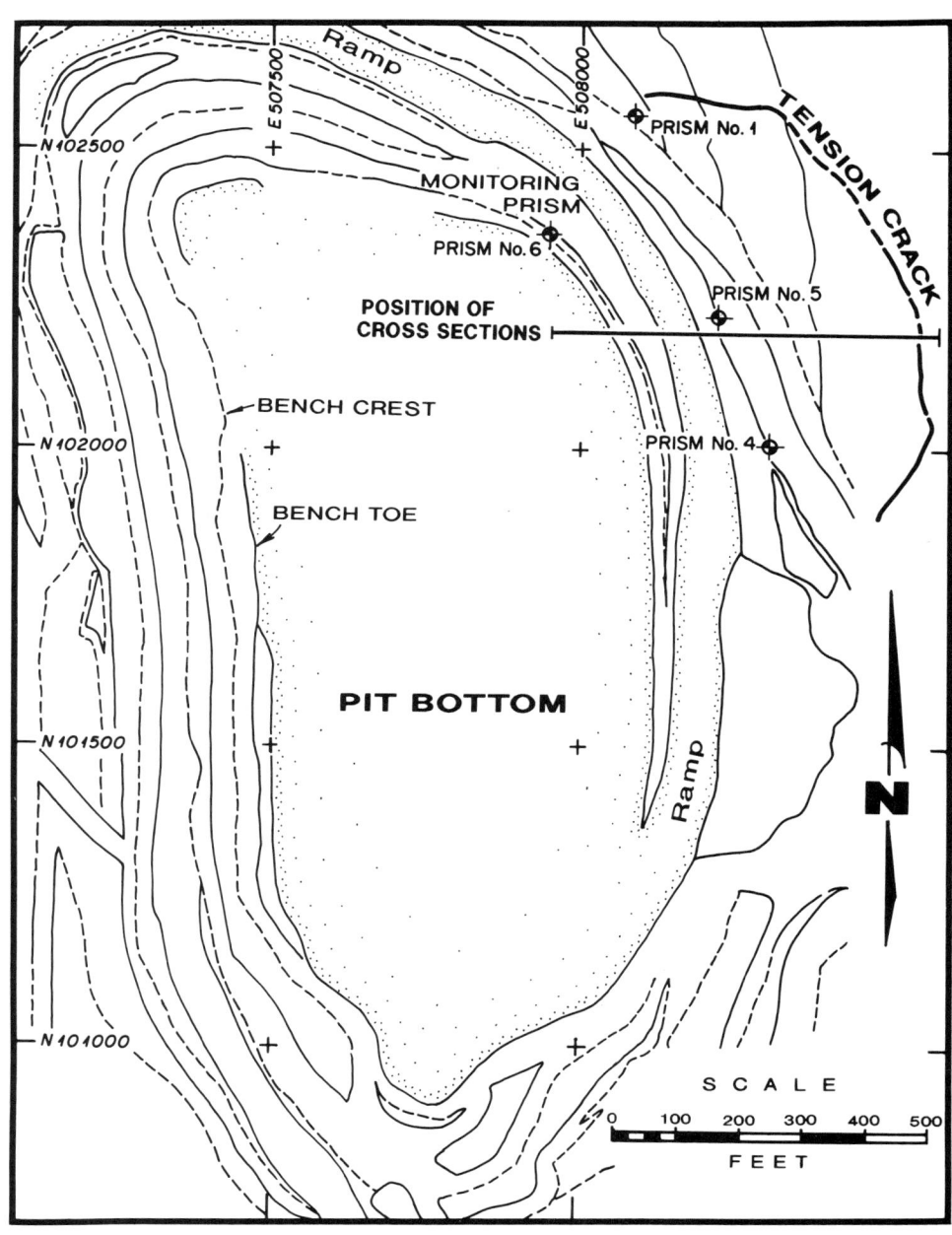

Figure 8. Pit plan with location of cross section, tension crack, and monitoring prisms.

Figure 9. Photographs of tension crack after (A) three months and (B) six months.

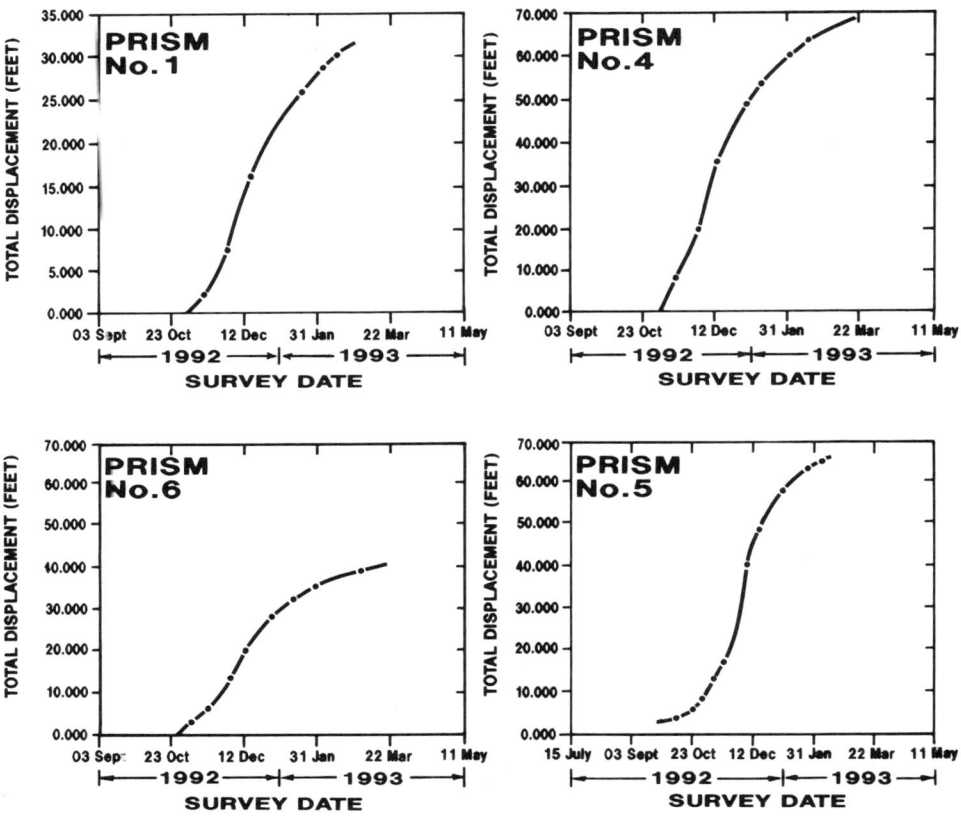

Figure 10. Slope monitoring information for total slope displacement.

Figure 11. Floor heave of 5–10 ft (2–3 m) 100–150 ft (30–45 m) from slope toe.

base of the slope, floor heave was noticed up to 175 ft (55 m) horizontally from the slope toe. The floor heave initially was in the order of 2–3 ft (1 m), but eventually reached a height of between 25–35 ft (8–10 m) above the pit floor (Fig. 11).

The landslide initiated a reappraisal of the geologic slope model. Further rock testing was performed and the stability recalculated. The geologic model was changed in response to recognition that the alteration was more intense than anticipated, with widespread development of low strength clay between the slope toe and a strong rhyolite rock mass some 200 ft (60 m) from the toe. The failure surface was determined to consist partly of a low angle shear/thrust containing thick clay gouge and the completely altered andesite. The strength of the failure surface was assessed from laboratory shear strength testing and from back analysis of the slide. Both the laboratory testing and back analysis results showed that the strength of the argillized andesite was lower than that assumed in the initial analysis. The altered andesite in the toe area more closely approximated a grade 4 as opposed to a grade 3 alteration. When slope height is plotted against slope angle with factor of safety curves for the two grades of alteration, it is apparent that the completely altered andesite could sustain stable slopes only with slope angles 10° lower than that for the grade 3 alteration (Fig. 12).

Modeling part of the failure surface as fault controlled and shear strengths similar to the grade 4 alteration, the Modified Bishop and Janbu circular analyses gave comparable factors of safety for the slowly moving landslide. A further analysis was performed using a block generated failure surface where the shear zone existed to calculate whether significantly

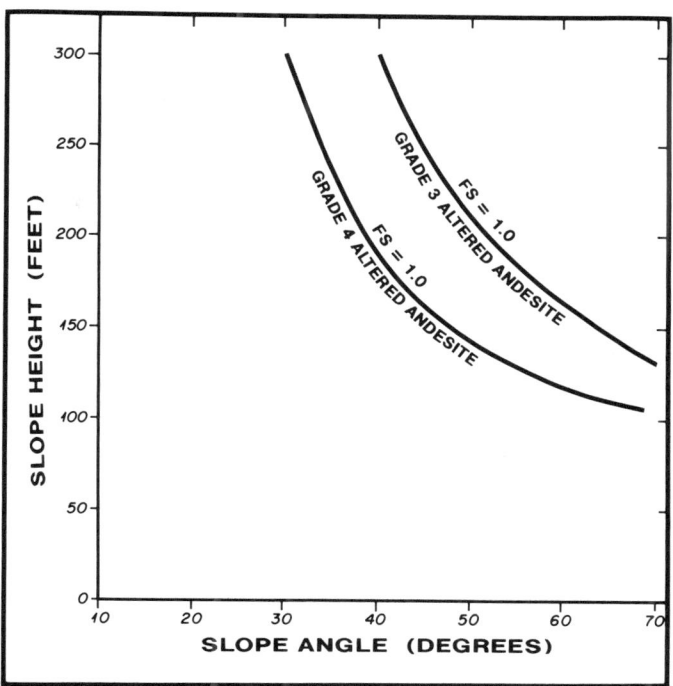

Figure 12. Factor of safety as a function of slope height and slope angle for alteration grades 3 and 4.

different factors of safety would be generated (Fig. 13). Both types of stability analyses showed that for the slope angle excavated, the factor of safety was below unity and that the slope would continue to fail unless the slope angle was lowered or

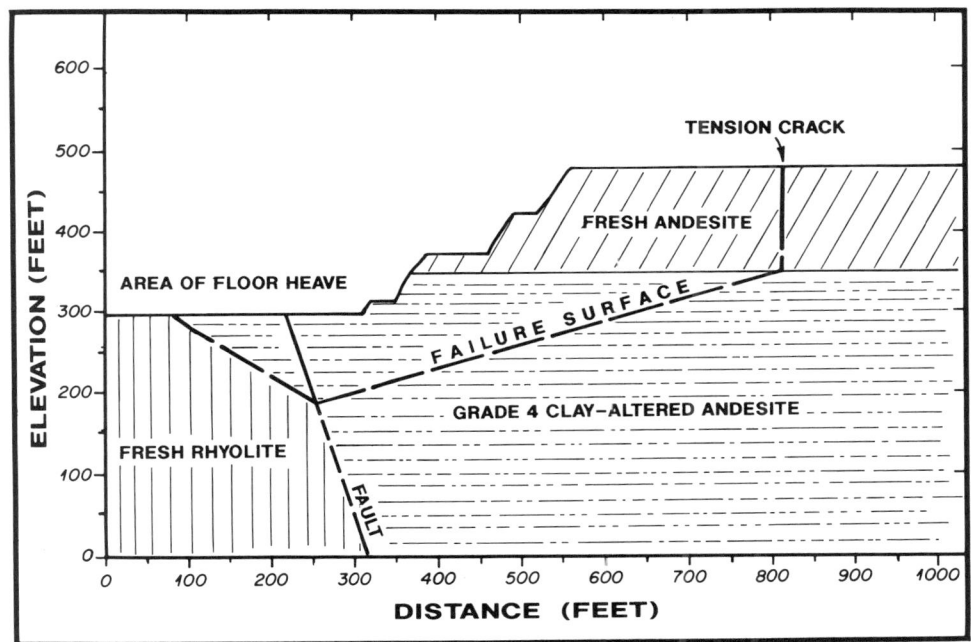

Figure 13. Revised failure surface and simplified geologic cross section through the landslide.

buttressed. A buttress was easily executed by leaving the floor heave in place, effectively acting as a toe surcharge, which increased the resisting force. The pit continued to be deepened outside the area of floor heave, so that the mine operation did not experience significant operational difficulties. Prisms were installed on the unstable slope, and a computerized electronic distance measurement (EDM) permitted continuous slope monitoring. The resulting data provided real-time total displacement, velocity, and acceleration information on the landslide. Any major movements that might develop automatically initiate safety warning signals. The pit continued to be safely excavated to the full design depth with the buttress in place. After reaching the design depth, the buttress was progressively removed in order to recover gold ore beneath the buttress. Buttress removal was discontinued when the operation became uneconomical.

CONCLUSIONS

Argillic alteration produced from hot hydrothermal fluids produces a dramatic reduction in rock mass strength as a result of the formation of clay-rich lithologies. Correct identification of the grade of alteration is crucial in recognizing where the rock mass has undergone significant change. The strength behavior of the argillic altered lithology can then be determined, and its effect incorporated into the slope design.

ACKNOWLEDGMENTS

Permission to publish the information in this paper is gratefully acknowledged from FMC Gold Corporation and Amax Gold, Inc. Manuscript review by Colleen Murray is appreciated.

REFERENCES CITED

Bienawski, Z. T., 1973, Engineering classification of jointed rock masses: Transactions of the South African Institute of Civil Engineers, v. 15, p. 335–344.

Barton, N. R., 1973, Review of a new shear strength criterion for rock joints: Engineering Geology, v. 7, p. 287–332.

Creasey, S. C., 1959, Some phase relations in hydrothermally altered rock of porphyry copper deposits: Economic Geology, v. 54, p. 351–373.

Guilbert, J. M., and Park, C. F., Jr., 1986, The geology of ore deposits: New York, W. H. Freeman and Co., 537 p.

Hemley, J. J., and Jones, W. R., 1964, Chemical aspects of hydrothermal alteration with emphasis on hydrogen metasomatism: Economic Geology, v. 59, p. 538–569.

Hoek, E., and Brown, E. T., 1980, Empirical strength criteria for rock masses: Journal of the Geotechnical Engineering Division, American Society of Civil Engineers, v. 106, p. 1013–1035.

International Society for Rock Mechanics (ISRM), 1981, Rock characterization, testing and monitoring: International Society for Rock Mechanics suggested methods: Pergamon, London, 211 p.

Krank, K. D., and Watters, R. J., 1983, Geotechnical properties of weathered Sierra Nevada Granodiorite: Bulletin of the Association of Engineering Geologists, v. 20, p. 173–184.

Lee, S. G., and de Freitas, M. H., 1989, A revision of the description and classification of weathered granite and its application to granites in Korea: Quarterly Journal of Engineering Geology, v. 22, p. 31–48.

Lumb, P., 1962, The residual soils of Hong Kong: Géotechnique, v. 12, p. 226–243.

McFeat-Smith, I., Workman, D. R., Burnett, A. D., and Chau, E. P. Y., 1989, The geology of Hong Kong: Bulletin of the Association of Engineering Geologists, v. 26, p. 17–108.

Meyer, C., and Hemley, J. J., 1967, Wall rock alteration, in Barnes, H. L., ed., Geochemistry of hydrothermal ore deposits: New York, Holt, Rinehart and Winston, 670 p.

Ruxton, B. P., and Berry, L., 1957, Weathering of granite and associated erosional features in Hong Kong: Bulletin of the Geological Society of America, v. 58, p. 1263–1292.

Sales, R. H., and Meyer, C., 1948, Wall alteration at Butte, Montana: AIME Transactions, v. 178, p. 9–35.

Sharma, S., 1992, Slope analysis with XSTABL: U.S. Department of Agriculture, Intermountain Research Station, Contract No. INT-89416-RJV, 149 p.

MANUSCRIPT ACCEPTED BY THE SOCIETY MAY 20, 1994

Index

[Italic page numbers indicate major references]

A

Abalone Cove landslide, 15, 19
  bentonite, 21
aerial photography, 80
African continental margin, 108
Aina Haina, Oahu, landslide, 79, 83
Alani-Paty landslide, Oahu, 79, *83*
  Alani element, 83, 102, 103
  geotechnical properties, *87*
  hydraulic properties, *88*
  hydrology, *90*
  Kahaloa element, 83, 85, 101, 102
  mechanics, *99*
  movement, 90, *99*
Altamira Shale, 15
alteration, *139, 141*
American Society for Testing and Materials standard, 4, 121
ancient landslide deposits, 17, 117
andesite, 141, 142
anisotropically consolidated-undrained (ACU) test, *4, 5, 8,* 10
Apennines, Italy, 108, 113, 117
argillic alteration, *139*
argillite, 131
argillization, 139
Athens, Ohio, 129
atomic absorption spectrometry, 21
Atterberg limit tests, 2, 13, 14, 19, 20, 43, 112, 114

B

basal slip surface, 83
  depth, 80
basalt, 141
bentonite, Abalone Cove landslide, 21
  Palos Verdes Peninsula, *13, 19, 21,* 35
  Portuguese Bend landslide, *13,* 29
bond strength, 7
Briones Formation, San Pablo Group, 2
Briones Park field site, California, 1
  soil characterization, *2*
  stability analysis, *8,* 10
  test results, *4, 6*
  testing procedures, *3*
Briones Park, California, 1
Briones Reservoir, California, 2
Bromhead torsional ring shear apparatus, 30

C

calcarenite, 108
calcium montmorillonite, 13, 21
Cascade Mountains, Washington, 66

Casentino, Italy, 108
Catalina Schist, 15
cementation, 7
Certaldo, Italy, unstable slopes, 116
Chianti Mountains, Italy, 113
Cincinnati, Ohio, 68
classification systems, alteration, *141,* 142
clay fraction, 14, 21, 32
claystone, 121, 122, 123, 131
  durability, *124*
Coast Range, northern California, 2
cohesion, 29, 34
collapse, 8
  mechanisms, 53, 60
Colorado, landslides, 79, 102
compression tests, 1
Conemaugh Group, 129
consolidated drained stress-controlled strain-rate (CDSR) test, 43, 45
consolidated drained triaxial (CDTX) test, 43
consolidated undrained (CU) triaxial test, 29
consolidation, 9
  coefficient, 29
constant-shear-drained (CSD) test, *4, 5, 6, 7,* 10
contractive behavior, 8, 9
conventional slice methods, 51
Coulomb failure potential, 68
Coulomb frictional failure, 66
critical friction angle, 107
critical height, 51, 53, 55, 57
critical state, 113
critical state soil mechanics, 7
Cucaracha shale, 20

D

debris apron, 86, 102
debris flows, 1, 8, 121, 126, *129*
decarbonization, 140
deformation, 29, 79, 83, 102
Delhi-Pike landslide, 103
deviator stress, 39
dilation, 6
direct shear
  conventional, 32
  long sample, 32
direct shear (DS) test, 1, 3, *4, 6,* 10, 29, 30
displacement, 102
disturbance, 3
drainage conditions, *8,* 19
Drucker-Präger yield criterion, 39, 40, 43
Dunkard Group, 128, 129
Dupuit assumptions, 65
Dupuit water table models, modified, *65*

E

effective normal stress, 14
effective yield friction angles, 45
Elsa basin, Italy, 108, 116
en echelon faults, 19
engineering analysis, *144, 145*
erosion, 6, 126
  excessive, *128*

F

factor of safety, 13, 14, 35, 51, 52, 53, 55, 68, 69
failure envelope, 29, 30, 34, 36
failure height, 53, 57
failure modes, *9*
failure patterns, 53
failure scars, 1
feldspar, 141
field velocity profiles, 48
fissures, 19
flow failure, 1
flow slides, San Francisco Bay area, 1
fluvial erosion, 108
Flying Triangle landslide, 14
Franciscan Terrane, northern California, 39
friction angle, 116

G

gabion wall, 19
gradation analysis, 2
grain-size analysis, 1, 6, 43, 112
granite, 141
Green-Ampt analysis, 97, 98
ground water, 19, 34, 48, 79, 80, 87, 89, 90, *91, 95,* 97, 100, 102, 103, 139
ground water flow, slope stability, *63, 64, 66*

H

half-grabens, discontinuous, 19
Hind Iuka landslide, Oahu, 83
Honolulu Volcanics, 86
Honolulu, Oahu
  landslide, *79*
  rainfall, 90
Hulu-Woolsey landslide, Oahu, 80, 103
human activity, 111
hydration-dehydration, 140
hydraulic conductivity, 63, 79, 83, 88, 89, 90, 102, 103
hydraulic properties, 63, *88*
hydrologic conditions, 39
hydrolysis, 140
hydrometer analyses, 21
hydrothermal alteration, *139,* 140
hydrothermal fluids, 139

Index

I
inclinometer survey, 39
infiltration, 80, *95*, 97
infinite slope method, 45
Inspiration Point, California, 15
irrigation, 1

K
Kamiloiki Valley, Oahu, 86
kinematically admissible failure mechanism, 52, 56
kinematics-based limit analysis, 51
Klondike Canyon landslide, 15
Kona storms, Oahu, 90
Koolau Basalt, 86, 91
Koolau Range, Oahu, 79, 82
  rainfall, 90
Kuliouou Valley, Oahu, landslide, 79

L
landslides
  factors of safety, 14
  failure, 13
  Palos Verdes Peninsula, *13*
  regrading, 19
  San Francisco Bay area, 1
lateral earth pressure, 80
limestone, 63, 108
limit analysis approach, *51*, *52*
limiting strain criterion, 42, 47, 48
liquefaction, static, 66
liquid limit, 14, 20, 21, 25, 32, 35, 36, 43
Long Sample Direct Shear Device, 30
longitudinal stretching, 79
lower bound theorem, 52

M
Malaga Mudstone, 15
Manoa Valley, Oahu, 86, 103
  landslide, 79, 80, 83
mica, 141
mineralization, 139
  economic, 142
mining, 140
Minor Creek landslide, California, 39, 103
  movement, *29*
  velocity profile, *48*
  viscoplastic slope movement, *39*
Mohr-Coulomb parameters, 40
Mohr-Coulomb yield condition, *54*
Monterey Formation, 14, 15
Montespertoli, Italy, unstable slopes, 116
movement, post-failure rates, 39
mudrock, *131*
  characterization, *121*, *122*
  durability, 121, *125*
  undercutting, 129, 131, *132*, *134*, 137
mudstone, 121, 122, 123, 131, 135
  durability, *124*
Mugello basin, Italy, 108, 113

O
Ohio
  landslides, 79, 102
  slope movements, 131
Ohio River valley, sliding, 63, 65, 66
Ontario shale, 122
ophiolite, 108
overburden stress, removal, 6
overconsolidation, 6
oxidation reduction, 140

P
paleoclimatic factors, 107, 118
Palolo Valley, Oahu, 86
  landslide, 79, 83
Palos Verdes anticline, 15
Palos Verdes fault, California, 15
Palos Verdes Peninsula, California
  landslides, *13*
  shear strength, *25*
Pennsylvania, slope movements, 131
perching, 102
permeability, 29
permeability tests, 88
phonolite, 142
Pierre Shale, 128
plane failures, *134*
plastic limit values, 21, 25, 43
plasticity index, 14, 20, 32, 35, 36, 43, 113, 114, 116, 122
plasticity test, 1
pore pressure, 39, 51, 52, *53*, 56, 57, 58, 64, 79, 93, 97, 98, 99, 100, 102, 103, 116, 118
  development, 29
  diffusion models, *66*
  excess, 9
  measurement, 1, 3, 6, 8
Portuguese Bend landslide, *13*
  bentonite, 29
  movement, *15*
  residual share strength testing, *19*
  shear strength, *29*
  slope stability analyses, *29*
  stabilization, 16, 17, 19
  testing procedures, *29*
Portuguese Point, California, 15
Portuguese Tuff, 15, 21
pressure waves, propagation, 80
principle of maximum work, 52

R
rainfall, 1, 48, 66, 79, 80, *90*, 92, 94, 95, 98, 107, 109, 111, 112
regolith, failures, 1
residual cohesions, 13
residual friction angle, 13, 14, 25, 29, 32, 34
residual shear strength, *13*, *19*, 21, 29, 30, 35, 36, *87*
residual strength parameters, 4, 7, 13, 113, 116
ring shear device, 29, 32, 36
Rio Costilla valley, New Mexico, 64
  landslide debris, 64
River Arno drainage basin, Italy, 109

rock alteration, 139
rock falls, *132*
Rock Mass Rating, 142
rock mass stability, *139*
rock mass strength, *142*, 150
rotational failure, 52, *55*
rotational mechanism, 60

S
safety factor. *See* factor of safety
San Francisco Bay area, 1, 68
  flow slides, 1
  landslides, 1
San Marcello Pistoeise, Italy, mudslide, 111
San Miniato, Italy, unstable slopes, 116
San Pablo Group, 2
sandstone, 2, 68, 108
scanning electron microscope (SEM) analysis, 142
secant modulus method, *41*
seepage, 1
shale, 121, 122, 123, 128, 131
  durability, *124*
shear direction, reversal, 30
shear force, 7
shear strain, 6
shear strength, 14, 29, 113, 117, 140
  Briones soil, 6, 7, 8, *9*, 10
  Palos Verdes Peninsula bentonites, *25*, 35
  Portuguese Bend landslide, *13*, 14, *19*, *29*
shear stress, 4, 14
shear tests, 35
shear thickening behavior, 40
shear thinning behavior, 40
shearing, 99
shortening, 79, 83, 99, 100
Siena basin, Italy, 108
silicification, 140
siltstone, 2, 68, 121, 131, 142
single well (slug and recovery) tests, 88
slake durability, 121, 122, 129
sliding potential, *68*
slope collapse, *54*
  rotational failure, *55*
  translational failure, *54*
slope failure, 10, 107
slope inclination angle, 57, 58
slope instability, 126
slope movement, 49
slope stability, *63*, *66*, 68, 80, 121, *131*
  groundwater flow, *63*, *64*, *66*
  limited analysis approach, *51*, *57*
  pore pressure effects, *53*
slope-stability analysis, 10, 29, 39, 51
slump failures, *128*
slumps, 121, 126
smectite, 3, 79, 102
soil behavior, 39
soil characterization, *6*
soil fabric, bonding, 7

soil structure, destruction, 3
South Shores landslide, 19
Spencer's Method, 32
stability analysis, 1, *8*, 51, 99, 103
stability factor, 57
stiffness, 3
strain parameters, 10
strain-rate parameters, *45*
strain-rate tests, 39, 49
strength loss, 4, 9
strength parameters, *9*, 10
strength tests, 4, 49
stress, history, 6
   intensity, 47
   normal, 34
stress ratio, 3
stress-strain behavior, 1, 9
stress-strain characteristics, 29
stretching, 79, 83, 99, 100
submarine gravitational processes, 108
surface drainage, 19
surface runoff, *94*

T

tectonic uplift, 108, 111
tectonization, 108
torsional ring shear apparatus, 36
translational failure, *54*
translational mechanism, 61

triaxial compression test, 3, 10, *39*
Tuscany, Italy, geomorphology, *108*
   landslides, *107*
   mudslides, *117*
two dimensional steady-state flow models, *65*
two-test method, *42*, 43, 49

U

undercutting, 121, 126, *129*, 131, *132*, *134*, 137
Unified Soil Classification system, 3
unloading test, 1, 3
upper bound theorem, 52, 54, 60
Utah, landslides, 79, 102

V

Val Marecchia, Italy, 113
Valdarno basin, Italy, 108, 113
Valmonte Diatomite, 15
velocity profile, 39
vertical hydraulic head diffusion, 66
vertical velocity profile, 40, 41
viscoplastic behavior, 48
viscoplastic constitutive model, 39, 41, 49
viscoplastic deformation, 39
viscoplastic flow, 46
viscoplastic model, 39, *40*, *43*

viscoplastic slope movement, *39*
viscous behavior, 39, 40, 49
Volterra basin, Italy, 108

W

Waahila Ridge, Oahu, 86
Waikiki, Oahu, rainfall, 90
Waiomao landslide, Oahu, 79, 83, 102
Walsenburg, Colorado, 128
water balance models, *65*
water pressure, subsurface, 80
water table, 19
wave erosion, Palos Verdes Peninsula, 15
weathering, 86, 131, *136*, 137, 139
wedge failures, *133*
West Virginia, slope movements, 131
Wierton, West Virginia, 68
Wood County, West Virginia, 128

X

x-ray diffraction (XRD) analysis, 3, 6, 87, 142

Y

yield behavior, 49
yield friction angles, 47, 48
yield strength, 3, *9*, 39, 41, *43*, 48

Typeset by WESType Publishing Services, Inc., Boulder, Colorado
Printed in U.S.A. by Malloy Lithographing, Inc., Ann Arbor, Michigan